Walter Franz | Peter Gabel | Stephan Gauss | Uwe Hempelmann | Rainer Henning
Wilhelm Kettler | Hans-Jörg Kremitzl | Gerhard Rösler/translated by Manfred Binder
Sandra Weixel | Gerhard Wilker

# *Colour Technology of Coatings*

Wilhelm Kettler et al.
Colour Technology of Coatings
Hanover: Vincentz Network, 2016
European Coatings library
ISBN 3-86630-699-7
ISBN 978-3-86630-699-8



Please ask for our book catalogue
Vincentz Network, Plathnerstr. 4c, 30175 Hanover, Germany
T +49 511 9910-033, F +49 511 9910-029
books@european-coatings.com, www.european-coatings.com

Layout: Danielsen Mediendesign, Hanover, Germany

ISBN 3-86630-699-7
ISBN 978-3-86630-699-8

European Coatings library

Walter Franz | Peter Gabel | Stephan Gauss | Uwe Hempelmann | Rainer Henning
Wilhelm Kettler | Hans-Jörg Kremitzl | Gerhard Rösler/translated by Manfred Binder
Sandra Weixel | Gerhard Wilker

# *Colour Technology of Coatings*

# *Preface*

Like many other textbooks, this one has its origins in the classroom and is the culmination of more than 10 years' experience of teaching courses on colorimetry to engineers and technicians from various branches of industry.

The *VDMI*, the VdL and the FPL asked various experts at pigments, paints, and instrument makers in 2000 to design various training modules for colourists working on industrial, plastics, paint, and print applications of colour.

The objective was to provide a technically correct and up-to-date introduction to those many aspects of colour and colour applications in industry. The target group was engineers deeply involved in colour applications in various industrial sectors. The first set of training modules was offered in 2001. Until retiring in 2014, *Dr. Tasso Bäurle* successfully steered the ship for almost 14 years, taking charge of the training modules. He continually refined and modified the underlying concept to meet the demands of the participants and participating industries and also acted as publisher of the German edition of this book. The latest version comprises just two modules covering **elementary concepts of colorimetry** and **in-depth insights in colorimetry**. Although the main focus is on paint applications, all the methods presented can be readily adapted to colour applications in other industries.

The main purpose of this book is to provide a comprehensive survey of relevant industrial colour applications and numerous concepts of physical and physiological pigment optics in order that a written record may be preserved of the specialist knowledge of all the lecturers involved in the coloristic training course. The colour problems discussed in this book include optics and chemistry of solid-colour and effect pigments, colourant formulation, optical microscopy of effect colour shades for pigment identification, methods of elementary and advanced colorimetry, measurement and visual assessment of solid and effect colour shades, colour tolerances and acceptability, and colour-order systems. Compared to the first German edition, this English edition contains a further chapter devoted to the newly emerging area of visual texture assessment of effect colour shades. Although technology for measuring visible texture is already available on the market, no mathematically rigorous definition of texture parameters and their dimensionality has been formulated so far that would be accepted throughout the colour community. However, a combination of multidimensional texture and colour information is the appropriate paradigm to adopt for proper physical characterisation of the visual appearance of effect colour shades. The contents of this book are a mix of objective detachment on one hand and a detailed first-hand knowledge and practical relevance on the other.

The mathematics throughout the book have been kept to a minimum, even though the technical treatment of colour problems is being driven more and more by mathematical models. Many references at the ends of chapters cite original papers. The reader is encouraged to consult these as further sources of information and to supplement the physical and physiological basics presented in this book with more general mathematics and rigor, as necessary.

As this English edition of the training course for colourists was being prepared, one of the authors, *Dr. Gerhard Rösler*, unexpectedly passed away in December 2012. I believe that Gerhard would be pleased with the expanded English edition of our joint project.

Wilhelm H. Kettler

Wülfrath, Germany, January 2016

# Contents

# I Fundamentals of colour perception

*Stephan Gauss*

## 1 Human colour vision

The term "**colour**" has different meanings. It is used variously to describe a characteristic of an object perceived by the eye, a paint or surface coating applied to a garden fence, and it can refer to a printing ink. Consider also the various materials, which give these products their colour. Such **colourants** are either pigments (insoluble in the paint or coating medium), or dyes (soluble in the medium), see Chapter IX. Colourants can be further ordered according to their "colour index" (C.I.), which more accurately should be called "colourant index."

A question that often arises during daily work in the paint and coatings industry is: "is today's production batch the same colour as the reference sample?" This book will show how we can obtain useful numbers from this perception in the brain to help us answer that question.

Any description of the term "colour" must distinguish between the physical processes that lead to a sensory stimulus in our eyes and our own subjective evaluation of that stimulus, which is transmitted from the eye to the brain. This is shown in Table I.1.

The left column in Table I.1 is more bio-physical in nature and is discussed in more detail in Sections I.1.1 and I.1.2. The right column is explained in Section I.1.3.

It is important to understand that the colour of an object exists only in our mind; different people may describe the colour of the same object differently. For this reason, it is common practice in the commercial world to talk instead about the colour **differences** from a given reference standard and not about the colour itself. Naturally, the coloristic attributes of the reference standard must be close to those of the sample being assessed. Because colour differences are usually small, people usually find it easier to agree on them. These colours are also known as related colours.

German standard DIN 5033 (Part 1)[1] defines the term colour in very sober terms:

*Colour is the sensation of a part of the visual field, which the eye perceives as having no structure and by which this part can be distinguished alone from another structureless and adjoining region when viewed with just one motionless eye.*

### Colour as a sensory perception

In this book the term colour will be used to describe the sensory perception which the brain associates with a given attribute of an object. This perception is affected

*Table I.1: Objective and subjective colour perception.*

| External, objective | Subjective |
|---|---|
| Physical stimulus, colour stimulus (ϕ) | Colour perception |
| Colorimetry spectrum | Colouring perceptions |

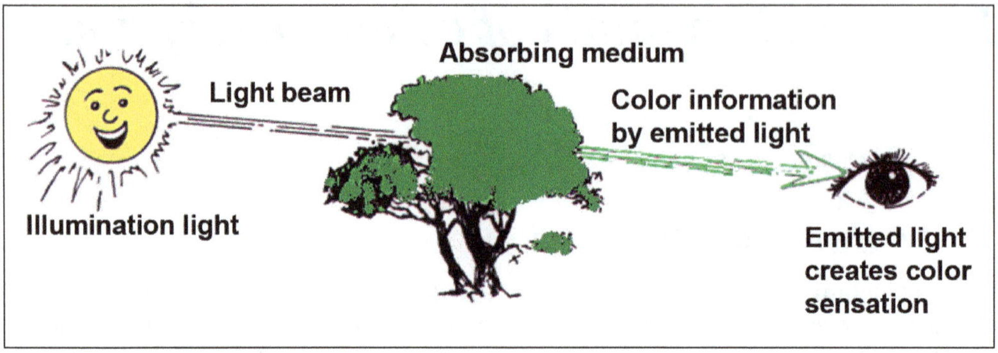


*Figure I.1: Our impression of the colour of an object is the outcome of interaction between a light source, the observed object, the human eye, and signal processing in our brain.* Source: Clariant Produkte (Deutschland) GmbH

by many other influences, in addition to the given attribute of the object. Thus, although the particular perception may vary from person to person, the environment and our own physical condition influence our colour perception, too. This is echoed in the phrase "looking at the world through rose-tinted glasses".

If we wish to perceive the colour of a non-luminous object, we must first illuminate it with a light source. Alternatively, we could observe the colour of a light source (luminous colour). These two ways are equivalent as regards the actual colour stimulus and the associated colour perception. Thus, our colour perception results from the interaction of three things:

- Light source
- Coloured object
- Detector (eye and brain)

This is illustrated in Figure I.1.

## 1.1 The human eye

In humans, the eye is the most important sensory organ [2, 3]. It is the origin of most of the signals processed in our brain. Because of this importance, nature has provided it with

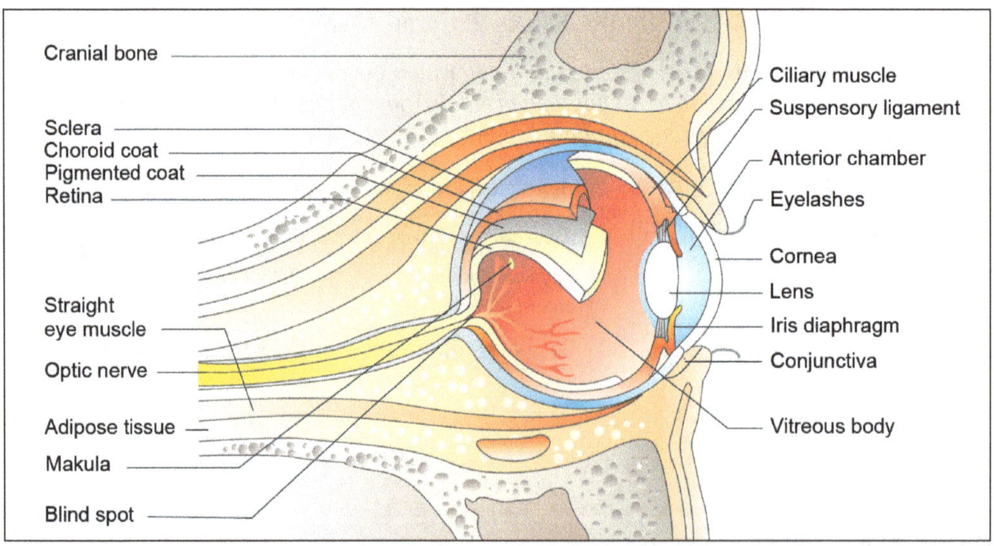


*Figure I.2: Schematic diagram of the human eye.* Source: Klett Verlag[4]

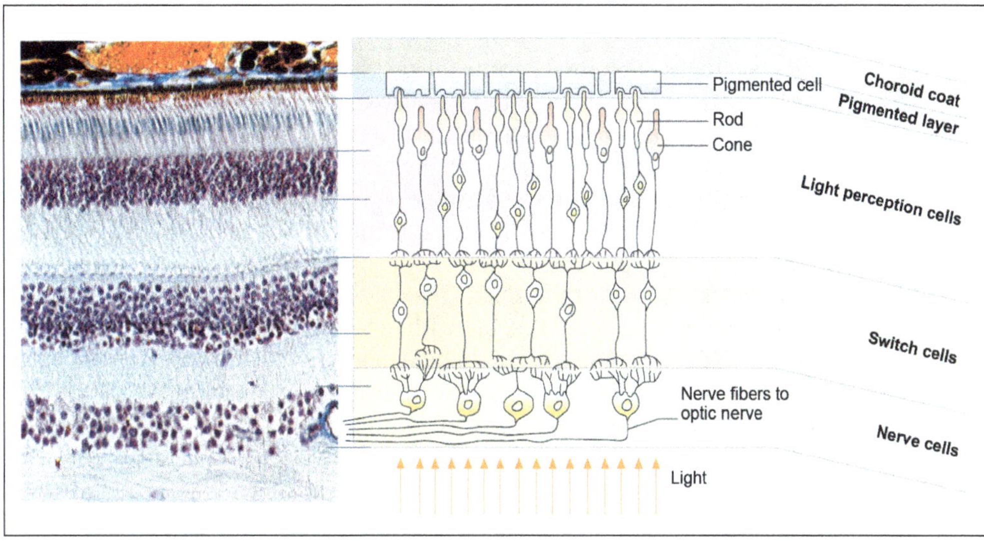


*Figure I.3: Schematic diagram of the human retina showing the arrangement of rods and cones. The light is incident from the bottom.* Source: Cornelsen Verlag[5]

several protective mechanisms. It is located in the eye-socket, where it is protected by the nasal bone, the cheekbone and the frontal bone. It has eyelashes to keep dust and dirt from entering and it features an eyelid-closure reflex for coping with an emergency. Its surface is continually cleaned by lachrymal fluid. Over the course of evolution, the eye as a means of providing vision has passed through a number of designs. A schematic diagram of a human eye is shown in Figure I.2.

### *1.1.1 Optical structure*

Light passes through the cornea and the jelly-like lens of the eye into the vitreous humour (vitreous body). In front of the lens is the iris, which acts as an aperture that enables the eye to adapt to varying lightness levels. The iris is capable of capturing three aperture stops, in the manner of a camera diaphragm. This corresponds to a correction factor of 1 : 8, but does not quite reflect the geometric conditions of 2 to 8 mm diameter for the pupil (aperture) (Stiles-Crawford effect). The ciliary muscle around the lens enables the latter to adapt to different focal distances in the manner of a telephoto lens by making the lens more spherical for focusing on near objects and flatter for distant objects. At the back of the vitreous humour is the retina, which contains the cells responsible for processing the sensory signals. Its structure is also shown in Figure I.3.

### *1.1.2 Signal processing and special features*

The retina is a light-sensitive, irregularly structured layer of tissue that lines the back of the vitreous humour[6]. Colour vision is only possible at angles of up to 40° around the optical axis. At greater angles, only monochromatic (black and white) vision is possible, while at smaller angles, colour vision and resolution increase. Close to the optical axis and located on the retina is the fovea, the area of highest resolution and greatest visual acuity. It has a visual angle of around 2° and is located within the macula, which contains pigments believed to provide the photoreceptors with additional protection against intense exposure to illumination. Surprisingly, the fovea is not on the optical axis, but is offset above it by about 4°. On the other side of the optical axis, offset by 10°, is another special feature of the human eye: the **blind spot.** This is an area where all the nerve fibres come together and leave the retina. Consequently, there are no photoreceptors in the blind spot.

### *Rods and cones*

There are essentially two types of **photoreceptor** cells in the retina, namely rods and cones, which derive their name from the shape of their sensitive areas. Rods and cones are not evenly distributed across the retina – for example, the macula contains a high concentration of just cones. The proportion of rods increases with distance from the optical axis, with the result that areas of the retina outside the visual angle of 40° contain only rods. Aside from this variation in rod and cone distribution, the concentration of all sensory cells decreases with increase in distance away from the retina and so too does the density of connections between these sensory cells and the brain. Inside the macula, each cell (here: cone) is connected to the visual centre by one nerve fibre, but with increase in distance from the centre more and more sensory cells are connected to a single nerve fibre. Ultimately, in the outer part of the retina, more than 100 rods and cones are connected to one nerve fibre. As a result, there are only 1 million nerve fibres for the roughly 6 million cones and 100 million rods in the human eye. Apart from photoreceptors and nerve fibres, the retina contains cells for processing the electrical signals, and fine blood vessels.

### *Individual images make up the overall picture*

One consequence of the structure of the human eye is that we mostly see using just 0.02% of the retina, i.e. that portion which is located in the macula. Recent studies have shown that the images of the viewed object are not created on the retina in the manner of a camera, but rather that small eye-movements cause the macula to keep refocusing on new areas. The brain composes all these individual images, which are captured every split second, to form a steady image. In other words, our vision is not due to our eyes fixating on an object, but instead is a composite of a large number of individual images.

## *1.2 The photoreceptor cells in the human eye*

Electromagnetic waves in the visible wavelength range trigger a chemical reaction in the photoreceptor cells, generating an electrical signal, which is transmitted by the nerve fibres to the visual centre in the brain (**colour stimulus**). Photoreceptor cells are around 40 μm in length and 2 μm in diameter. Surprisingly, they are located in the retina in such a way that the light ray from the lens has to pass through the entire cell to reach the photosensitive area. All the nerve fibres stretch between the retina and the vitreous humour and

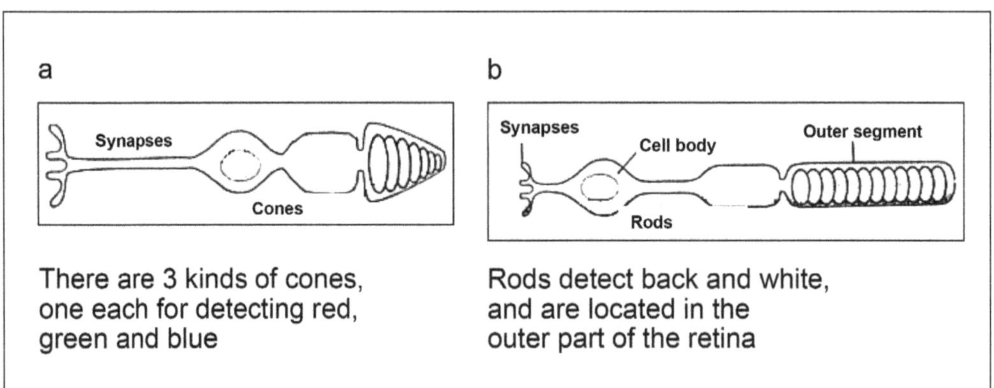


*Figure I.4: Schematic diagram of the (a) rods and (b) cones in the human retina. The light is incident from the left.* *Source: Clariant Produkte (Deutschland) GmbH*

exit to the brain via the blind spot. A schematic diagram of the structure of rods and cones is shown in Figure I.4.

Rods and cones have different functions. **Rods** do not convey any sense of colour, but rather are responsible for our ability to distinguish between light and dark in low light conditions (scotopic vision). They are sensitive to luminance levels down to below 0.1 cd/cm$^2$, and are responsible for our night vision. A single photon is enough to make them fire. As the light intensity increases, the rods become less responsive and the **cones** start to take over. Cones are two orders of magnitude less photosensitive and are responsible for our day vision. The stimuli experienced by the cones ultimately give rise to our colour vision (photopic vision). There are three types of cones in the eye, which sense light in different bands of the visible spectrum. The photosensitive protein molecules (opsins) located in the outer parts of the cones (and the rods) differ in their maximum sensitivity. The stimuli experienced by the three types of cones are processed separately. It is the different stimulation of the three types that creates the perception of colour in our brain.

### *1.2.1 Spectral sensitivity of the receptors*

The advent of micro-spectrometers after 1960 permitted the spectral sensitivity of the rods and the three types of cones to be analysed directly. The cones are named according to their sensitivity to wavelength ranges: short (S), medium (M) and long (L). In the literature, they are also referred to as R (red), G (green) and B (blue). They exhibit maximum sensitivity at 420 nm (S), 530 nm (M) and 560 nm (L). These spectral sensitivities are shown in Figure I.5, and are relative to the overall sensitivity of the fovea. It should be noted that this fundamental data, which will later be used in colorimetric calculations (see Section I.7), was determined on a small number of test persons (17) in the 1920s [7, 8]. The chosen persons were deliberately younger than 30 years old in order that incipient yellowing of the eye-lens and age-related macular degeneration could be ruled out. This shows that everyone's eyes are different and that everyone's colour perception is subjective. The eye sensitivities discussed here are therefore average values for a fairly small test sample (people with normal vision; emmetropia). New tests to broaden the data base are planned.

#### *Lightness and colourfulness*

The three sensitivity ranges, which are shown in Figure I.5, overlap extensively. Rather than being a drawback, this allows us to see colour. The stimuli to which the photoreceptors respond need to be differentiated on the basis of **light intensity** and **wavelength**. A strong signal fired by an L photoreceptor may be caused by light of high intensity at 500 nm (range of low sensitivity) or of low intensity at 560 nm (range of high sensitivity). At each individual wavelength, the signals emitted from the three types of cones are in a fixed ratio. Any signal-processing system can derive information about the light intensity from the absolute signal height while the spectral composition can be determined from the ratio of the different signals. This is how we are able to perceive **lightness** and **colour** separately. Details on human signal processing during vision can be found in *Kaiser* and *Lee* [3, 9]. Our current understanding is that the three cones and the rods generate three types of signal (pairs of parameters) that are transmitted to the brain. These three signals together are called **colour stimulus** and can be represented mathematically as a vector.

It should also be noted that there are substantial differences in the number of S, L and M cones in the retina. That is to say, there is no defined pattern equivalent to the red, green and blue pixels on TV screens. Recent studies assume that the frequency distribution

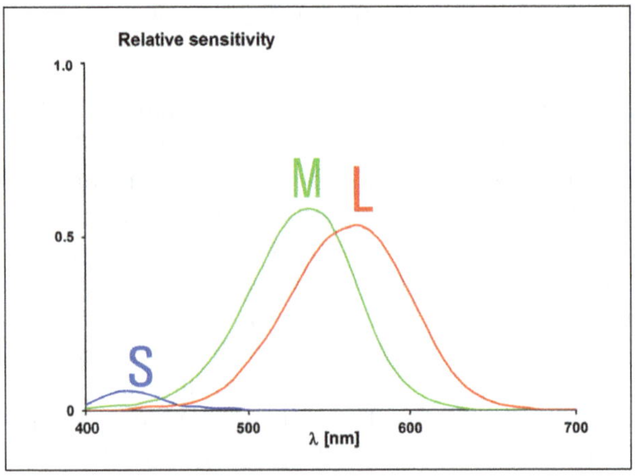


*Figure I.5: Sensitivity of the three types of cones S (blue), M (green) and L (red) as a function of wavelength. Types M and L overlap extensively, with their maxima occurring at 540 nm (M) and 650 nm (L).*

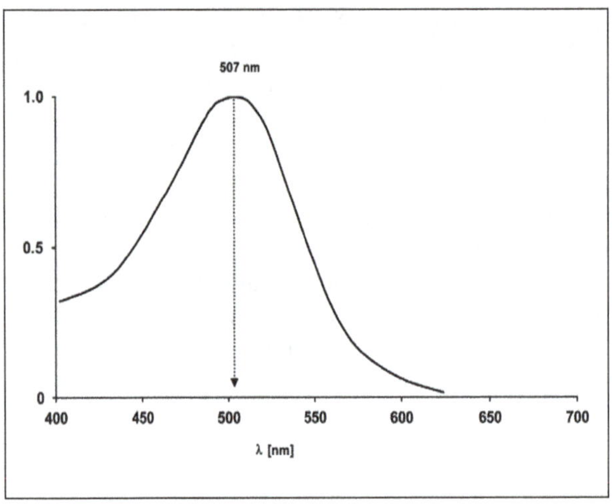


*Figure I.6: Spectral sensitivity of human photoreceptor rods. Their maximum sensitivity occurs at 498 nm, i.e. between those of the S cones and the M cones.*

for the S, M und L receptors is 1:3:6 [10]. One reason for the relatively small number of S cones might be chromatic aberration of the eye-lens. As its focal point varies with the wavelength, it cannot accurately focus the entire visible spectrum on the retina. The eye is likely optimised for the M and L cones whose maxima are closer together while a blurry image will reach the S cones. Consequently, a fine distribution is not needed for this wavelength range.

Consider now the rods, which fire even at low light levels. Their light-sensitive pigment is rhodopsin, which mainly absorbs in the blue-green range. Its sensitivity (response) curve is shown in Figure I.6. The colour stimulus of the rods cannot distinguish between light of high intensity in the low-sensitivity region (e.g. above 600 nm) and of low intensity at the sensitivity maximum at 507 nm. As a result, humans cannot see colour differences in low light, which is why “at night all cats are grey”. In the transition region from rod vision to cone vision (twilight), the greater sensitivity of the rods in the blue range gives rise to the belief that blue objects become brighter than in broad daylight. This is called the Purkinje effect.

While humans possess trichromatic vision, there are animals which have just 2-cone and even 4-cone sensors (tetrachromatic vision). Some birds, reptiles and fish have cones, which are sensitive in the UV range.

### *1.2.2 Visual defects*

Our vision changes as we get older: the eye-lens tends to lose its elasticity and become yellowish. Again, some 8% of males and 0.5% of females are born with visual defects (ametropia). Their cones may exhibit different sensitivity or even none at all. This kind of colour blindness can be diagnosed with the aid of special colour displays called Ishihara plates that exploit the effect of colour confusion. Another way to test for colour blindness is to use an anomaloscope. With this instrument, the patient uses a set of three primary-colour light sources to match a given reference colour (additive colour mixing, see Section I.3.1).

# 1.3 Colour perception

## 1.3.1 Chromophoric attributes

The names we give to colours and our perceptions of them are mainly determined by our cultural history. This would explain why different cultures have different views on colour.

As early as 1850, *Grassmann* and *von Helmholtz* described three distinct characteristics of human colour perception. This **three-colour theory** or **trichromatic model**[1] was posited around one hundred years before experimental proof of the existence of three different cones in the retina was found. Some years later, in 1878, *K. Hering* published his **opponent colour theory** according to which our colour perception is based on pairs of hues, namely red/green and blue/yellow, as well as on black and white. This model has since been verified experimentally.

Our perception of colour is based upon three fundamental attributes:

- **Brightness** or, even more important, lightness
- **Hue**
- **Chroma** or **colourfulness** (chromaticness), and **saturation**

Current definitions of all these terms are presented in more detail in ISO 11664 Part 1 and DIN 5033 Part 1 [1, 12]

**Chroma** is a measure of how much a colour stimulus differs from that of an uncoloured surface of the same lightness. As in the case of **lightness,** the human eye evaluates sensory stimuli on a relative basis. Chroma can therefore be thought of as relative colourfulness. Finally, by expressing chroma as a ratio of lightness, we obtain the **saturation**.

## 1.3.2 Colour constancy

The human brain does not convert transmitted colour stimuli directly into a colour perception, but instead automatically takes account of the conditions in which the observations are made. The raw colour signals themselves do not penetrate into our consciousness.

Colour vision incorporates the concept of colour constancy. Our visual interpretation of a hue always takes ambient lightness into account. For example, the **illumination level** outdoors in direct sunlight is several orders of magnitude greater than it is inside. The stimuli reaching our eyes outdoors should lead to white-out of any object, which would only appear white. The combined input from the eyes and the brain leads to a recognition that the different stimulus triggered by a green apple, say, stems from a change in lighting conditions and not from a change in the hue of the apple itself. Colour constancy is therefore based on the notion that the lightness of an object is determined by the lightness of the environment or of a reference object, and our colour perception changes accordingly ("related colours", see also Section I.1). For a good example, consider our visual perception of the moon. If we look at it in the afternoon, it appears pale yellow against the blue of the sky. Some hours later, in the dark, we perceive it as being bright yellow even though its lightness, which is the result of illumination by the sun, has not changed much during this short time. A further special effect is that we can make out the details of the moon's surface at night.

[1] *The basic idea behind this concept was published in 1807 by Thomas Young*

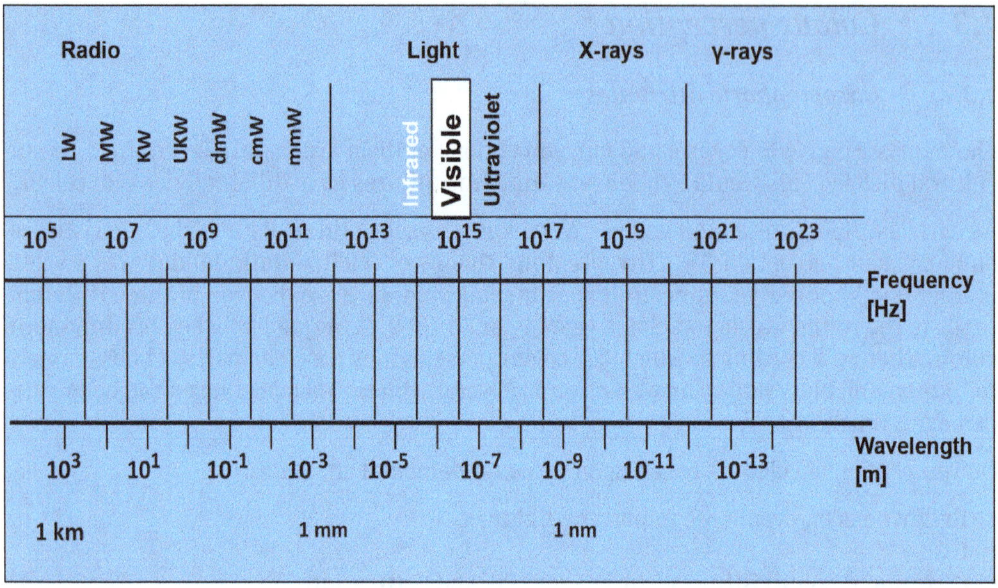


*Figure I.7: Wavelength and frequency of electromagnetic radiation on a logarithmic scale, along with different applications. Only wavelengths in the narrow range from 400 to 700 nm are detectable by the human eye.* Source: Clariant Produkte (Deutschland) GmbH

## 2 Light as Electromagnetic Radiation

The light perceived by the human eye consists of electromagnetic radiation. Figure I.7 shows the different wavelength and frequency ranges of electromagnetic radiation on a logarithmic scale, along with several applications in these ranges. The human eye can only detect radiation in the tiny wavelength range of 370 to 700 nm (1 nm = $10^{-9}$ m). Due to the low sensitivity of our eyes outside this range, colour measurements are made over the range 400 to 700 nm by international agreement.

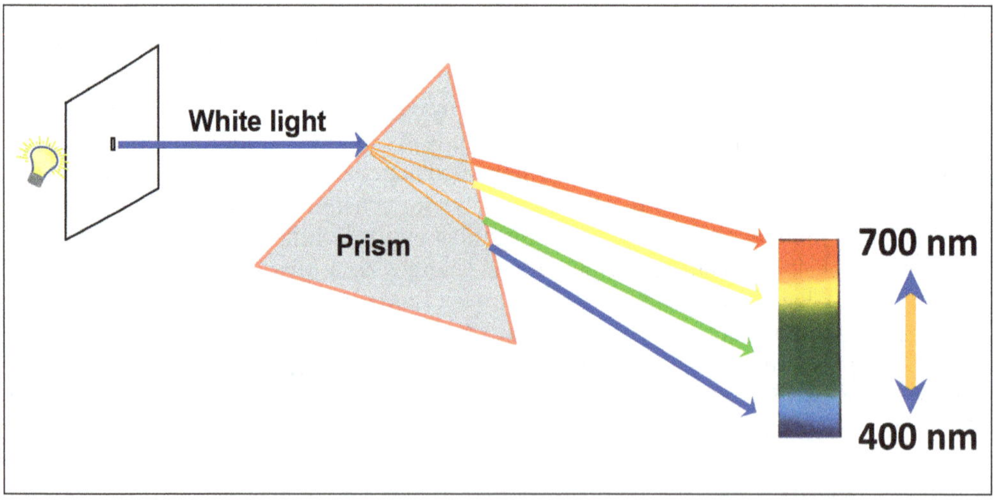


*Figure I.8: A prism disperses the white light into its components: light of different colours and wavelengths (Newton, 1766). A diffraction grating could be used instead to achieve the same dispersion.* Source: Clariant Produkte (Deutschland) GmbH

The relative spectral power S (λ) of the radiation usually differs from wavelength to wavelength, and gives rise to different colour perceptions. In colorimetry, the relative spectral power is usually referenced to 560 nm. Back in 1766, ***Isaac Newton*** demonstrated in his prism experiment (see Figure I.8) that white sunlight is a combination of different colours. In other words, there is no wavelength for white light. Light of different wavelengths creates different colour perceptions in humans. Radiation with a wavelength ranging from 450 to 490nm appears blue, while that in the range 490 to 560nm is seen as green, and wavelengths longer than 630nm are perceived as red. The adjacent wavelength ranges are ultraviolet (<400nm) and infrared (>700nm). Electromagnetic radiation itself is colourless; it is human sensory perception, which converts the various spectral power distributions into the concept of colour in our brains. The spectral power distribution emitted by an object carries the information about the colour of the object.

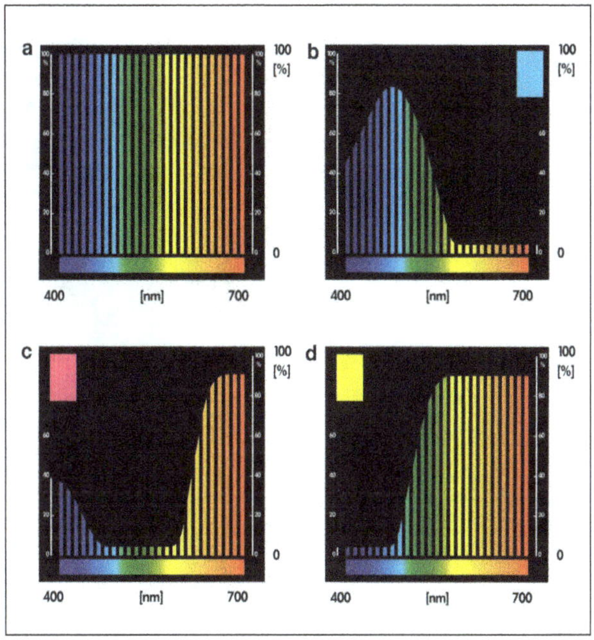


*Figure I.9: Schematic diagram of a spectrum of ideal white (a), blue (b) and yellow(d). Mixing blue and red creates violet (c). A bar chart symbolizes the colorimetric measurement of the radiation distribution at discrete sampling points. Measurements, e.g., every 20 nm yield 16 sampling points in the range 400 to 700nm.* Source: Datacolor AG

The intensity distribution of the radiation is referred to as a **spectrum.** The curve of this distribution is called a **remission curve** or ***reflectance curve R(λ)***. Further details on reflection and remission are presented in SectionsI.4 and VII.2. Only the term reflectance curve is used in daily life. Figure I.9 shows the intensity distributions of a blue (Figure I.9b) and a yellow (Figure I.9d) hue. The colour distribution in a rainbow or in Isaac Newton's prism experiment is created by shifting an intensity maximum along the visible wavelength range, whereby the colour curve changes from blue to green to yellow and finally red. The steeper or greater the slope of the intensity maximum, the purer is the hue. In theory, a spectrum could be based on an abrupt, rectangular distribution where the intensity jumps from 0% to 100%. Such (theoretical) colours are called **ideal colours**.

A constant intensity distribution from 400 to 700nm would be perceived as achromatic. Depending on the absolute intensity value, an ideal white, ideal grey or black would be created (Figure I.9a). A single intensity maximum is not the only way to create a colour - the radiation intensity in the spectrum can be any shape. That is how other colours are created in addition to the colours of the rainbow. For example, violet is not seen in the rainbow but is created by a spectrum having maxima in both the blue and the red wavelength ranges. Figure I.9c shows an example of such a spectral power distribution containing a mixture of blue and red. This intensity distribution is perceived as violet. The two methods for mixing colours are presented in Section I.3.

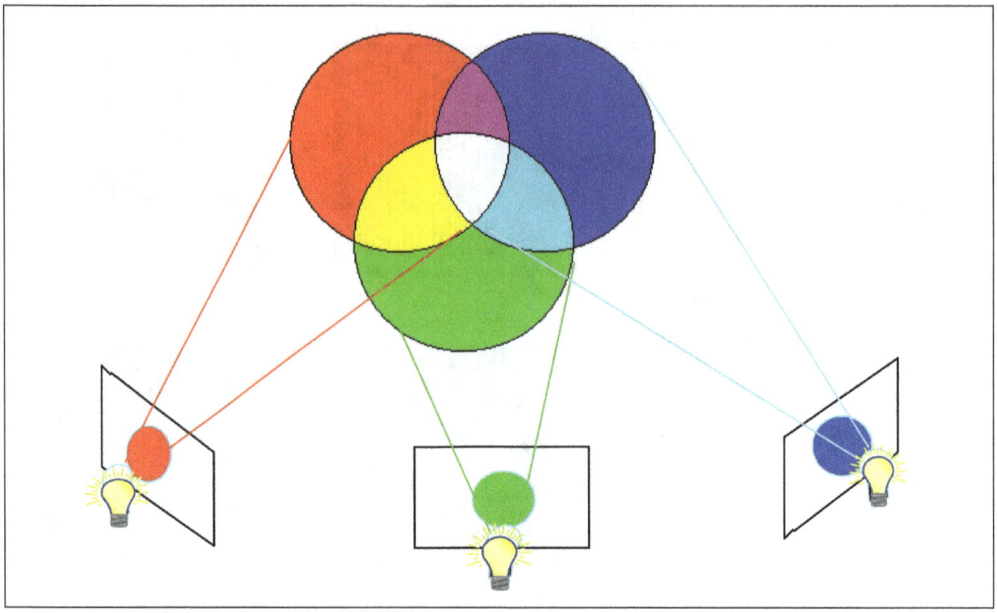

*Figure I.10: In additive colour mixing, a combination of all three primaries yields white. This type of colour mixing occurs most often with coloured lights.* Source: Clariant Produkte (Deutschland) GmbHa

# 3 Colour mixing

Colour mixing is important in daily life because commercial colourants (pigments and dyes) do not cover the full range of all perceptible colours. The technical background to matching colours to a reference sample (recipe calculation) is discussed in detail in Section XI. The basics are presented below.

## 3.1 Additive colour mixing

There are basically two kinds of colour mixing. In additive colour mixing, different hues add up to eventually yield white. This is shown schematically in Figure I.10. Probably the easiest example of this type of mixing occurs in every TV set. The screen consists of separate red, green and blue pixels. Varying the intensity of these three pixels allows nearly every colour perception to be created in the viewer's brain. Adding red and green together, for example, produces yellow. Additive colour mixing is thus the concept behind the mixing of coloured light. Where all three pixels have the same intensity, white is created. Red (R), green (G) and blue (B) are called the primary colours (primaries) of additive colour mixing. Complementary colours are two coloured lights that add up to white (see also Grassmann's 1st law).

## 3.2 Subtractive colour mixing

When we experimented with paints as children, however, we had a different experience. Mixing red and green produces a grey-brown colour – not yellow as described above. Mixing of pigments, i.e. colourants, usually follows the rules of subtractive colour mixing. Consider the fundamentals of colour creation. The red colour of a pigment is based on the absorption of blue and green wavelengths whereas green is created by the absorption of red and blue wavelengths. Mixing of red and green colourants will lead to absorption of nearly

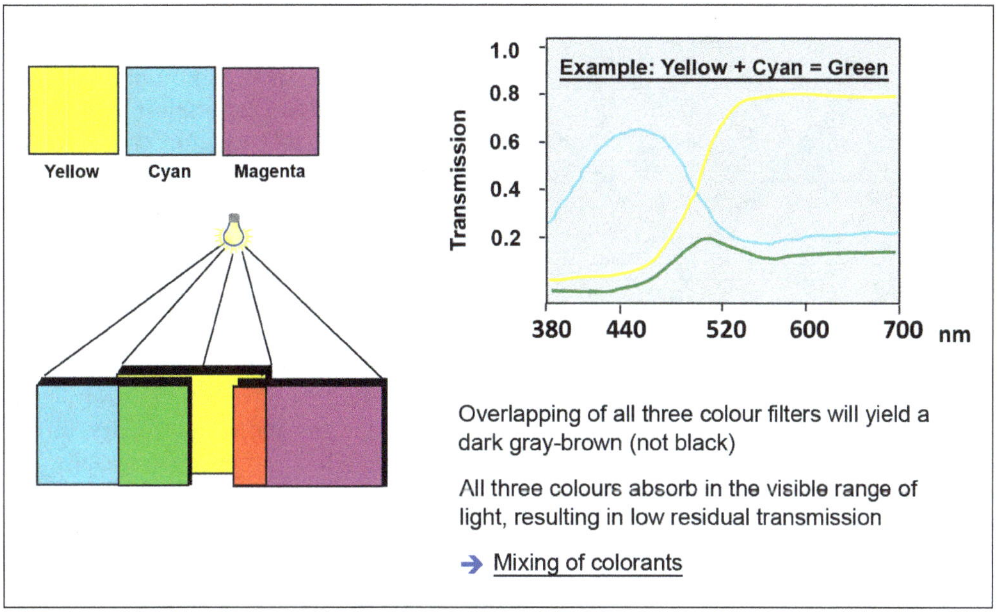


*Figure I.11: Subtractive mixing of cyan and yellow pigments yields green (a). The spectra (b) illustrate how the green reflectance curve is generated by the subtractive mixing of cyan and yellow reflectance curves.* Source: Clariant Produkte (Deutschland) GmbH

the whole visible spectrum, leaving only grey-brown. The term "subtractive" is somewhat misleading. A better term would be "the addition of the absorption effect of colourants." The primaries for subtractive colour mixing are cyan (C), magenta (M) und yellow (Y), and are especially important in the printing industry. They add up to yield a kind of black (in the printing industry a really good black (K [for key]) is included by way of a fourth colour component). Subtractive mixing of cyan und yellow is illustrated in Figure I.11. This reveals another important aspect of this type of colour mixing, namely that subtractive mixing mostly yields spectra of low intensity (darker colours) and flatter slope. As a result, the mixed colour has less chroma (and less brilliance) than the original colours.

Both kinds of colour mixing can take place simultaneously. For example, coloured pictures can be created by printing several colours on top of each other (subtractive mixing) and by printing pixels of several colours beside each other. We must remember that the process behind colour perception in the human brain is different from the technique used in colour reproduction. In daily technical communications involving colour, the RGB/sRGB and CMYK colour spaces spanned by their primaries play an important role. They are not the same as the general colour spaces presented in Chapter V. The range of colours that can be produced by a colourant system is called the colour **gamut**.

# 4 *Interaction of light and matter*

When light strikes matter (material), the interaction that occurs depends on the nature of the matter. And this will also influence the colour of the light.

**Transmission**: Transmission occurs when the light beam passes through the material without modification. Such a material is said to be transparent. In addition, if the material is also colourless, the light beam will exit the material with the same intensity distribution (spectrum). Only a small quantity is reflected at the point where it enters and exits the material (interface of material and air or vacuum).

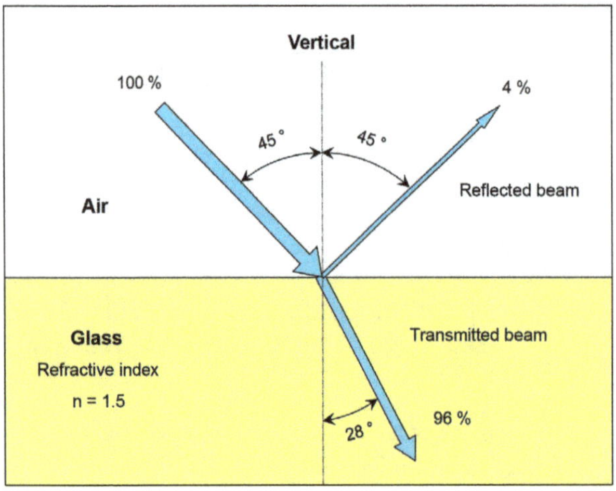


*Figure I.12: At the interface between air or a vacuum and glass, some of the light is reflected, but most is refracted towards the normal to the interface surface. Refraction is caused by the lower speed of light in matter.* Source: Clariant Produkte (Deutschland) GmbH

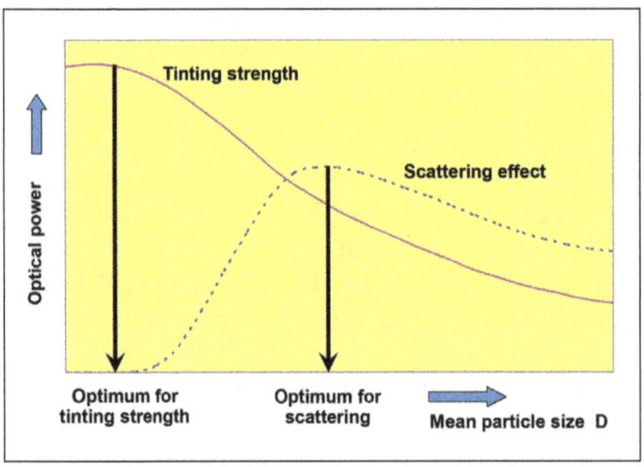


*Figure I.13: Relationship between the scattering effect of a pigment and its particle size. Scattering is at a maximum when the particle size is identical with the wavelength. Tinting strength exhibits a different dependence on particle size. On account of this, both properties can be varied during pigment production.* Source: after L. Gall

**Reflection:** The amount reflected depends on the difference in the speeds at which the light travels in the two materials. Light travels faster in a vacuum or air than it does in matter. The ratio of these two speeds is called the refractive index. Normal glass has a refractive index of 1.5 and reflects around 4% of the light incident on its surface. This type of reflection occurs at the interface of materials that have different refractive indices and, as a corollary, different light-propagation speeds.

**Absorption:** Electromagnetic radiation (light) can be partly or wholly absorbed as it passes through a material. If all of it is absorbed, the material is said to be **opaque**. The amount of radiation absorbed depends on the electronic structure of the material and whether the wavelength of the radiation matches the excitation energy needed to cause an electronic transition. For this reason, different wavelengths within the radiation will be absorbed to different extents. Back in 1760, *Lambert* established that the extent of absorption by a transparent medium varies with the thickness of the medium. The dependence of absorption on the concentration of the absorbing material is known as *Beer's* law. The *Lambert-Beer law* is described in more detail in Section VII.3.

**Scattering:** When a light beam penetrates a material, it can also be scattered. Although the light changes direction in this case, its wavelength remains the same. As a result of this kind of interaction, the light beam may even be partially scattered out of the material (i.e. re-emitted). Scattering can take place repeatedly. These processes all occur in the first few micrometers beneath the surface of a material. The white colour of milk, clouds and white pigments are all familiar examples of scattering. The blue colour of the sky is also the outcome of light scattering by air molecules, although refractive index also plays an important role in this interaction.

The greater the difference in the refractive indices of two materials when a light beam passes from one to the other, the more the light is scattered.

The amount of scattering is also influenced by particle size, with particles the same size as the wavelength causing the most, and larger particles the least [13]. As the particle size of pigments can be manipulated during production and further processing, it is possible to produce pigments of the same colour index for use in different kinds of applications, pigments which scatter more and thus have greater hiding power or are more transparent, e.g. for printing inks. Varying the particle size of the pigments therefore enables both the hiding power and the hue to be modified. Consequently, pigment manufacturers can produce various pigments of different **hiding power** and different **hues** under the one **Colour Index** (see Chapter X). For the same reason, pigments from different suppliers having an identical colour index can exhibit different hues in application.

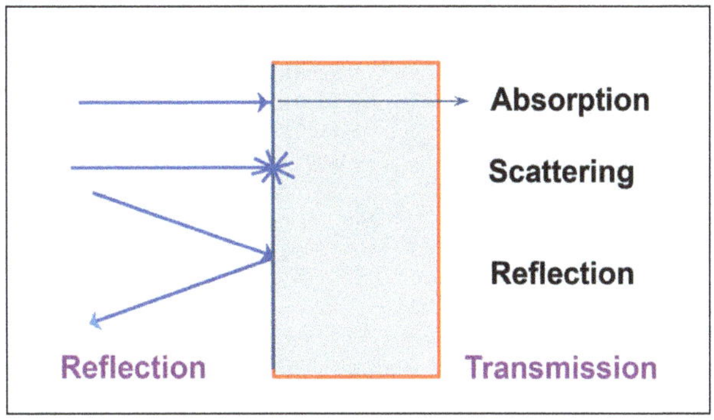


*Figure I.14: Three different kinds of interaction are possible when light strikes matter. The absorption effect is described by the coefficient K(λ) and the scattering effect by S(λ). These are discussed in more detail in Section VII. 3 ff.*
Source: Clariant Produkte (Deutschland) GmbH

The scattering properties of a material determine whether it is transparent, translucent or opaque. An opaque material does not transmit any light at all. The various types of interaction that occur when light strikes matter is illustrated opposite.

# 5 Standard illuminants and light sources

Everybody knows that the perceived colour of an object depends on the nature of the illumination. On one hand, the amount of illumination entering the human eye can be varied extensively by the "human aperture", i.e. the pupil. However, our ability to see colour vanishes if the radiation intensity is too low (night vision) or too high, in which case we are blinded. On the other hand, it is known that the perceived colour also depends on the nature of the light source. An object is perceived to have a different colour at sunset than under the full mid-day sun; the colour perception created by a light bulb is different from that created by sunlight - there is also the complication that an object may be illuminated with coloured light. This means that it is essential for us to state the type of illuminant when we report the results of a colour assessment, whether visual or colorimetric (see DIN 6174). This is a consequence of the fact that different kinds of illumination are described as being white illumination. The reason for this is that we humans perceive the various spectra emitted by different light sources all to be white. This is also known as metamerism and is discussed in more detail in Section I.9.

## 5.1 Standard illuminants

**Standard illuminants** are defined and published by the International Commission on Illumination (usually abbreviated CIE for its French name). These are theoretical spectral power distributions, irrespective of whether the actual light sources for them are technically available. The first standard illuminants were introduced in 1931 and 1964. At that time, even the spectral power distributions of natural sources were standardised. Direct sunlight was represented by **standard illuminant B** and average daylight by **standard**

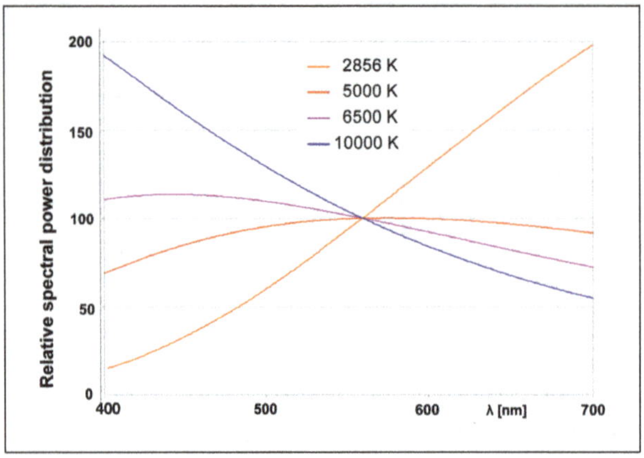


*Figure I.15: Schematic diagram of the spectral power distributions of blackbody radiators at 4 different colour temperatures.* Source: Clariant Produkte (Deutschland) GmbH

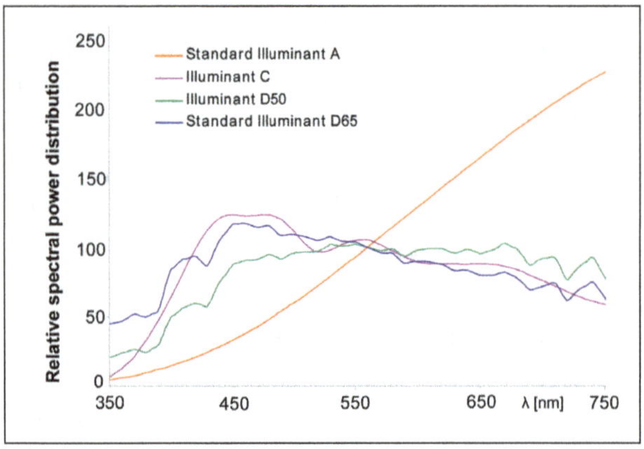


*Figure I.16: Relative spectral power distributions of CIE standard illuminants A, $D_{65}$ and $D_{50}$.* Source: Clariant Produkte (Deutschland) GmbH

**illuminant C**. Since then, CIE has focused on standard illuminants based on theoretical spectral power distributions of blackbody radiators. Illuminants B and C are no longer standardised. Blackbody radiators have a continuous spectrum without any peaks or gaps. The relationship between temperature and radiant power is given by Planck's radiation law.

It states that the spectrum of an ideal blackbody radiator is directly related to its temperature – we have a colour temperature. As the temperature is stated in Kelvin (K), the absolute temperature scale, we must add 273degrees to a Celsius temperature to arrive at its equivalent Kelvin temperature.

A spectral power distribution S (λ) shows the wavelength dependence of the radiation. For the standard illuminants, the spectral power distributions are astated relative to the value at 560nm. Comparison of different power distributions shows that they all intersect at 560nm. Figure I.15 presents the relative power distributions S (λ) for blackbody radiators at 2856K, 5000K, 6500K and 10,000K over the range 400 to 700nm. As the colour temperature rises, the proportion of blue, short-wavelength power rises.

The current versions of CIE Report 15 [11] and ISO 11664 Part 2 [12] define standard illuminants A, $D_{65}$ und $D_{50}$ on the basis of blackbody radiators. They contain tables of the spectral power distributions, which are shown in Figure I.16. Note that the standard illuminants are defined in the range of 300 to 780 nm, which is wider than the sensitivity range of the human eye. The wider range allows, e.g., specific effects in the ultra-violet range that affect the visible range (e.g. fluorescence, optical brighteners) to be investigated with a standardised illuminant. Standard illuminant A represents a light bulb having a filament with a temperature of 2856K. D65 describes average natural daylight on the Earth – the surface temperature of the sun is 5800K, but absorption effects in the atmosphere raise the colour temperature to 6504 K. D50 is an illuminant mainly used in the graphics industry.

The other type of CIE standardised illuminant is the fluorescence emitted by fluorescent lamps. Officially, CIE has defined the two illuminants F2 (cool white fluorescent/CWF) and F11 (three-band fluorescent source/TL84). These illuminants consist of a small continuous background portion overlaid with a line or a band spectrum. The relative spectral power distributions of these two standard illuminants are shown in Figure I.17. Further information on these kinds of illuminants is presented in the next section.

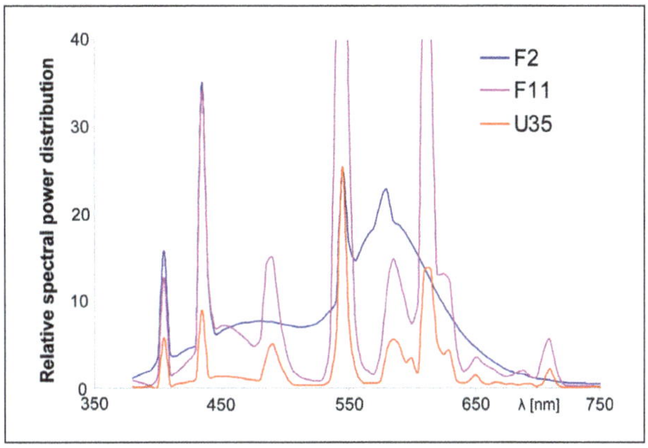


*Figure I.17: Relative spectral power distribution of the CIE standard fluorescent illuminants F2 and F11 and the energy-saving lamp U35.* Source: Clariant Produkte (Deutschland) GmbH

CIE Report 15 also lists tables of spectral power distributions for further illuminants, but these have not been officially standardised. They include, for example, the aforementioned natural illuminants B and C. New illuminants will be developed, launched onto the market and standardised in the future. In the European Union, incandescent light bulbs are banned, and so illuminant A will disappear from our daily lives, to be replaced by energy-saving lamps, such as type U35. Standardisation of the LED illumination spectra is a more recent topic. LEDs are a high-growth area and the standardisation process is underway.

All colorimetric measurements based on light sources actually available must be converted to the standard illuminants before they can be used in further calculations. This is done by calibrating the spectrometer by comparing the readings for a calibration tile with the saved theoretical values for that tile. The software then calculates the correction factor for each value on the spectrum. Thus, a single measurement allows the data for all standard illuminants to be determined simultaneously. Further details of this calibration process are given in Section II.2.

## 5.2 *Light sources*

Light sources used in colorimetry can be classified into two main groups: **thermal radiators** and **luminescence radiators**. The latter can be further sub-classified into gas-discharge lamps and electroluminescence lamps.

### *Thermal radiators*

In a thermal radiator, a single-coil filament is electrically heated to a temperature at which it emits a continuous spectrum of electromagnetic waves. The spectrum varies with the temperature, i.e. with the applied electric current. As the temperature rises, the colour of the emitted light shifts from dark red to yellow to white. Such light sources approximate a blackbody radiator. The best known source for this is the tungsten filament coil which, when heated to 2856K, can replicate standard illuminant A. Apart from the metal used, the emitted spectrum is defined by the gas inside the glass bulb (krypton, argon or halogen (bromine)). This gives rise to a certain amount of deviation from the theoretical blackbody radiator. To compensate, the concept of the correlated colour temperature has been introduced. A (correlated) colour temperature can be assigned to luminescence radiators in the same way.

### *Gas-discharge lamps*

In gas-discharge lamps, free charge carriers are accelerated by an electric field. This excites the atoms as well, inducing them to emit light of characteristic wavelength. Increasing the gas pressure in the system can trigger other kinds of interactions, giving rise to

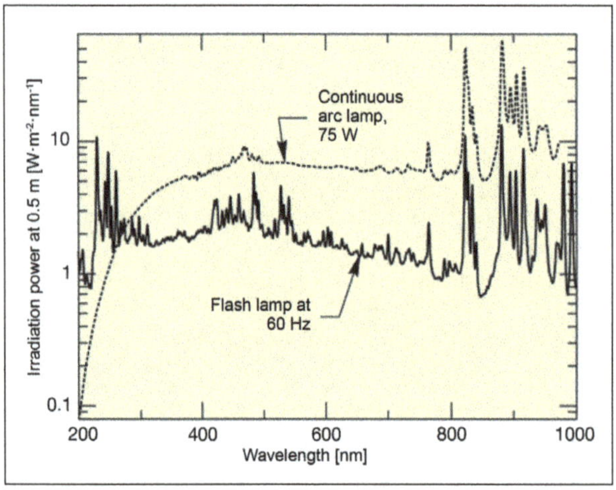


*Figure I.18: Comparison of the spectral power distributions of a continuous xenon arc lamp and a Xe flash lamp. The latter is more common in large spectrophotometers.* Source: Datacolor AG

a continuous underground to the line or band spectrum. Thus, gas-discharge lamps can also be used in colorimetry.

### *Standard illuminant $D_{65}$*

Standard illuminant D65 is usually replicated with a xenon discharge lamp because the emission spectrum of the high-pressure type largely coincides with that of standard illuminant D65. A few extraneous peaks are eliminated with filters. Nowadays most devices are based on xenon flash lamps because these do not need to be warmed up for 30 to 60 minutes before measuring starts. They are also safer because high pressure is not used. However, the flash lamps have a smaller proportion of continuous spectrum and more extensive filtering of the line spectrum is required.

### *Fluorescent sources of light*

Low-pressure mercury vapour lamps usually serve as the light source for fluorescent illuminants. They emit a line spectrum in both the short-wavelength visible range and the UV range. The UV emission is used to stimulate fluorescence in the phosphors on the glass-wall coating. The phosphors employed determine whether standard illuminant F2 (CWF) or F11 (three-band fluorescence/TL84) is emitted. The same process is also used to illuminate light booths. The popular name “neon lamp” to describe these light sources is a misnomer.

### *LED light sources*

Soon after light-emitting diodes (LEDs) of various wavelengths were developed, it was proposed that they be used to replicate the standard illuminants. Since short-wavelength, i.e. blue, LEDs already existed, this idea was soon adopted. The spectrum emitted by LEDs is not continuous, but merely an overlapping of numerous line spectra. Thus, the calibration software has to use correction factors to smooth the peaks and troughs to replicate the standard illuminants. For this reason, LED light sources cannot be used to visually assess samples under standard illuminant conditions. The advantages of LED light sources are their low energy consumption and small size. They are therefore chiefly used in handheld spectrometers.

Several technologies have been developed for producing LEDs that emit white light for use around the house. However, the differences in the technologies have hampered standardisation. These “white” LEDs have a high content of blue radiation and thus a very high colour temperature.

## 6 Standard observer

In Section I.1, we discussed the structure of the human eye. We talked about the rods and the cones, their relative numbers and their spatial distribution about the retina. We also looked at the various wavelength-dependent sensitivities of these “sensors”. In colorimetry,

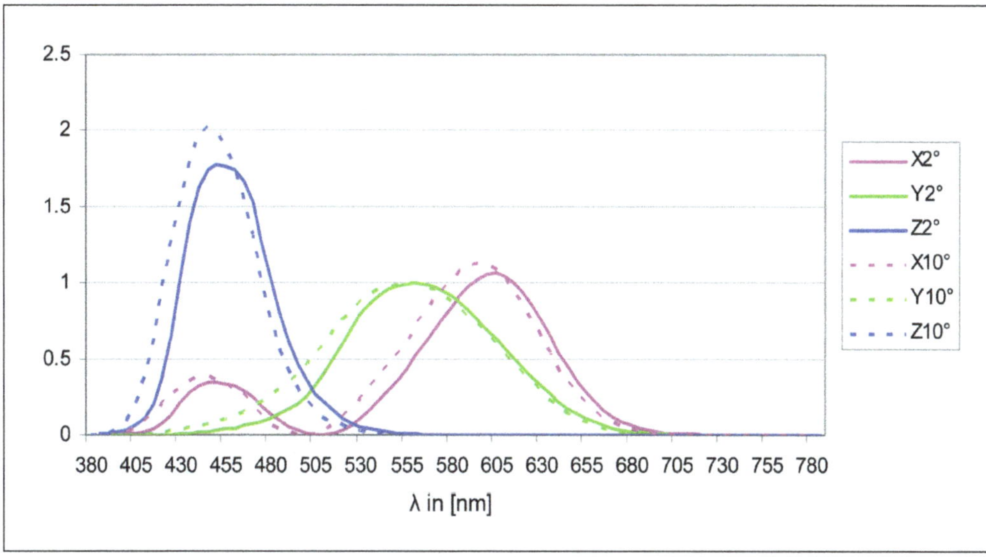


*Figure. I.19: The colour-matching functions $\bar{x}(\lambda)$, $\bar{y}(\lambda)$ and $\bar{z}(\lambda)$ are a mathematical characterisation of the visual sensation of the human eye. Their values were defined by the CIE. Two variants of the functions exist, i.e. the 2° and 10° standard observer, and the choice between them depends on the size of the assessed sample. The dashed lines indicate the data for the 10° observer. The three interlinked functions are scaled so that the maximum for $\bar{y}(\lambda)$ is a value of 1.* Source: Clariant Produkte (Deutschland) GmbH

the human eye is replaced by a mechanical sensor and accordingly the eye's sensitivity must be transformed into a mathematical model. In this model, human colour perception is replaced by the concept of response curves. This was introduced in 1931 by the CIE. As in human signal processing, the number of signals (three cone types + one rod) is reduced to three.

### *2° standard observer/1931 CIE standard observer*

The concept of the 2° standard colorimetric observer draws on the experiments of Wright and Guild[7, 8]. It is the sensory perception which a small observed object triggers in the fovea only (see Section I.2). The three response curves $\bar{x}_2(\lambda)$, $\bar{y}_2(\lambda)$ and $\bar{z}_2(\lambda)$ are called colour-matching functions. Their values are defined in CIE Report 15[11] and were officially standardised in ISO 11664-1[12] in 2008. Figure I.19 shows a plot of these functions. They are defined such that the $\bar{y}2(\lambda)$ curve represents the lightness perceived by humans. In addition, the maximum of this function is chosen so as to have a value of 1 at 550 nm. A visual angle of 2° roughly corresponds to an object the width of a thumbnail, held at arm's length.

### *10° standard observer/1964 CIE standard observer*

However, the samples we usually assess in daily work are larger. So, in 1964, the 10° standard observer was introduced (also known as the 1964 standard observer). The 10° visual angle in this case corresponds to the width of the palm of the hand, held at an arm's length. This matches most paint drawdowns prepared in a coatings laboratory. The colour-matching functions for the 10° observer $\bar{x}_{10}(\lambda)$, $\bar{y}_{10}(\lambda)$ and $\bar{z}_{10}(\lambda)$ are illustrated by the broken lines in Figure I.19. The small variations in response are a consequence of the different distributions of S, M and L cones and the inclusion of rods for the 10° observer. At a 10° visual angle, areas of the retina beyond the fovea are stimulated. The 10° experiments performed back then excluded the 2° area of the central fovea. In daily practice, the 10° observer is normally implied, unless the 2° observer is specifically mentioned.

These response curves illustrate the vanishingly small response of the human eye to wavelengths shorter than 400 nm and longer than 700 nm. Accordingly, colorimetric calculations are restricted to the range 400 to 700 nm. It should be mentioned that the tests performed in 1930 collected data from just 17 persons [7, 8] (see Section I.3.2) and the results exhibit considerable scatter. Thus the currently defined colour-matching functions represent an average and the differences between the 2° observer and the 10° observer are smaller than the scatter of the raw data upon which these functions are based.

# 7 CIE 31 system

Let us review what we have learned so far. We have seen that the triple play of light source - object - observer (see Figure I.1) is necessary for colour perception. We discussed light sources and their standardisation to standard illuminants (Section I.5), examined the different types of interactions that occur when light strikes matter (Section I.4) and explained the structure of the human eye and the basics of colour perception (Section I.1). This flow of information is summarised on the left side of Figure I.20. The right side illustrates the analogous situation in colorimetry. Here, we use numbers as a substitute for colour perception in the human eye and signal processing in the brain. The colour-matching functions $\bar{x}(\lambda)$, $\bar{y}(\lambda)$ and $\bar{z}(\lambda)$ proposed by CIE in 1931 and 1964 and introduced in Section I.6 are the first combination of mathematical functions for representing colour perception in humans. Together with the measured reflectance curve $R(\lambda)$, a triplet of numbers can now be assigned to each colour. These numbers (X, Y and Z) are called **tristimulus values**.

## 7.1 Calculation of tristimulus values

The various parameters needed for calculating tristimulus values were introduced in the previous sections. Thus, $S(\lambda)$ of the standard illuminant represents the spectral power distribution of the incident radiation from the theoretical light source illuminating the object. $R(\lambda)$, the reflectance factor at wavelength $\lambda$, represents the reflected light that catches the eye. The eye itself is represented by the colour-matching functions. These functions either are defined by the CIE and listed in tables at specified wavelength increments (e.g. every nm, every 5 nm or every 10 nm) or $R(\lambda)$ is measured by the spectrophotometer in those nm increments. Modern spectrophotometers mostly measure in 10 nm increments, and so yield 31 data sets for the range 400 to 700 nm. The visual perception of the human eye is therefore based on the sum of each of the products $S(\lambda) \cdot R(\lambda)$ for all wavelengths. Note that the more mathematically correct integration is replaced by the sum of the 31 products. Multiplication of this result by the colour-matching functions yields the **tristimulus values**. The product $S(\lambda) \cdot R(\lambda)$ is called the stimulus.

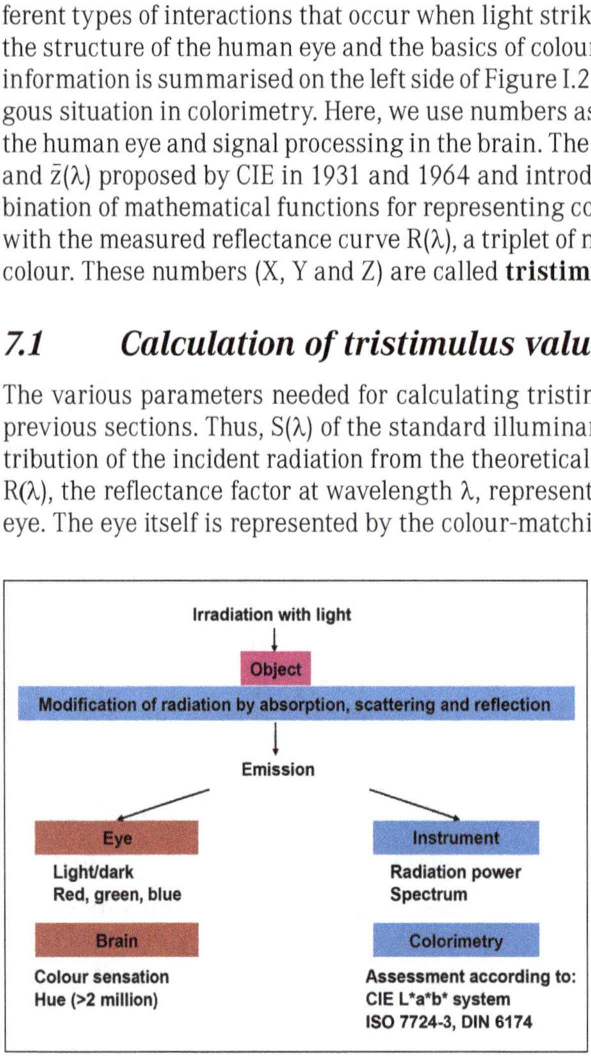


*Figure I.20: Sketch of the information flow involved in colour perception and colour measurement. For the latter, visual perception and signal processing in the human brain must be replaced by mathematical functions. Tristimulus functions form the basis for these functions.* Source: Clariant Produkte (Deutschland) GmbH

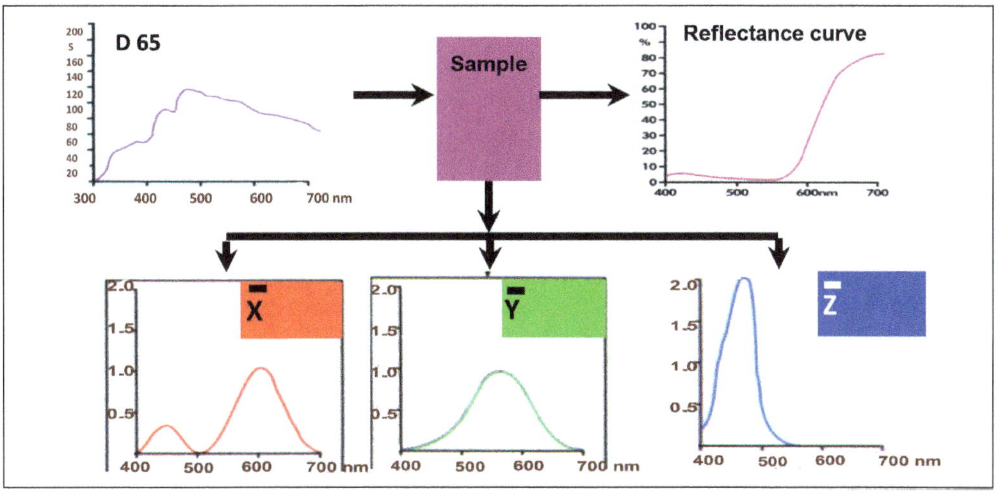


*Figure I.21: Schematic diagram of the process for calculating tristimulus values obtained by summing the product of the spectral power distribution S(λ) of the illuminant, the reflectance factor R(λ), and the colour-matching functions in the range 400 to 700nm. Given 10nm increments, this means that 31 products must be added up for each of X, Y and Z.* Source: Clariant Produkte (Deutschland) GmbH

*Equation I.1*

$$X = k\,\Sigma\,S(\lambda)\cdot R(\lambda)\cdot \bar{x}(\lambda)$$
$$Y = k\,\Sigma\,S(\lambda)\cdot R(\lambda)\cdot \bar{y}(\lambda)$$
$$Z = k\,\Sigma\,S(\lambda)\cdot R(\lambda)\cdot \bar{z}(\lambda)$$

Summation is applied over the range 400 to 700 nm in each case, with the factor *k* introduced for the purpose of standardisation; for a pure (ideal) white, therefore, Y = 100. In Section I.6, it was pointed out that colour-matching function $\bar{y}(\lambda)$ also serves as a measure of the lightness perceived by humans.

*Equation I.2*

$$k = 100/\Sigma\,S(\lambda)\cdot \bar{y}(\lambda)$$

For this reason, Y is sometimes also called luminance. This type of standardisation officially applies only to the 2° observer (1931), but is also used for the 10° observer because the deviations are very small. Since S(λ) is different for each standard illuminant, *k* also varies for each illuminant. Figure I.21 illustrates the process for calculating the tristimulus values X, Y and Z.

Ideal white represents a sample that reflects 100% of each incident light wavelength. $Y_n$ is the notation employed for all standard illuminants as they are also white. Accordingly, Y = 100 cannot be exceeded. Exceptions to this rule are fluorescent samples that absorb energy at wavelengths <400nm and re-emit it in the visible range. More precisely, it should be mentioned here that for the 10° observer the official designations are $X_{10}$, $Y_{10}$ and $Z_{10}$, in correspondence with the colour-matching functions $\bar{x}_{10}(\lambda)$, $\bar{y}_{10}(\lambda)$ and $\bar{z}_{10}(\lambda)$. The standard illumination used in the measurement must also be cited.

From a mathematical point of view, integration would be more precise and X, Y and Z would be represented by the area under the curves. For X, this would take the form:

*Equation I.3*

$$X = k\int S(\lambda)\cdot R(\lambda)\cdot \bar{x}(\lambda)\,d\lambda$$

Integration is done over the range 400 to 700nm in increments of d λ (e.g. 10nm). In the literature, an integration range of 380 to 760nm is sometimes mentioned, but the colour-matching functions are virtually zero outside the 400 to 700nm range. The CIE 31 system has since standardised in ISO 11664-3[12].

## 7.2 Chromaticity coordinates

We would not impress anyone by boasting that "the colour of my new car is X = 35, Y = 65 and Z = 15". Values of X, Y and Z can therefore only serve as an intermediate step toward further calculations. Along with the X, Y and Z system, CIE introduced the concept of **chromaticity coordinates** x, y (both lower case) in 1931. These are intended to provide a convenient two-dimensional representation of colours. They are calculated as follows:

*Equation I.4*

$$x = X/(X + Y + Z)$$
$$y = Y/(X + Y + Z)$$
$$z = Z/(X + Y + Z)$$

As a consequence, "lower case z" automatically substitutes to z = 1-x-y.

### Spectrum locus

An x-y plot of the chromaticity coordinates of all spectral colours from 380 to 780 nm yields the spectrum locus, along with all the intervening mixed colours created to fill the space bounded by the spectrum locus. This diagram is called a **chromaticity diagram** and is shown in Figure I.22. It is also known as the CIE colour triangle or in laboratory parlance "the shoe-sole" or "the horseshoe". The curve is nearly identical for the 10° observer from 1964. Table I.2 lists the x, y and z values for selected hues.

With this display format we now have a two-dimensional diagram of colour – but it does not contain any lightness information. So we add luminance Y to yield a 3-dimensional Yxy colour system. However, an actual three-dimensional diagram of this system does not exist. The x-y chromaticity diagram is therefore always a reduced, incomplete colour diagram for *one* lightness value. The values x and y define a **chromaticity** consisting of **hue** and **chroma** (saturation) and thus represent a link to visual colour assessment.

As can be seen in Figure I.22, the "shoe-sole" contains white and the different kinds of white. The tristimulus values for ideal white are labelled $X_{n,}$ $Y_n$ and $Z_n$. Examples of the values for standard illuminants and the two types of observers are shown in Table 1.3.

Starting from the achromatic white, the saturation (chroma) of a hue increases outwards, reaching its theoretical maximum at the boundary of the spectrum locus. The outcome of mixing two colours (additive mixing) can be seen on the line linking the two colour locations of the pure colours. The link between the final points of the spectral locus from red to blue represents a special case. This link is called the **purple line**. It enables us to obtain purple colours that would otherwise be absent from the chroma-

*Table I.2: x, y and z values for selected hues.*

| Color coordinates for ideal spectral colors | | | | |
|---|---|---|---|---|
| Shade | nm | x | y | z |
| Red | 670 | 0.74 | 0.26 | 0 |
| Orange | 620 | 0.69 | 0.31 | 0 |
| Yellow | 585 | 0.54 | 0.46 | 0 |
| Green | 525 | 0.08 | 0.84 | 0.08 |
| Blue | 470 | 0.12 | 0.07 | 0.81 |
| Violet | 400 | 0.18 | 0 | 0.82 |

*Table I.3: Examples of the tristimulus values for ideal white for standard illuminants A and D65, for both the 2° observer and the 10° observer.*

| Illuminant | $X_n$ | $Y_n$ | $Z_n$ |
|---|---|---|---|
| A(2°) | 109.85 | 100.00 | 35.59 |
| A(10°) | 111.14 | 100.00 | 35.20 |
| $D_{65}$(2°) | 95.05 | 100.00 | 108.88 |
| $D_{65}$(10°) | 94.81 | 100.00 | 107.30 |

ticity diagram. If we now extend this to three colours, we create a triangle that contains all the different colours obtainable from the three starting colours. *Burns* [13] makes an interesting observation about such chromaticity diagrams (CIE triangles): "whenever you see a coloured representation of a chromaticity diagram it must be wrong". This also applies to Figure I.22. The reason is that the limiting spectrum locus as the outer boundary of the diagram is based on ideally brilliant colours. However, the pigments and dyes that would be needed to produce the spectrum locus in print or on a computer monitor do not exist in practice. The colourants that actually are available cover no more than half of the theoretically available area of the triangle.

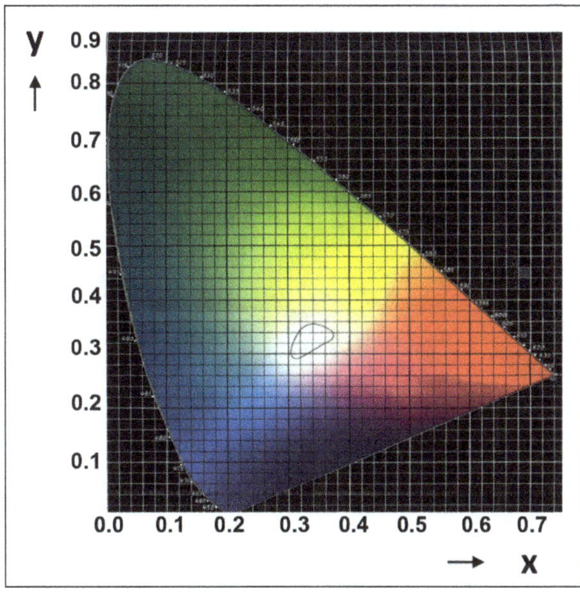


*Figure I.22: The x-y chromaticity diagram based on the 1931 CIE publication. The pure spectral colours define the outer boundary. The space inside the boundary represents the totality of all renderable colours, including the achromatic point. Lightness is another attribute that leads to different sets of "shoe-soles/horseshoes" at different lightness values.*

The main problem with the CIE **Yxy colour system** from 1931 is the poor equidistance. It is standard practice in industry to measure colour distances or colour tolerances against a reference colour. Where the human eye determines identical colour differences for two colours, the x-y diagram will, to an extent depending on the colour location in question, exhibit very different colour differences. As a result, it is not possible to communicate meaningful colour differences based on this colour system. Yet precisely that is essential for relations between customer-supplier and designer-producer, etc. This lack of **equidistance** in the chromaticity diagram can be seen in *MacAdam* ellipses (see Section I.8.4). As an example: a given colour difference in the green area will give rise to very large coordinate differences, while in the blue-violet area the same colour difference will be perceived at much smaller coordinate differences. Not only the size but also the orientation of the ellipses varies with the colour location in question. Whereas they are stretched parallel to the y axis for green-yellow colours, principal axes for blue and red colours are parallel with the purple line. Consequently, nowadays the X, Y and Z chromaticity coordinates are only an intermediate step in the calculation of further colour metrics. That said, the x-y chromaticity diagram is still used to define colours in certain application areas (e.g. traffic lights).

# 8 CIELAB system

The problems with the XYZ system encouraged CIE to develop a more extensive colour system that featured improved equidistance. Development work started in 1973, but it was not until 1976 that the CIE 1976 (L*a*b*) colour system was introduced. To this day, it is the most widely employed colour system. In 2008, this CIE guideline was incorporated into ISO 11664 Part 4. Before 1973, several other colour systems had been proposed, but they will not be discussed in further detail here. One of them was a 'Lab' system proposed by *Hunter*, which

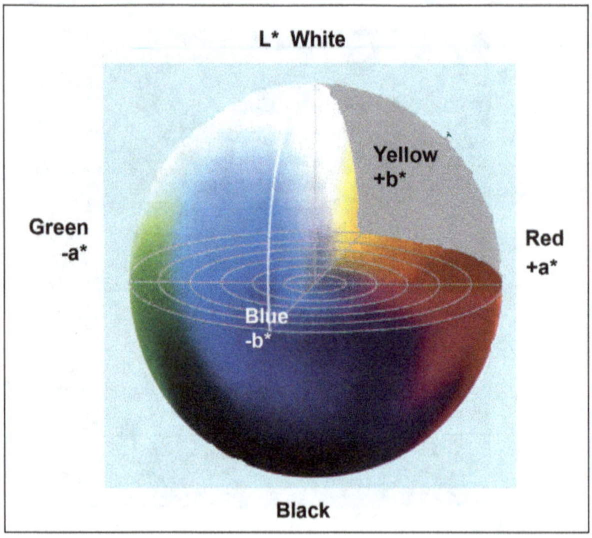


*Figure I.23: Plot of CIELAB colour space based on a red-green axis a*, a yellow-blue axis b* and the lightness axis L*.*
Source: Datacolor

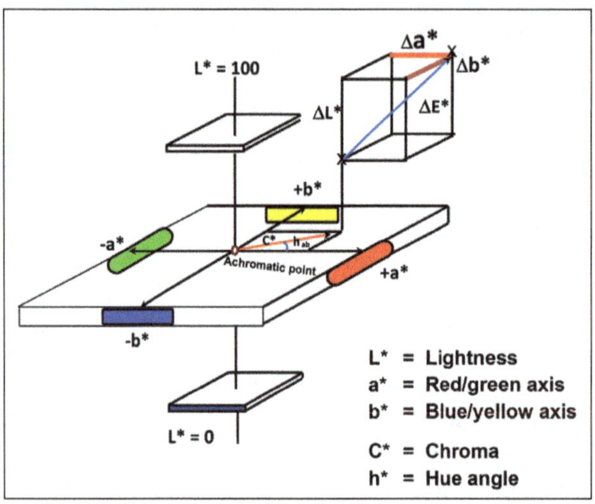


*Figure I.24: Schematic diagram of the CIEL*a*b* coordinate system for visualizing the coordinates of the colour location of both the reference **R** and the sample **P**. In a second step, it is possible to derive the total colour difference $\Delta E^*_{ab}$ and its different parts $\Delta L^*$, $\Delta a^*$ and $\Delta b^*$ between these two colour locations. Switching from the Cartesian coordinate system L*a*b* to the cylindrical coordinate system L*C*h, we obtain the new coordinates C* (chroma, saturation) and h (hue angle). The latter directly correlates with visual colour assessment and should be used for highly chromatic (brilliant) colours. The colour difference between P and R remains the same; it is merely expressed by $\Delta L^*$, $\Delta C^*$ and $\Delta h$.* Source: Clariant Produkte (Deutschland) GmbH

is why the CIE coordinates introduced later are officially marked with an asterisk (*).

In daily documentation, this convention is usually ignored since CIE itself calls its colour space **CIELAB**.

For completeness, it should be mentioned that, at around the same time, another CIE committee defined a more-or-less equivalent system called CIELUV[1] (ISO 11664 Part 3[12]). However, the CIELAB system is the most popular system. Historically, CIELUV has tended to find greater application in additive colour mixing, i.e. coloured lights or illuminants.

At that time, tests on human colour vision based on mixing light in light booths led to the development of a concept based on opponent colour theory. A similar concept was presented by *Hering* back in 1878 (see Section I.1.4.1). The outcome was a Cartesian coordinate system based on a **red-green axis a***, a **yellow-blue axis b*** and a **black-white axis L*** (lightness), with red and yellow extending in the positive coordinate direction. This system is based on the human experience of colour vision that, for example, yellow can tend to be greenish or reddish (i.e. orange), but never bluish. On the other hand, red can tend to be yellowish (orange) or bluish (violet), but never greenish. This is the basis of the CIE76 (L*a*b*) colour space (CIELAB) as shown in Figure I.23.

Unlike the case for the CIE tristimulus (XYZ) system, a three-dimensional coordinate system is

[1] *The CIELUV system is based on the coordinates L*, u* and v*. These can be obtained from the XYZ system by slightly modifying the equations described in Section I.8.1 for the CIELAB system. A cylindrical coordinate system employing C*uv and huv representing the same space can also be developed.*

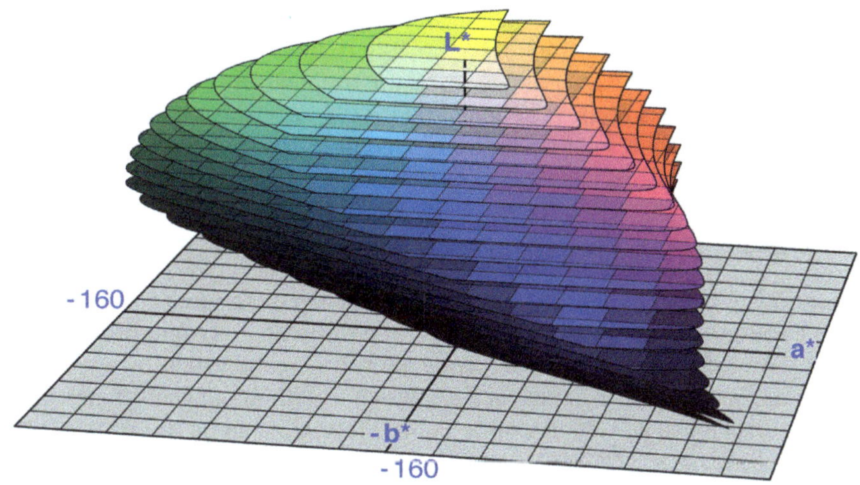


*Figure I.25: Starting from ideal colours, which can only range from 0 to 100 % in intensity at one wavelength, we arrive at this shape for CIELAB colour. It is neither a sphere nor infinitely large* [15]. Source: Prof. B. Hill

created at once, with each colour represented by a specific point in colour space. Figure I.24 shows a schematic diagram of CIELAB colour space and its coordinates. The $L^*$ axis of lightness is restricted between $L^* = 0$ for black and $L^* = 100$ for white. All colours on this axis are achromatic, consisting of different levels of grey. In a cross-section through CIELAB space at fixed lightness, the coordinates of $a^*$ and $b^*$ represent the coloured portion of the stimulus. The two axes $a^*$ and $b^*$ are limited by the ideal colours of 100% spectral intensity at a single wavelength. At its maximum extension, CIELAB colour space is somewhat shapeless, as shown in Figure I.25 [15]. So-called "**CIELAB units**" were introduced to describe their lengths, but they are not related to other units of measure.

The systems underlying CIELAB are all derived from the *Munsell* colour system, which only defined an illuminant C and the 2° observer. CIE introduced the 10° observer and other illuminants.

## 8.1 The L*, a*, b* coordinates

As established by CIE, the new coordinates are calculated from the previous X, Y and Z coordinates as follows [11]:

*Equations I.5 to I.7*

$$L^* = 116\, f(Y/Y_n) - 16$$
$$a^* = 500\, [\, f(X/X_n) - f(Y/Y_n)\, ]$$
$$b^* = 200\, [\, f(Y/Y_n) - f(Z/Z_n)\, ] \qquad \text{where:}$$

*Equations I.8 to I.13*

$$f(X/X_n) = (X/X_n)^{1/3} \qquad \text{for } X/X_n > (24/116)^3$$
$$f(X/X_n) = (841/\,108)(X/X_n) + 16/116 \qquad \text{for } X/X_n \leq (24/116)^3$$
$$f(Y/Y_n) = (Y/Y_n)^{1/3} \qquad \text{for } Y/Y_n > (24/116)^3$$
$$f(Y/Y_n) = (841/\,108)(Y/Y_n) + 16/116 \qquad \text{for } Y/Y_n \leq (24/116)^3$$
$$f(Z/Z_n) = (Y/Y_n)^{1/3} \qquad \text{for } Z/Z_n > (24/116)^3$$
$$f(Z/Z_n) = (841/\,108)(Z/Z_n) + 16/116 \qquad \text{for } Z/Z_n \leq (24/116)^3$$

In the literature, the fraction $(24/116)^3$ is often written as 0.008856; it is written as a full fraction here for the sake of clarity.

In the equations above, X, Y and Z are the tristimulus values of the measured sample and $X_n$, $Y_n$ and $Z_n$ are those of the corresponding white reference for the chosen illuminant. As described in Section I.7.2, ideal white has a $Y_n = 100$. In the CIELAB system, normalization to the corresponding white $X_n$, $Y_n$ and $Z_n$, which is implicit in the equations, causes all achromatic colours (shades of grey) to lie on the L axis – unlike the case for the CIE 1931 chromaticity diagram (shoe-sole) for the XYZ system. CIE has defined equations for the reverse calculation from CIELAB to the XYZ system.

## 8.2 *The L*, C*, h coordinates*

There is no equivalent in the CIELAB system for the colour assessment employed by humans to describe colours in terms of characteristics called **hue** and **chroma**. It can adopt only **lightness** $L_{ab}$ directly. However, the equivalence can be established by switching from the rectangular Cartesian coordinates of $L^*a^*b^*$-space to cylindrical coordinates. These are already shown in Figure I.24 as **distance coordinate** $\mathbf{C^*_{ab}}$ and as **angular coordinate** $\mathbf{h_{ab}}$ (h does not have an asterisk). $C^*_{ab}$ defines the distance of the colour location from the lightness axis $L_{ab}$ and represents chroma. The $L_{ab}$ axis yields information about variation in lightness only and nothing about colour (achromatic axis). The further the colour location is away from the axis, i.e. $C^*_{ab}$ is increasing, the more brilliant and more saturated the colour appears. The hue angle $h_{ab}$ defines the hue of the colour location. It is measured counter-clockwise, starting from the positive $a^*$-axis. As a consequence, the $L^*C^*h$ system is equivalent to the $L^*a^*b^*$ system and simultaneously has a direct link to the principle underlying human colour assessment.

The $L^*$-axis is identical in both systems; the other two coordinates are defined by the following equations:

*Equations I.14 and I.15*

$$C^*_{ab} = (a^{*2} + b^{*2})^{1/2}$$
$$h_{ab} = \arctan(b^*/a^*)$$

The subscript $_{ab}$ represents the link to $L^*a^*b^*$ space – as opposed to LUV space. For the most part, this subscript is not used either.

## 8.3 *Colour differences*

Usually, human colour assessment of a sample is based on comparison with a reference. This can be achieved in CIELAB space with the aid of coordinates for a sample **P** and a reference **R**. Further details are described in the related ISO standards [11, 15, 16].

The colour differences "sample **P** – reference **R**" are obtained as follows:

*Equations I.16 to I.18*

$$\Delta L^* = L^*_P - L^*_R$$
$$\Delta a^* = a^*_P - a^*_R$$
$$\Delta b^* = b^*_P - b^*_R \qquad \text{in LAB space and, similarly,}$$

*Equations I.19 and I.20*

$$\Delta C^*_{ab} = C^*_{ab,P} - C^*_{ab,R}$$
$$\Delta h_{ab} = h_{ab,P} - h_{ab,R} \qquad \text{in LCh space.}$$

In accordance with the equations above, a positive value of $\Delta a^*$ signifies that sample **P** has a higher red content than reference **R** whereas a positive $\Delta C^*_{ab}$ signifies that the sample has higher chroma or purity.

The **total colour difference**, also called colour distance, between sample and reference corresponds to the 3-dimensional difference in CIELAB space. It is defined by the connecting

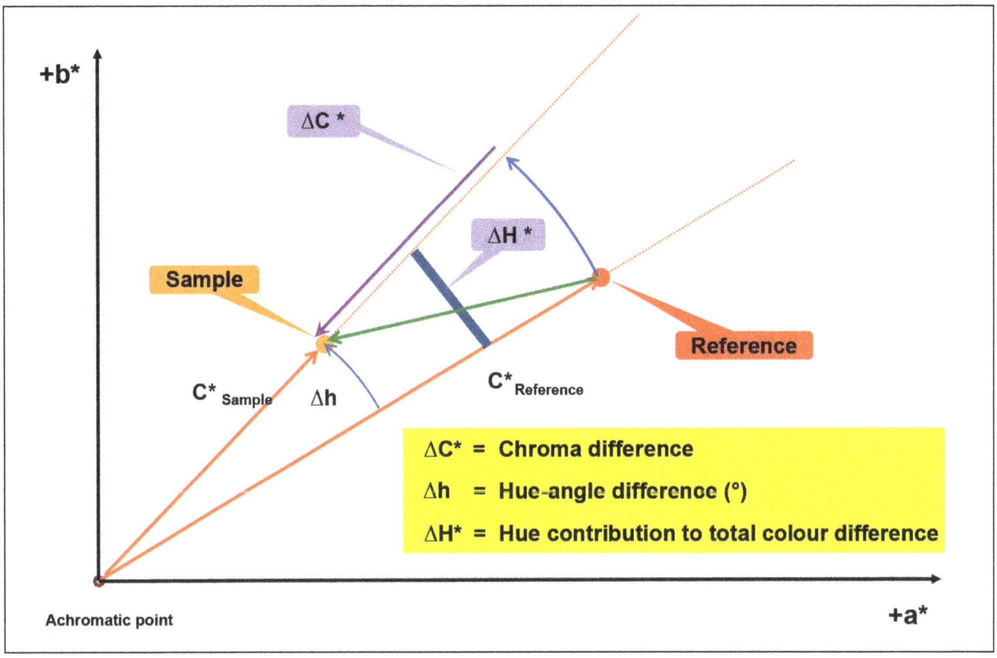


*Figure I.26: Schematic diagram showing the splitting of the total colour difference $\Delta E^*_{ab}$ into the hue component $\Delta H^*_{ab}$ and the chroma component $\Delta C^*_{ab}$. The lightness $\Delta L^*_{ab}$ is not shown. $\Delta H^*_{ab}$ is defined as the distance between both chords at half the length of $\Delta C^*$.*

vector ΔE*ab between **P** and **R**. This is also shown in Figure I.24. In the CIELAB-system, this is readily obtained by vector addition of the colour differences ΔL*, Δa* and Δb*.

*Equation I.21* $$\Delta E^*_{ab} = [(\Delta L^*)^2 + (\Delta a^*)^2 + (\Delta b^*)^2]^{1/2}$$

This does not work as directly in the LCh system because there h is measured in degrees. Consequently, $\Delta H^*_{ab}$ was introduced to account for the hue component in the colour difference. Figure I.26 illustrates how this parameter is defined. Half the distance between P and R $\Delta H^*_{ab}$ is defined as the interval between both chords through **P** and **R**. Thus, we can now establish

*Equation I.22* $$\Delta E^*_{ab} = [(\Delta L^*)^2 + (\Delta C^*_{ab})^2 + (\Delta H^*_{ab})^2]^{1/2}$$

as a second method of calculating the colour difference. In addition, this is the equation for splitting the colour difference into its three parts or colour difference components.

It should be noted that $\Delta H^*_{ab}$ defines the **hue component** of the colour difference $\Delta E^*_{ab}$. Although 'H' on its own does not exist and hue is defined by the hue angle h (ISO 11664 Part 3), nevertheless, it is common practice to talk about the LCH (uppercase H) diagram of CIELAB colour space. A look at Figure I.26 reveals another difference between $\Delta h_{ab}$ and $\Delta H^*_{ab}$. Two pairs of samples **P** and reference **R** may have identical hue differences $\Delta h_{ab}$ but, if $\Delta C^*_{ab}$ is different at the same time, then both pairs will have different values for $\Delta H^*_{ab}$.

It may now be appreciated why the $L^*a^*b^*$-system is preferred for **achromatic colours**, i.e. colours having small values of $C^*_{ab}$. For such colours close to the achromatic $L^*$ axis, even a small variation in hue could give rise to a large value in $\Delta h_{ab}$, even though this colour difference might not be perceptible. The reverse is true for colours of high saturation and chroma, that is, the LCH-system is preferable.

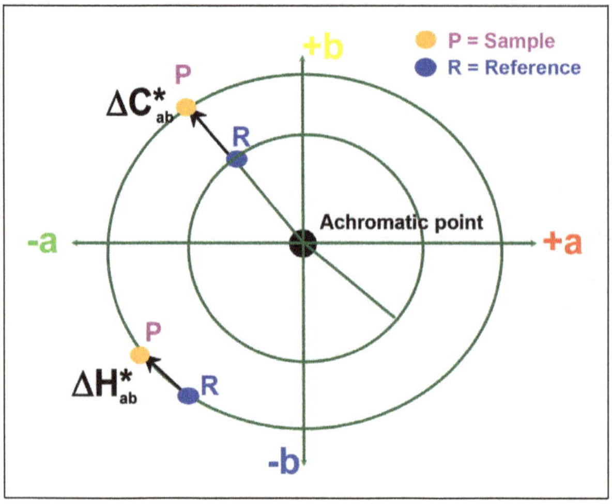


*Figure I.27: This plan view of the a*-b* plane compares the Lab colour differences and the LCH differences based on two pairs of sample **P** and reference **R**. Both pairs have by way of colour difference a negative $\Delta a^*$ and a positive $\Delta b^*$. While the pair at top left differ in chroma $\Delta C^*_{ab}$ only, the pair at bottom left differ in hue $\Delta h_{ab}$ only. Thus, the LCH system is a much better match for human colour perception when it comes to saturated colours.* Source: Clariant Produkte (Deutschland) GmbH

| - ← | ΔH* Hue/Shade | → + |
|---|---|---|
| Redder ← | Yellow | → Greener |
| Redder ← | Orange | → Yellower |
| Bluer ← | Red | → Yellower |
| Bluer ← | Violet | → Redder |
| Greener ← | Blue | → Redder |
| Yellower ← | Green | → Bluer |

**Tolerances for ΔH* and ΔL***

| | | |
|---|---|---|
| 0.0 ... 0.1 | /=/ | = corresponding |
| 0.1 ... 0.2 | 1 | = trace |
| 0.2 ... 0.5 | 2 | = slightly |
| 0.5 ... 0.8 | 3 | = noticeably |
| 0.8 ... 1.4 | 4 | = distinctly |
| 1.4 ... 2.0 | 5 | = considerably |

*Figure I.28: Comparison of the visual colour assessment of different colours and the algebraic sign of $\Delta H^*_{ab}$ and $\Delta h_{ab}$. For example, if a sample of yellow is greener than the reference, $\Delta H^*_{ab}$ has a positive algebraic sign.* Source: Clariant Produkte (Deutschland) GmbH

As already mentioned, the LCH system has the advantage of providing a direct link to visual colour assessment and allows colourists to gain an idea of the actual colour difference from the readings for $\Delta C^*_{ab}$ und $\Delta H^*_{ab}$. Figure I.27 illustrates the advantage of the LCH system over the Lab-system for colours that are close to the achromatic L* axis. It shows a plan view of the a*-b* plane, with two pairs of sample and reference **P** and **R** on the left. Both pairs have by way of colour difference a negative $\Delta a^*$ and a positive $\Delta b^*$. Nevertheless, one pair differs only in chroma $\Delta C^*_{ab}$ (top left) while the other differs only in hue $\Delta h_{ab}$ (bottom left). On the basis of the $\Delta a^*$ and $\Delta b^*$ values alone, it is not possible to draw a direct conclusion about the colour difference between **P** and **R**. That would require taking into account the relevant quadrant in which the sample pair is located.

The link between $\Delta h_{ab}$ and $\Delta H^*_{ab}$ on one hand and visual colour assessment on the other can be clarified by looking at the a*-b* plane. A yellow hue may be either greener or redder (more orange) than a reference sample. Accordingly, $\Delta h_{ab}$ or $\Delta H^*_{ab}$ will have a positive algebraic sign (greener) or a negative one. The corresponding relationships for other hues are shown in Figure I.28.

A corresponding comparison for chroma difference $\Delta C^*_{ab}$ and lightness difference $\Delta L^*_{ab}$ is shown in Figure I.29.

While colorimetric results can vary continuously, visual assessments are discrete steps, such as 2 for yellower, 5 for duller etc. These steps can be reported either in the form of numbers or in words:

/=/ = corresponding
1 = trace
2 = slightly
3 = noticeably
4 = distinctly
5 = considerably
6 = substantially

**Lightness and chroma/purity**

**Darker ←– $\Delta L^*$ +→ Lighter**

**Duller ←– $\Delta C^*$ +→ Purer**

*Figure I.29: Equivalents for visual assessments of chroma difference $\Delta C^*_{ab}$ and lightness difference $\Delta L^*_{ab}$.*

Source: Clariant Produkte (Deutschland) GmbH

Only in the case of samples based on white colour reductions having a pigment: white ratio of 1 : 10 or 1 : 20 have direct links been established between the visual assessment steps and the CIELCH values. They are shown in Table I.4. The step height for $\Delta C^*_{ab}$ is slightly larger than for $\Delta H^*_{ab}$ and $\Delta L^*_{ab}$; for example, the upper visual step limit of 2 = slightly corresponds to $\Delta C^* = 0.8$.

Finally, it should be mentioned that the total colour difference $\Delta E^*_{ab}$ can only be split into its colour difference components for both the Lab system and the LCH-system in cases where colour differences $\Delta E^*_{ab} < 5$. For larger differences, the linear relationship assumed between the characteristics used cannot be applied. The basic definitions given in equations I.5 to I.15 are obviously non-linear.

## 8.4 Colour tolerances and MacAdam ellipses

In daily practice, we are interested not so much in the absolute chromaticity coordinates or colour values as in the colour differences between the actual sample and the reference. Accordingly, the magnitude of the colour differences is the criterion for approving or rejecting the sample tested. Tolerance ranges for acceptance of a product must be agreed between supplier and customer. These tolerances are influenced on one hand by the perceptibility threshold of the human eye and on the other by variations in sample preparation (see Section II.8). In addition, the human perceptibility threshold must be seen as distinct from the colour tolerance that, for example, is acceptable to pigment manufacturer (supplier) and paint manufacturer (customer). Whereas the perceptibility threshold is $\Delta E^*_{ab} \geq 0.15$ in the achromatic and dark region on account of lightness differences, it is around $\Delta E^*_{ab} \approx 0.6$ for saturated colours with high chroma. In the following, this difference in perceptibility threshold is discussed as a problem not of the human eye but of the colour system.

The CIE chromaticity diagram (colour triangle/"shoe-sole") was published in 1931 but it soon became clear that the colour differences in this system did not match the

*Table I.4: Comparison of CIELCH values with the corresponding visual colour assessment steps.*

| Relative tinting strength (100 +/- ....%) | ΔC (CIELAB units) | ΔH und ΔL (CIELAB units) | Visual assessment stage | |
|---|---|---|---|---|
| from 0 % to 1 % | 0 to 0,10 | 0 to 0,10 | Corresponding to | /=/ |
| from 1 % to 2 % | 0,11 to 0,30 | 0,11 to 0,20 | trace | 1 |
| from 2 % to 5 % | 0,31 to 0,80 | 0,21 to 0,50 | slightly | 2 |
| from 5 % to 10 % | 0,81 to 1,40 | 0,51 to 0,80 | noticeably | 3 |
| from 10 % to 20 % | 1,41 to 2,20 | 0,81 to 1,40 | distinctly | 4 |
| from 20 % to 40 % | 2,21 to 3,00 | 1,41 to 2,00 | considerably | 5 |
| from 40 % | from 3,01 | from 2,01 | substantially | 6 |

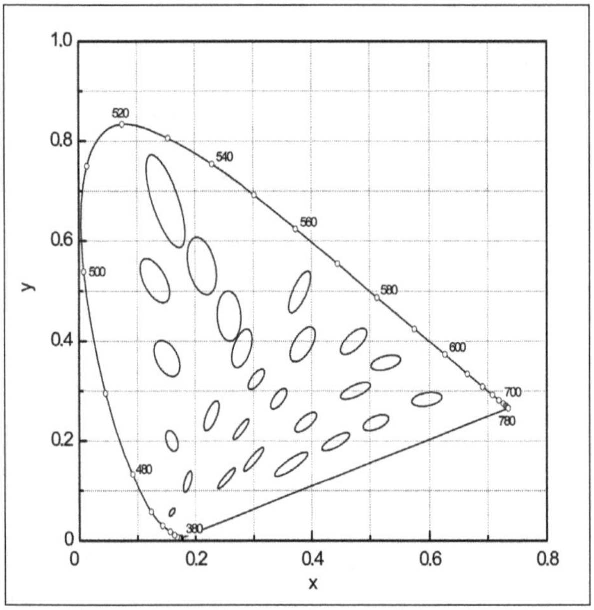


*Figure I.30: The lack of equidistance in the CIE 31 system is revealed by MacAdam ellipses[17]. These ellipses visualize colour differences that humans see as having identical colour distances. To illustrate the effect more clearly, the ellipses have been enlarged by a factor of 10. Using the CIELAB system instead of CIE 31 reduces but does not fully eliminate this effect.* Source: W. Kettler

colour differences experienced by humans. What, for humans, was a small colour difference in the green range represented quite large differences in the CIE colour triangle. In contrast, in the violet range, small differences in the CIE system were perceived by humans as large colour differences. In the early 1940s, *MacAdam* and *Brown* [17, 18] studied the perceptibility threshold of colour variations in the chromaticity diagram. Colour differences felt to be identical around a reference colour are located on an ellipse – not a circle – around the reference. The length and orientation of the long axis of these ellipses depend on both the hue and the chroma (or saturation). Figure I.30 shows a plot of the ellipses on the CIE chromaticity diagram. To illustrate the effect more clearly, the scale of the ellipses has been enlarged by a factor of 10. The ellipses are quite large for green hues and small for violet hues. Their orientation rotates through nearly 90° between green and red. These results in the CIE 31 system can be transformed into the CIELAB system, where they remain (small) ellipses, or three-dimensional ellipsoids.

Even though the **lack of equidistance** in CIE 31 and CIELAB colour spaces is widely known, industry continues to agree standardised delivery specifications for all colours. This can lead to discussions between supplier and customer where a reading does not seem to match the visually perceived difference [19]. For automotive colours, the permissible tolerances between paint manufacturer and automotive industry are defined in DIN 6175 (Part 1 for unicolour, Part 2 for effect paints). For further information, see Sections VI.7 and VI.8. However, the complicated concept employed in that standard is not transferable to other applications, and is mainly founded on the potentially high damages for which suppliers and insurance companies may be liable.

Since human perception should serve as *the* criterion for colour assessment, the MacAdam ellipses document a problem of the present-day status of colorimetry. The search began for solutions as far back as 1931. This led to the CIELAB system in 1976 as the first step; other colour systems have since followed. These newer systems are described in more detail in Chapter V "Instrumental colour difference assessment".

# 9 Metamerism

We encounter metamerism not only when we conduct colour measurements but also in our daily lives. It is the familiar experience that a jacket and trousers which match in colour in the shop no longer match in colour when we get them home. Naturally, this is only one example of the phenomenon, albeit a negative one. But if we want a better understanding of colour matching, we need to understand the background to metamerism.

## 9.1 Colour constancy

As explained in Section I.1.4., our perception of colour also depends on the illuminant. A body, which is illuminated with different coloured lights, will elicit different colour stimuli in our vision system. The various (standardised) illuminants presented in Section I.5.1 appear white to our eye, even though they have quite different spectral power distributions. "White fluorescent" tubes in department stores emit different spectra from sunlight or domestic light-bulbs. This explains why the colour perception of an object changes somewhat when the illuminant is changed (as in the jacket-and-trousers example). This phenomenon is not metamerism. The variations in colour location brought about by changing the illuminant are a type of colour inconstancy of a surface colour. The colour signal processing which occurs in our brains is prone to chromatic adaptation. Discussed in Section I.1.2, it causes the eye to adapt to or to correct differences in lightness or hue following a change in illumination.

Adequate colour-matching enables hues to be optimised for good colour constancy. Further details on this are presented in Section XI. A colour constancy index can be established by determining the change in the colour location for one illuminant relative to a reference illuminant as a type of total colour difference. Allowance is made for chromatic adaptation by introducing empirical correction factors into the calculation [20].

## 9.2 Metameric pairs

The concept of metamerism always relates to a colour assessment of a pair of samples - as in the jacket-and-trousers example - where a change of illumination occurs. As shown in Section I.9.1, the colour location of an object usually changes when the illuminant changes. **Metamerism** is said to occur when a pair of samples, e.g. a pair of colours, has the same colour location in one illuminant but different colour locations in another. More precisely, the tristimulus values and the colour locations of the pair are identical for the first illuminant while the colour locations are different for the second illuminant. This is said to be a **metameric pair** of samples. This effect is based on the fact that, in human vision, there is no unique link between the colour perception and the spectrum or the reflectance curve. One and the same colour stimulus can be generated by different reflectance curves. Signal processing by human photoreceptors is discussed in Section I.1.3. It is the reason why we recognize the photograph of an object as the object itself. The pigments used for printing a colour picture of an apple, say, have different reflectance curves from those of the dyes in the apple's skin. In addition, the photographic reproduction of the apple can be accomplished with a three-colour or six-colour printer, for example. Thus, as we have seen, metamerism can have both positive (e.g. photography) and negative impacts (jacket-and-trousers problem). It is this phenomenon that enabled companies to reproduce unknown paint compositions, i.e. from competitors, in the first place. It can also be used to obtain identical colour perceptions, even though it might not be possible to use the exact same pigment because the application may have been created in a powder coating, a waterborne coating or a solventborne system. Further reasons can be different fastness specifications (e.g. interior application, exterior application) or the task of obtaining identical colours on different-coloured materials (coating, plastics, printing) or of replacing pigments containing heavy metals by ones which are environmentally sound and less harmful to health. Where different reflectance curves lead to the same colour stimulus, a change of illuminant will have different impacts on the perception of those reflectance curves. The various (standard) illuminants are presented in Section I.5.1. Although all of them create the impression of

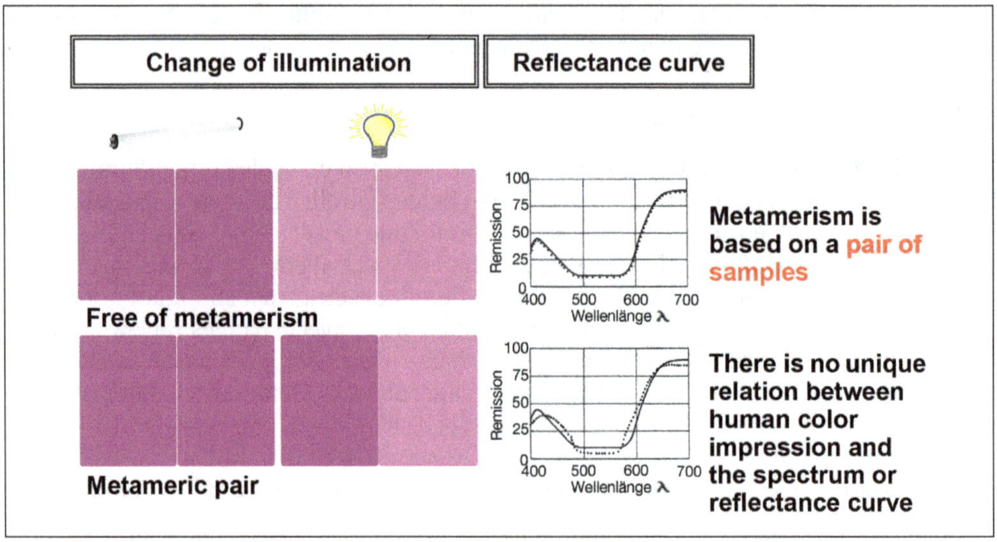


*Figure I.31: Metamerism can occur when pairs of samples are assessed under different illuminants. For human vision, there is no strict link between colour perception and spectrum (or reflectance curve). Different reflectance curves may yield identical colour perceptions. Metamerism is expected when the reflectance curves for a pair of samples overlap three times or more.* Source: Clariant Produkte (Deutschland) GmbH

white to our eyes, illuminant D65 has a much higher energy content in the blue range while illuminant A has a substantially higher content in the red range. Changing the illuminant changes the weighting of different parts of the spectrum, and this may create a different perception under the other illuminant. Figure I.31 illustrates the background to metamerism. One criterion governing the appearance of metamerism in a sample pair is three-fold or multiple overlapping of the two reflectance curves.

The phenomenon of metamerism has been discussed here as an effect of visual colour assessment. In daily practice, it is crucial when it comes to colour matching based on colorimetric measurements. Good matching during colour reproduction will reduce or even prevent metamerism. Further details are presented in Section XI. Such formulations match the original not only as far as human colour vision is concerned, but also in terms of spectral curve.

So far, we have looked at illuminant metamerism. But there are also

- Observer-geometry metamerism: when changing from a 2° observer to a 10° observer
- Observer metamerism: when two persons have different colour perceptions [1]
- Geometry metamerism: when the illumination geometry or observer direction is changed (especially in the case of effect pigments)

However, these usually have much less impact and can be corrected by appropriate countermeasures. A typical example is the coloristic design of a vehicle interior. Here, various materials (e.g. textiles, plastics, leather) with different surface structures (smooth, grained, profiled) are in the line of sight. In this case, the illumination will vary from the light in the show-room, to sunlight to courtesy lights, and the observers will be designers, sales persons and customers. All these aspects pose challenges when it comes to matching the various materials.

[1] *This phenomenon is more commonly interpreted as a normal colour difference rather than a metameric effect.*

## 9.3 Special metameric index

The magnitude of the metamerism effect is indicated by the specific metameric index M. This is calculated by determining the colour difference for a given pair of samples when switching from reference illuminant B to a test illuminant T. If the CIELAB-system is used, the equations given in Section I.8.3 yield the following relationships:

*Equation I.23* $M_T = [(\Delta L^*_T)^2 + (\Delta a^*_T)^2 + (\Delta b^*_T)^2]^{1/2}$

Similarly

*Equation I.24* $M_T = [(\Delta L^*_T)^2 + (\Delta C^*_{Tab})^2 + (\Delta H^*_{Tab})^2]^{1/2}$ where

*Equations I.25 to I.28*

$\Delta L^*_T = L^*_T - L^*_B$

$\Delta a^*_T = a^*_T - a^*_B$

$\Delta b^*_T = b^*_T - b^*_B$

$\Delta C^*_{Tab} = C^*_{Tab} - C^*_{Bab}$ and

*Equation I.29* $\Delta H^*_{tab} = (a^*_B b^*_T - a^*_T b^*_B) \cdot [0{,}5(C^*_{Tab} C^*_{Bab} + a^*_T a^*_B + b^*_T b^*_B)]^{1/2}$

(for the definition of dH, see also Figure I.26)

When stating the metameric index, the reference illuminant (most often $D_{65}$), the test illuminant, the colour difference equation employed (in this case CIELAB or CIELCH) and the observer (most often the 10° observer) must be reported as well.

This simple definition of $M_T$ only holds when there is no colour difference between the two originals (reference and sample) under illumination with reference illuminant B. However, in practice this is quite unrealistic. Thus, in this case, the chromaticity coordinates given above must first be corrected. Different types of corrections are published in the literature, e.g. DIN 6172 [21], such as multiplicative correction, additive correction and the Cohen-Kappauf model [22], but these will not be discussed any further here. For Euclidean colour spaces like CIELAB (1967) or DIN 99, these corrections are quite straightforward, but for those colour spaces with correction terms, such as CIE 94 and CIEDE 2000, they are additionally dependent on the colour location.

# 10 Literature

## 10.1 General literature

Society of Dyers and Colourists: Colour Physics for Industry, 2nd Edition, Edited by Roderick McDonald

R. W.G. Hunt, M. R. Pointer, Measuring Colour, 4th Edition, John Wiley & Sons Inc.

G. A. Klein: Farbenphysik für industrielle Anwendungen, Springer, Berlin (2004)

R. S. Berns, Billmayer and Saltzmann's Principles in Color Technology, 3rd Edition, Wiley-Interscience, New York (2000)

## 10.2 References

[1] DIN 5033-1: Farbmessung -Teil 1: Grundbegriffe der Farbmetrik

[2] W. G. K. Backhaus, R. Kriegel, J. S. Werner, Eds, Color Vision: Perspectives from Different Disciplines, Walter de Gruyter, Berlin (1998)

[3] P. K. Kaiser, R. M. Boynton, Human Color Vision 2nd Ed., Optical Society of America, Washington D.C. (1996)

[4] Klett Natura 3, 18, Biologie für Gymnasien, Klett Verlag Stuttgart

[5] Biologie 7 - 10 Hessen, 21831, Cornelsen Verlag, Berlin

[6] S. E. Palmer, Vision Science: Photons in Phenomenology, MIT Press, Cambridge, MA (1999)

[7] W. D. Wright, A re-determination of the trichromatic coefficients of the spectral colors, Trans. Opt. Soc. London 30, 141 - 164 (1928)

[8] J. Guild, The Colorimetric Properties of the Spectrum, Philos. Trans. R. Soc. London Ser. A 230, 149 - 187 (1931)

[9] B. B. Lee, Receptive Field Structure in the Primate Retina, Vision Res. 36, 631 -644 (1996)

[10] S. A. Hagstrom, J. Neitz, M. Neitz, Variation in cone population for red-green color vision examined by analysis of mRNA, Neuroreport 9, 1963 - 1967 (1998)

[11] CIE-Technical Report 15, 3rd Edition (2004), ISBN 3 901 906 33 9, CIE.co.at

[12] DIN EN ISO 11664, Teile 1-4

[13] R. S. Berns, Billmayer and Saltzmann's Principles in Color Technology, 3rd Edition, Wiley-Interscience, New York (2000)w

[14] Private communication: Prof. B. Hill, Aachen

[15] DIN 6174: Farbmetrische Bestimmung von Farbmaßzahlen und Farbabständen im annähernd gleichförmigen CIELAB-Farbenraum

[16] DIN 6176: Farbmetrische Bestimmung von Farbabständen bei Körperfarben nach der DIN99-Formel

[17] D. L. McAdam, Visual sensitivities to color differences in daylight, J. Opt. Soc. Am. 32, 247 (1942)

[18] W. R. J. Brown, D. L. McAdam, Visual sensitivities to combinet chromaticityand luminance differences; J. Opt. Soc. Am 39, 808 (1949)

[19] DIN Fachbericht 49, Verfahren zur Vereinbarung von Farbtoleranzen (1996)

[20] L. M. Changjun, M. R. Luo, B. Rigg, R. W. G. Hunt, CMC 2000 Chromatic Adaption Transform: CMCCAT 2000, Col. Res. Appl. 27, 49 (2002)

[21] DIN 6172:1973, DIN 6172:93, DIN 6172:2010 Metamerie-Index von Probenpaaren bei Lichtartwechsel

[22] J. B. Cohen, Visual Color and Color Mixture. The Fundamental Color Space, Univ. of Illinois Press, Urbana and Chicago (2001)

# II Colour measurement, colour measurement systems and visual colour assessment

## 1 Principles behind measuring coloured surfaces

*Gerhard Rösler, translated by Manfred Binder*

In industry, it is very often necessary to check that today's production does not vary in colour from yesterday's production or that today's production is up to specification. Of course, our eyes are capable of doing this job, but colour measurement systems afford a way to obtain objective assessments, to quantify differences numerically, and to set tolerances with a view to aiding decision-making. Furthermore, visual assessments of minor deviations are only possible if the two samples can be compared side by side at the same time. Measuring instruments on the other hand can store absolute readings so that different samples can be compared at different places and at different times. They not only furnish objective results, but also express the difference between two samples as a number.

### 1.1 Analytical and visual characterisation of colour

A colour measurement instrument can define the colour of physical samples in terms of numbers. It thus supports effective communication about colour within the industry. However, the human eye still remains the ultimate arbiter. Our brain "knows" what colour is, has no absolute colour memory and it is difficult to describe colour differences through language. This problem is exacerbated by differences in ambient light and our expectations.

A colour measurement system works by simulating human colour vision. It needs a light source to illuminate the samples, a receptor to collect the scattered light (equivalent to the human eye) and software to calculate the colour (the human brain).

Colours can basically be measured in two ways:

- the tristimulus method
- the spectrophotometric method

The basic setup of a colour measurement system is shown in Figure II.1.

#### 1.1.1 Method A: Tristimulus colorimeter

The tristimulus method is inspired by colour vision: the sample is illuminated with polychromatic light, and the light in the visible range that is reflected by the sample passes through three filters to three detectors, which measure the signals. This combination of physical illumination and spectral transmission by the three filters produces the best possible match for meeting the requirements of the standard observer and the chosen standard illuminant. The readings correspond directly to the tristimulus values X, Y and Z and can be transformed into chromaticity coordinates.

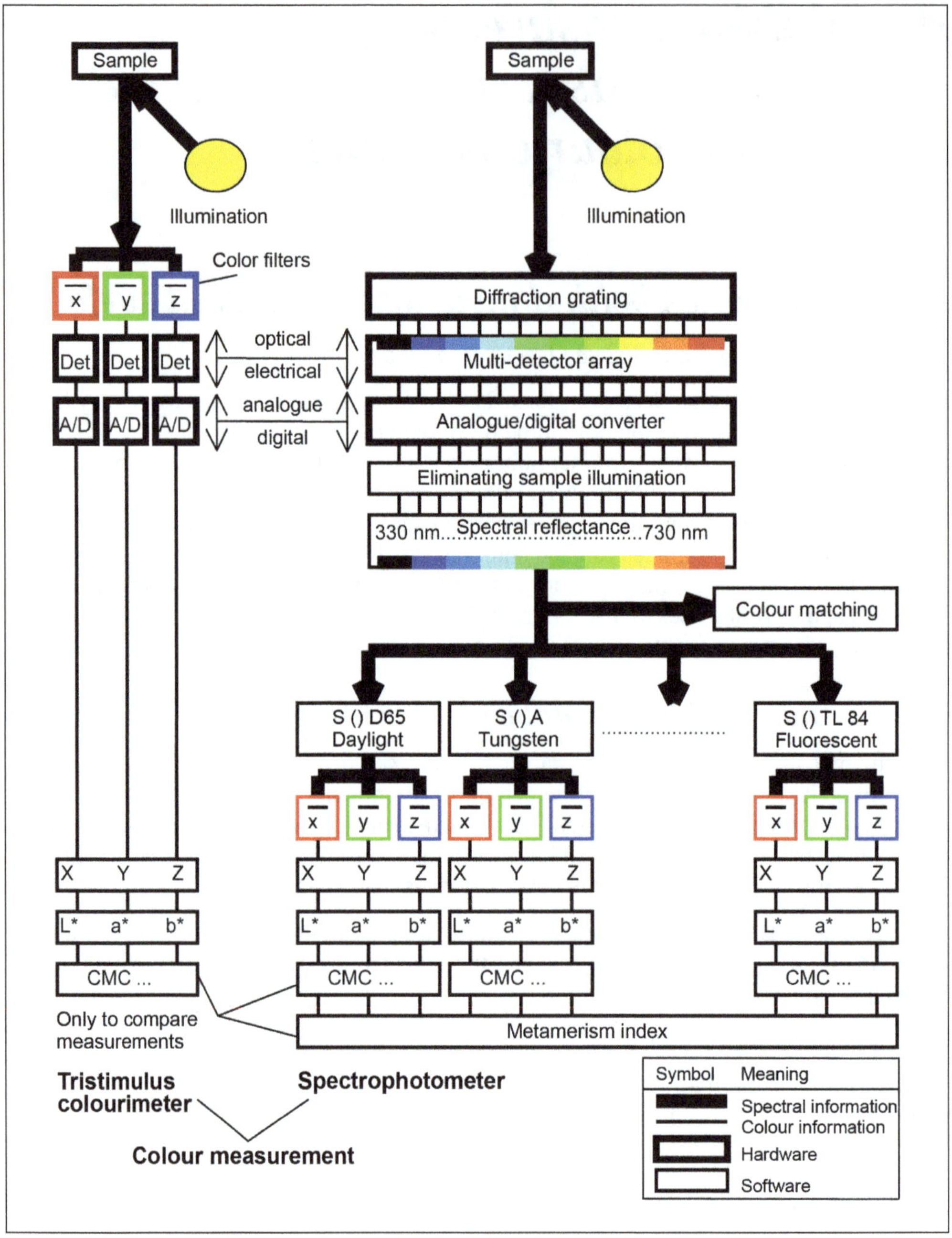


*Figure II.1: Method A - tristimulus colorimeter (left), and Method B - spectrophotometer with polychromatic illumination (right).*

**Advantage:** Simple in theory, but difficult in practice. It is difficult to produce the right combination of colour filters and photodetectors and so they must be produced separately as a pair.

**Disadvantage**: Only one light source, so metamerism cannot be determined. Poor inter-instrument agreement due to the filters and detectors chosen.

On account of these disadvantages, tristimulus colorimeters are seldom used nowadays.

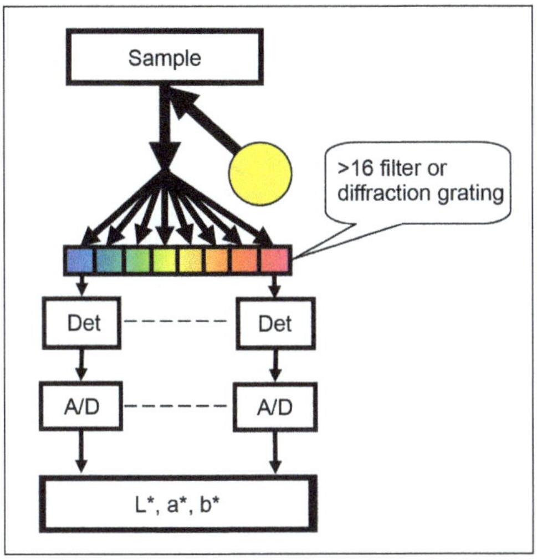


*Figure II.2a: Method B - parallel spectral method with polychromatic illumination.*

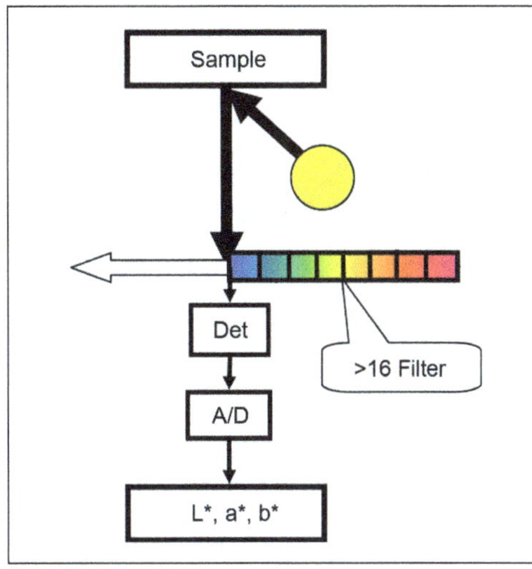


*Figure II.2b: Method C - sequential spectral method with polychromatic illumination.*

### *1.1.2 Methods B and C: Spectrophotometer with polychromatic illumination*

In methods B and C, the samples are illuminated with polychromatic light. The light reflected by the sample is dispersed into the different wavelengths of the visible range inside the spectrophotometer. The intensity of each wavelength or wavelength range (e.g. 5 nm, 10 nm, 20 nm) is measured in one or more detectors. If only one detector is used, the wavelengths are measured sequentially. This is a technically simpler solution, but measurements take longer (see Figures II.2a and II.2b).

Where several detectors are employed, all wavelengths are measured in parallel. Although the technical outlay is a little bit higher in this case, the results are obtained very quickly.

The intensity of the light $R(\lambda)$ reflected by the sample is measured relative to that of the illumination. It is therefore independent of the spectral power distribution of the source illuminant. Each colour has a characteristic reflectance curve $R(\lambda)$, which acts as its fingerprint.

One way to stabilise the readings is to utilise dual beams to measure not only the light reflected by the sample, but also the source light itself. Measuring the ratio of the source light to the reflected light compensates for fluctuations in the source light. Although this increases the measuring accuracy, it also makes the instrument design more complex.

The readings for $R(\lambda)$ and the tristimulus values are used to calculate the X, Y and Z values in accordance with standardised mathematical principles. Different standard illuminants and observers can be calculated because this is a relative spectral measurement.

### *1.1.3 Method D: Spectrophotometer with monochromatic illumination*

Method D consists in sequential illumination of the sample with light of different wavelengths in the visible spectrum (see Figure II.3). In this case, the reflected light is measured.

As described in Section II.1.1.2, the X, Y, Z values are calculated from the reflectance readings, the standard observer and the standard illuminant. Different illumination/observer scenarios can be determined.

**Disadvantage:** Error-free measurement of fluorescent samples is not possible; see Section II.1.1.4.

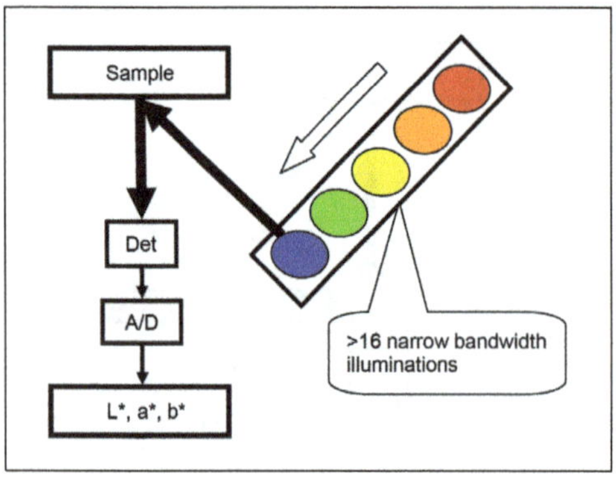


Figure II.3: Method D – sequential spectral measurement with monochromatic illumination.

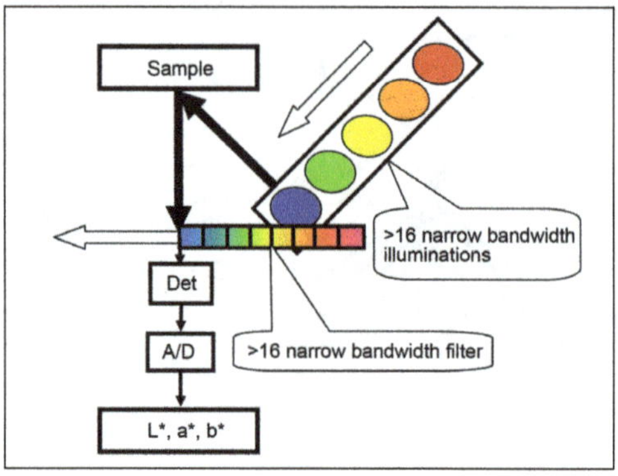


Figure II.4: Method E – bi-spectral measurement.

### *1.1.4 Method E: Bi-spectral measurement*

Method E utilises sequential monochromatic illumination, with sequential or parallel measurement of the light reflected by the sample. A full spectrum of the sample reflectance is

*Table II.1: Principles behind colourimeters.*

| | A | B | C | D | E |
|---|---|---|---|---|---|
| **Brief description** | Tristimulus colourimeter | Spectral measurement with parallel detection; many detectors | Spectral measurement with sequential detection; only one detector | Spectral measurement with sequential illumination; one detector | Bi-spectral measurement with sequential illumination and parallel or sequential detection |
| **Illumination** | Polychromatic | Polychromatic | Polychromatic | Monochromatic | Monochromatic |
| **Applications** | | | | | |
| **Spectral information for colour matching** | No | Yes | Yes | Yes | Yes |
| **Spectral resolution** | | 1 nm to 20 nm | Typically 10 nm to 20 nm | Typically 10 nm to 20 nm | 1 nm to 20 nm |
| **Conversion to different illuminants; metamerism index** | No | Yes | Yes | Yes | Yes |
| **Fluorescence measurement** | Yes | Yes, using the measuring illuminant | Yes, using the measuring illuminant | No | Yes, can be determined using all illuminants |
| **Time needed for measurement** | Short | Short | Long | Long | Very long |
| **Distribution** | Decreasing | Widespread | Widespread | Moderate | Slight |

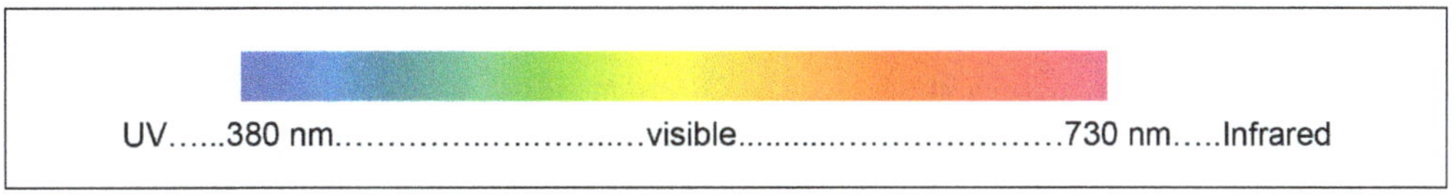


*Figure II.5: Spectral measuring range for colour measurement*

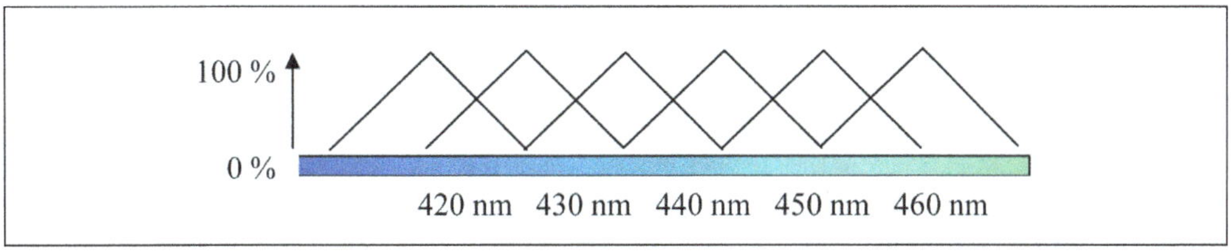


*Figure II.6: Ideal spectral sensitivities of the various receiver channels at 10 nm spectral resolution.*

recorded for each wavelength (see Figure II.4). In this instance, both passive reflection by the sample and fluorescence are measured separately. This means it is possible to determine the colour of a fluorescing sample in any illuminant.

### 1.1.5 Spectral measuring range, resolution and illumination

#### Spectral measuring range

The spectral measuring range is the range over which the human eye experiences colour vision. It extends from at least 400 to 700 nm, and preferably 380 to 730 nm. Some scientific analyses extending beyond visual correlations require wavelengths below 380 nm in the ultraviolet (UV) range and above 730 nm in the infrared (IR) range.

#### Spectral resolution

The spectral measuring range is divided into sub-ranges, in which 20 nm increments (full width at half maximum = wavelength width at 50 % sensitivity) represent the coarsest sub-range that is still useful. Typically, state-of-the-art instruments have resolutions of 10 nm, 5 nm and even 1 nm. Instruments with a full-width-at-half-maximum resolution of at least 5 to 10 nm are recommended. The crucial factor is the resolving power of the internal optics. Dividing the output values by interpolation cannot improve poor optical resolution.

#### Dynamic measuring range

The spectral reflectance factors are 100 % for absolute white and 0 % for absolute black. Two types of samples require a dynamic measuring range above 100 %: fluorescent samples and effect samples, such as metallic paints. The spectral reflectance of fluorescent samples can be significantly higher than 100 % (up to approximately 150 %). For metallic samples, reflectance at small aspecular angles (e.g. 45°:as15°) can well be several hundred percent.

#### Illumination

The tristimulus colorimeter (Method A) requires an accurate light source. The light must be real and lend itself to measurement. Any additionally required illumination must be real and not calculated - unlike the case for the spectral measurements.

For spectral methods B, C, D, the light source is not critical for measurement of non-fluorescent samples, as the light source is eliminated (as described in Section II.1.1.2). Correct measurement of fluorescent samples by methods B and C requires a real light source.

*Table II.2: Light sources for illuminating samples in colour measurement systems.*

| | Advantages | Advantages | Distribution |
|---|---|---|---|
| **Tungsten lamp** | Simple | Low efficiency, very little UV radiation | Widespread |
| **Xenon short arc lamp** | Close to daylight | Expensive, control gear required | Rare nowadays |
| **Xenon flash lamp** | Close to daylight, illuminates only during measurement, very bright compared to ambient light | Power supply, electromagnetic compatibility | Widespread |
| **White LED (blue LED with phosphorus)** | Simple, good durability | No energy below 420 nm | Slight |
| **White light made from several LEDs with different colours** | Good durability | Temperature drift must be compensated | Slight, but growing |
| **Sequential LEDs of different colours** | Simple, good durability | Only limited number of wavelengths available | Slight, but growing |

### *1.1.6 Spectrometers, monochromators and detectors*

As described earlier, it is essential to separate (disperse) the light reflected by the sample so that the intensity of the wavelengths can be measured individually.

The dispersive element (monochromator) in a spectrometer is a grating (often a reflectance grating or a transmission grating nowadays), a set of narrow band filters, or a prism (seldom nowadays).

There are two types of detection:

- Parallel, i.e. simultaneous measurement of all wavelengths by multiple detectors (quick, but high outlay)
- Sequential, with all wavelengths being measured by just one detector (slower, but less outlay)

An alternative to sequential detection is sequential illumination (e.g. by means of narrow band filters or LEDs).

In most modern colour measurement systems, the reflected light is measured with the aid of silicon semiconductors. These may be single detectors or integrated arrays, such as CCD, CMOS and NMOS. In the past, photomultipliers served as single detectors.

# *2 Measuring geometries*

*Gerhard Rösler, translated by Manfred Binder*

Different illumination conditions and viewing angles can yield different impressions of colour and/or colour differences. As with visual assessment, it is important to choose the right measuring geometry when it comes to instrumental colour measurement. Measuring geometries have been specified in international standards for many years now. However, with all applications, it is essential to find the geometry that will yield the right results as it can critically affect the usefulness of the colourimetric readings. New measuring geom-

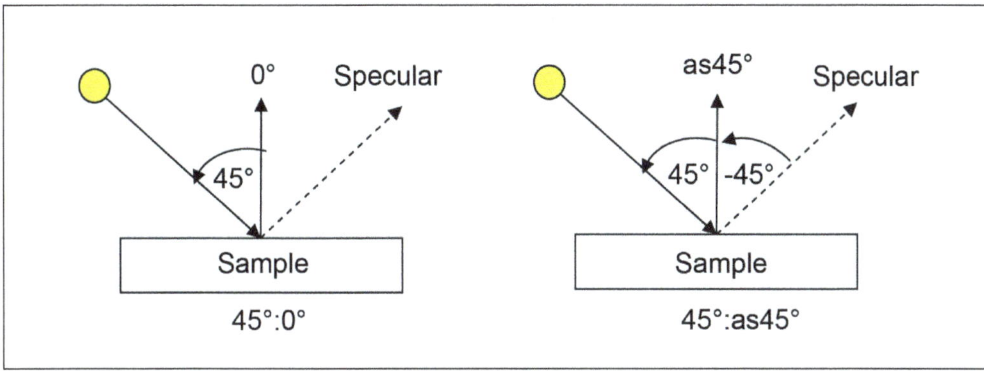


*Figure II.7: Angel definitions - conventional (calculated from the surface normal) and aspecular.*

etrics have been successfully employed in multi-angle instruments for quite some time. Such multi-angle measurements will soon find applications beyond that of effect paints.

The convention adopted for specifying measuring geometries is as follows: *xx*°:*yy*° or *xx*°:as*zz*°

The first term (*xx*) specifies the illuminating angle, always measured from the surface normal, while the second term (*yy*) indicates the viewing angle, measured from the surface normal. Alternatively, the term (as*zz*) signifies the viewing angle, as measured from direct specular reflection. ("as" stands for aspecular). For effect paints, the "as" term is highly descriptive of the measuring geometry, because it is the difference between the viewing angle and the specular reflection angle.

**Example A:** 45°:-30° is equivalent to 45°:as15°; the measurement is very close to specular reflection; the orientation of the aluminium or other effect flake will exert a strong influence here.

**Example B:** 45°:30° is equivalent to 45°:as75°; far away from specular reflection; scattering by the colourants will exert a strong influence.

Instead of having to deal with two angles and to do some calculations, the "as" notation makes it possible to immediately understand the distance from specular reflection. When discussing colour, it is highly advisable to add the "as" (e.g. as25°) in order that confusion may be avoided. In the past, it was unnecessary to indicate the illuminating angle, because all multi-angle instruments had a constant angle of 45° (illuminating or viewing, e.g. 45°:as15° or the reverse: as15°:45°). Now, though, new measuring instruments have a second illumination source for characterising interference effects. As a result, it has become necessary to define the illuminating angle, e.g. to distinguish 15°:as15° from 45°:as15°. For practical work, fully specifying just the 15°:as geometries is sufficient, because in most cases the conventional 45°:as geometries remain the same as before and so the addendum "45°:" is superfluous.

→ **Note:**
When it comes to measuring interference pigments, only the geometry close to specular reflection (as ±15°) is important, as that is where these pigments exhibit their specific characteristics.

The perceived colour of an object is the sum of the surface reflectance (also called *Fresnel* reflectance, first surface reflectance or specular component) and the typically diffuse

*Table II.3: The most common geometries of systems for reflectance colour measurement.*

| **Typical name** | **Sphere geometry "Specular included"** | **Sphere geometry "Specular excluded"** | **45°:0° 0°:45°** | **Coaxial illumination and measurement x°:x°** | **Colour measurement in specular reflectance for non-scattering samples x°:x°** | **Multi-angle measurement, 1 to 1 illumination** | **Multi-angle measurement, 2 to 2 illumination** |
|---|---|---|---|---|---|---|---|
| Geometry | di:8° | de:8° | | | | | |
| Geometry | | | | | | | 45°:as-15° |
| Geometry | | | | | | 45°:as15° | 45°:as15° |
| Geometry | | | | | | 45°:as25° | 45°:as25° |
| Geometry | | | 45°:as45° 0°:as45° | | | 45°:as45° | 45°:as45° |
| Geometry | | | | | | 45°:as75° | 45°:as75° |
| Geometry | | | | | | 45°:as110° | 45°:as110° |
| Geometry | | | | | | | 15°:as-15° |
| Geometry | | | | | | | 15°:as15° |
| Geometry | | | | E.g. 22.5°:22.5° | E.g. 15°:as0° and / or 45°:as0°, 60°:as0° | | |
| Variation | | de:0° in the paper industry | Special version for non-contact measurement | Non-contact measurement for large distances | Non-contact measurement | Special versions for non-contact measurement | |
| Standard | DIN 5033 | DIN 5033 | DIN 5033 | | | DIN 6175-2, ASTM E2194 | ASTM E2539 |
| Introduced | <1940 | <1940 | <1940 | 1985 | 1985 | 1989 (1982 preliminary version) | 2008 |
| Distribution | Very widespread | Widespread | Widespread | Slight | Slight | Slight | New on the market |
| Industry | A great many | A great many | A great many | Some | Glass coaters | Automotive and others | Automotive and others |
| Sample types | A great many | A great many | A great many | A great many | Low-E and sun-blocking glass | Metal effect and grained samples | Interference and metallic pigments |

reflectance from the body of the object. Surface reflectance (typically approx. 4% of the incident light) depends only on the refractive index, **not** on the colour of the body. Thus, surface reflectance always has the same colour as the illumination. The texture of the surface determines the spatial distribution of the first surface reflectance. This ranges from high-gloss to matte, with infinite variations. All of these create different visual impressions.

The di:8° ("specular included", see Section II.2.1) geometry always captures the total reflectance of the specimen, irrespective of the surface texture.

The 45°:0° geometry only captures the surface reflectance normal (0°) to the sample surface.

→ **Note:**
The phrase "specular excluded" only holds for high-gloss samples measured at 45°:0° and de:8° (see Section II.2.1). These geometries exclude first surface reflectance (gloss) completely only in the case of high-gloss samples. With di:8° measurements, the specular component must be subtracted by the software if it is not needed for further calculations like colour matching. The exact value depends on the refractive index of the material measured and therefore needs to be adjusted. See also ISO 7724-5 (Saunderson correction).

## 2.1 Sphere geometries for reflectance measurements

### *Diffuse:8° geometry, specular included (di:8°)*

The following conditions must be met for the di:8° geometry:

- The sample is illuminated evenly from all directions.
- The sample is in contact with the port of the Ulbricht or integrating sphere, the interior of which is covered with a white reflective coating.
- The measuring area must be smaller than the illuminated area and must be uniformly evaluated.
- The axis of the optical path is offset by 8° from the sample normal.
- The readings can be affected by the following factors:
  - Size of the port
  - Size and uniformity of the illumination
  - Size and uniformity of the optical path
  - The ratio of the area of the sphere ports to the internal area of the sphere (should be less than 10%)

### *Diffuse:8° geometry, specular excluded (de:8°)*

Here, the basic conditions are the same as for di:8°, but with the following additions:

- Unlike the case for di:8° geometry, there must be no illumination of the sample at -8° from the sample normal.
- Complete exclusion of specular gloss is only possible for high-gloss surfaces.

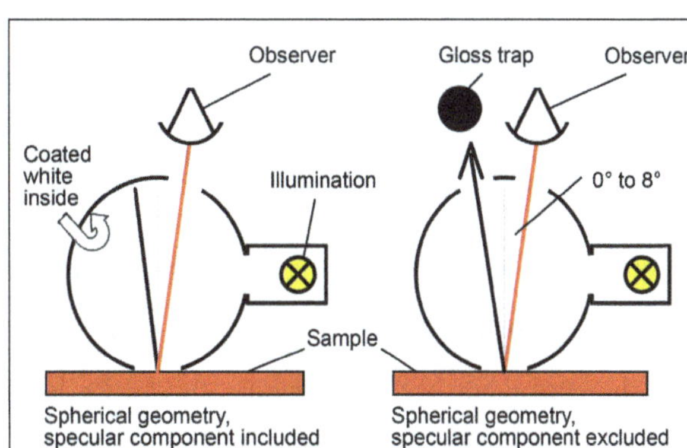


*Figure II.8: Sphere geometry di:8°(left) and de:8° (right).*

### *8°:diffuse geometry, specular included (8°:di)*

The same geometric conditions as for di:8 geometry, but with a reversed illumination path. Thus, the sample is illuminated at 8°, and the light reflected hemi-spherically from the sample is measured at all angles. The illuminated area of the port must be smaller than the port itself.

### *8°:diffuse geometry, specular excluded (8°:de)*

The same geometric conditions as for de:8° geometry, but with a reversed illumination path. Thus, the sample is illuminated at 8°, and the light reflected hemi-spherically from the sample is measured at all angles, but not at -8°. The illuminated area of the port must be smaller than the port itself.

### *Diffuse:0° geometry (de:0°), specular excluded*

The following conditions must be met for de:0° geometry:

- The sample is illuminated evenly from all directions, except at the sample normal.
- The sample is in contact with the port of the Ulbricht or integrating sphere, the interior of which is covered with a white reflective coating.
- The measuring area must be smaller than the illuminated area and must be uniformly evaluated.
- The axis of the optical path is normal to the sample.
- Complete exclusion of specular gloss is only possible for high-gloss surfaces.
- The readings can be affected by the following factors:
  - Size of the port
  - Size and uniformity of the illumination
  - Size and uniformity of the optical path

## 2.2 Directional geometries for reflectance measurements

For the directional geometries, namely 45°:0° or 0°:45°, the method for 45° illumination or 45° viewing can be sub-categorized as follows:

- a = from all 45° directions (annular),
- c = from several 45° directions (circumferential)
- x = from only one 45° direction.

### *45° annular:0° geometry (45°a:0° = 45°a:as45°), specular excluded*

The following conditions must be met for the 45°a:0° geometry:

- The sample is illuminated from all 45° directions.
- The axis of the optical path lies exactly on the sample normal.
- The measuring area must be smaller than the illuminated surface and must be uniformly evaluated.
- Complete exclusion of specular gloss is only possible for high-gloss surfaces.

### *0°:45° annular (0°:45°a = 0°:as45°a), specular excluded*

The following conditions must be met for 0°:45°a geometry:

- The same geometric conditions as for 45°a:0° geometry, but with illumination and viewing reversed. Thus, the sample is illuminated at the normal, and the light reflected from the sample is measured in all 45° directions.
- The illuminated surface must be smaller than the measuring area.

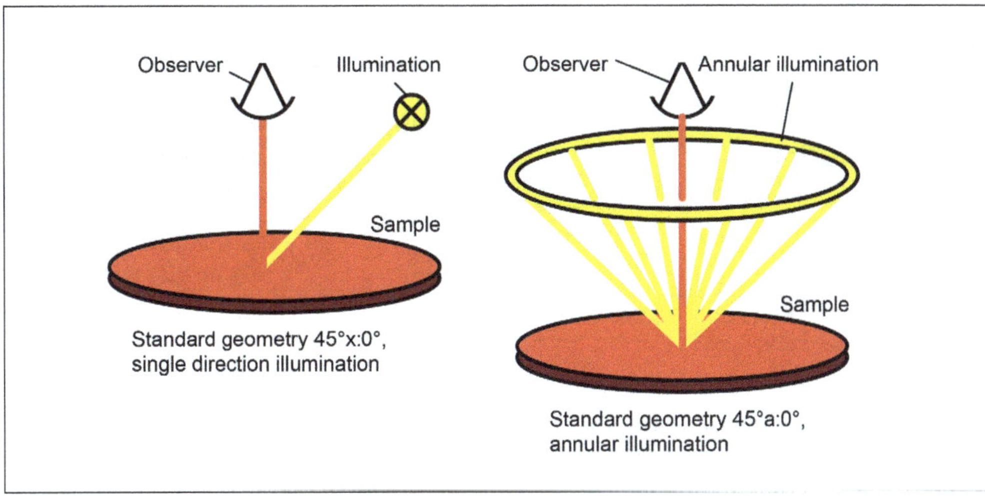


*Figure II.9: 45°x:0° Geometry (left) and 45°a:0° (right).*

### *45° circumferential:0° geometry (45°c:0° = 45°c:as45°), specular excluded*

The following conditions must be met for 45°c:0° geometry:

- The sample is illuminated from several directions at 45° to the sample normal.
- The axis of the optical path is exactly on the sample normal.
- The measuring area must be smaller than the illuminated surface and must be uniformly evaluated.
- Complete exclusion of specular gloss is only possible for high-gloss surfaces.

### *0°:45° circumferential geometry (0°:45°c = 0°:as45°c), specular excluded*

The following conditions must be met for 0°:45°c geometry:

- The same geometric conditions as for 45°c:0° geometry, but with reversed illumination path. Thus, the sample is illuminated on the normal, and the light reflected from the sample is measured at several 45° directions.
- The illuminated surface must be smaller than the measuring area.

### *45° single source:0° geometry (45°x:0° = 45°x:as45°), specular excluded*

The following conditions must be met for 45°x:0° geometry:

- The sample is illuminated from just one direction at 45° to the sample normal.
- The axis of the optical path lies exactly on the sample normal.
- The measuring area must be smaller than the illuminated surface and must be uniformly evaluated.
- Complete exclusion of specular gloss is only possible for high-gloss surfaces.

### *0°:45° single source (0°:45°x = 0°:as45°x), specular excluded*

The following conditions must be met for 0°:45°x geometry:

- The same geometric conditions as for 45°x:0° geometry, but with reversed illumination path. Thus, the sample is illuminated on the normal, and the light reflected from the sample is measured in just one 45° direction.
- The illuminated surface must be smaller than the measuring area.

*Table II.4: Table for selecting reflectance measuring geometries for different sample types and sample properties.*

| Property of sample | | | Goal of colour measurement | |
|---|---|---|---|---|
| **Sample body** | **Sample surface** | **Sample material** | **Best geometry for good visual correlation, e.g. final control** | **Best geometry for measurements requiring maximum independence from sample surface, e.g. colour matching** |
| Scattering, translucent to opaque (not transparent) | Matte or silky-gloss | Coating | 45°:0°, x°:x° | di:8° |
| Scattering, translucent to opaque (not transparent) | High gloss | Coating | 45°:0°, de:8°, x°:x° | di:8° |
| Metallic | Any | Coating | Multi-angle geometry 1 | Multi-angle geometry 1 |
| Interference | Any | Coating | Multi-angle geometry 2 (2 illumination directions) | Multi-angle geometry 2 (2 illumination directions) |
| Transparent over a non-scattering substrate (e.g. polished metal) | High gloss | Coating | di:8°, x°:as0° | di:8° |
| Transparent over a scattering substrate | High gloss | Coating | 45°:0°, x°:x°, multi-angle geometry 1 | di:8° |
| Scattering | Matte | Powder | 45°:0°, de:8°, di:8°, x°:x° | di:8° |
| Scattering, measurement through a glossy glass pane | High-gloss glass cover | Powder | 45°:0°, de:8°, x°:x° | 45°:0°, de:8° |

## 2.3 Measuring geometries for different sample types and sample properties

Table II.4 lists the various geometries for measuring matte to high-gloss surfaces.

## 2.4 Recommended geometries for transmission measurements

### *Direct:direct geometry (0°:180°)*

For the direct/direct geometry (0°:180°), the illumination is positioned on one side of the sample (direct) and therefore is normal to the surface. The measuring port is on the opposite side (direct).

→ **Note:**
These requirements are more or less met by measuring instruments described in Section II.2.1 (sphere geometries) in which there is sufficient distance between the light-emitting port of the sphere and the measuring optics.

### *Diffuse:direct measuring geometry (di:180°), total transmission*

For the diffuse:direct measuring geometry, the sample is illuminated in a closed Ulbricht sphere with a highly reflective white coating on the inside; this illuminates the sample evenly from all directions. For di:180° geometry, the sample has to be in contact with the measuring port. The measurement optics (direct measurement) are located on the opposite side of the sample. Total transmission includes the diffuse and the direct components of light transmission.

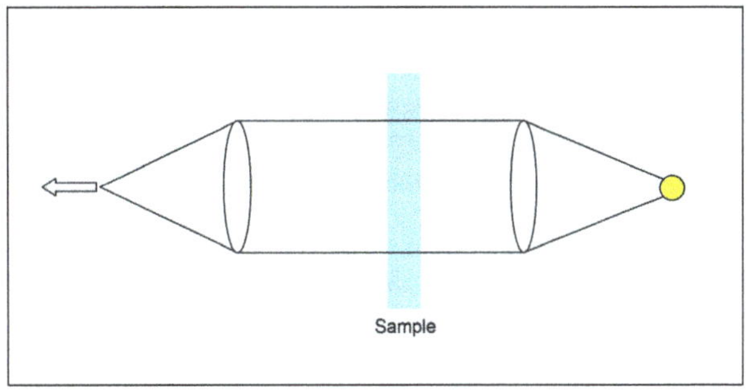


*Figure II.10: Geometry for transmission measurement 0°:180°.*

### *Diffuse:direct measuring geometry (de:180°), diffuse transmission*

The sample is placed directly against the sphere with a highly reflective white coating on the inside. On the opposite side of the sphere, there is an open port (often used for reflectance measurement). Thus, the sample is illuminated diffusely, without the direct component. The measuring optics are on the opposite side of the sample. Diffuse transmission includes only the diffuse component (not the direct component).

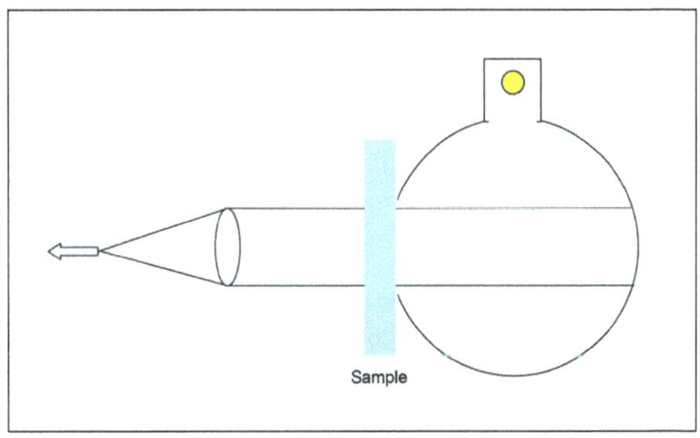


*Figure II.11: Total transmission di:180°.*

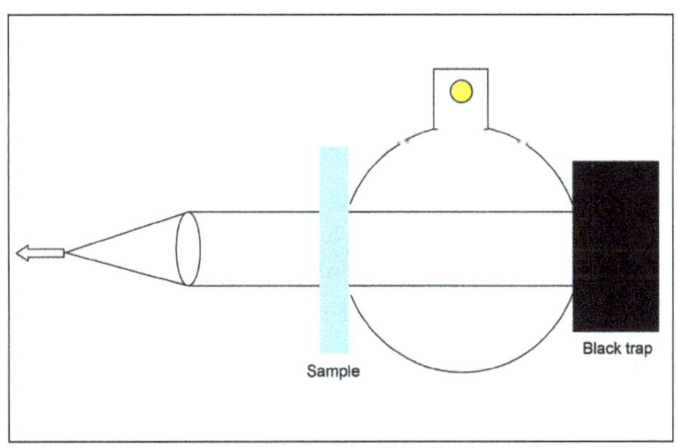


*Figure II.12: Diffuse transmission de:180°.*

### *Direct:diffuse measuring geometry (180°:di), total transmission*

The same geometric conditions as the di:180° geometry, but with reversed illumination path. Thus, the sample is illuminated at 180°, and the light reflected hemi-spherically from the sample is measured at all angles.

### *Direct:diffuse measuring geometry, (180°:de) diffuse transmission*

The same geometric conditions as for de:180° geometry, but with a reversed illumination path.

## 2.5 Notes on choosing the right geometry

→ **Note 1:**
It is important to specify the illumination and measuring conditions actually employed, even if they fall within the range of one of the aforementioned standard conditions. Some types of sample (e.g. retro-reflective materials) might need to be measured with other geometries and tolerances.

→ **Note 2:**
For the instrument manufacturer: If an integrating sphere (Ulbricht sphere) is used, steps must be taken to ensure that light from the light source does not directly illuminate the sample.

If the directed reflectance component (gloss) is to be taken into account, the sphere efficiency for that part of the sphere which absorbs the directed reflected light must have the same reflectance value as the rest of the sphere. Non-reflecting surfaces inside the integrating sphere must comprise less than 10 % of the sphere's total surface area.

→ **Note 3:**
When translucent samples are being measured, it should be borne in mind that light loss occurs, as some light will leave the translucent sample on the side of the measuring port. This will lead to measurement error and must be allowed for or minimised by appropriate steps during measurement.

→ **Note 4:**
When an integrating sphere is used to measure luminescent samples, the entire illumination system is influenced by the sample. Certain wavelengths absorbed and emitted by the sample will additionally illuminate the sphere and influence the radiation characteristics of the entire system (see CIE 76-1988: Intercomparison on measurement of (total) spectral radiance factor of luminescent specimens, 1988). For this reason, the area of the measuring port in a sphere should be less than 0.3 % of the total sphere surface area. An even better alternative is to use a 45°:0° or 0°:45° instrument.

## 2.6 Multi-angle geometries

Multi-angle measurements are needed for assessing directional colour properties of materials with surface effects (e.g. paints without high gloss surfaces, leather and plastic films with macro structures) as well as effects due to the volume of the material (e.g. metallic and interference).

### *Description of multi-angle geometries*

- The illumination is made at a certain angle, typically 45°, and sometimes 15° as well.
- For 45° illumination, the specular angle is -45° to the sample normal while it is -15° for 15° illumination.
- This specular reflectance angle is the reference angle for specifying the instrument geometry, and is called "as" for "aspecular". Aspecular angles are positive in the direction of (towards) the illumination; otherwise, they are negative.

### *Characterisation of metallic effects and surface effects*

The following angles are particularly useful for metallic and other surface effects:

- 45°:as15°, 45°:as25°, 45°:as45°, 45°:as75°, 45°:as110°

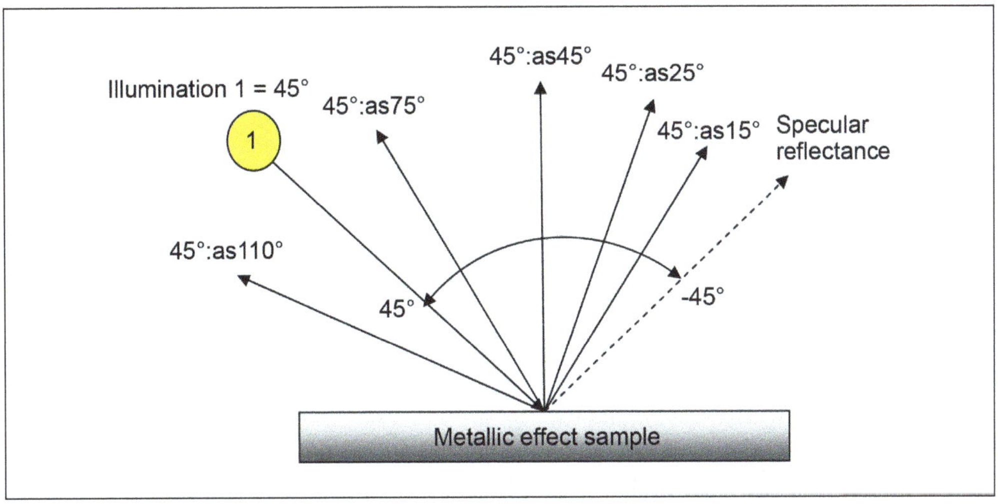


*Figure II.13: Important multi-angle geometry for measuring metallic and other effects.*

# 3 Measuring geometries for special effect pigments

*Peter Gabel*

The ongoing development of interference or special effect pigments (see Section X.4) poses a big challenge for the supplier of colour measurement systems. While metallic paints' characteristics are well defined under direct 45° illumination and a set of 3 or 5 aspecular viewing angles (see Figure II.13) published in DIN 6157-2 [1] and ASTM E2194 [2], instruments with fixed 45° illumination are unable to measure the extraordinary colour shift of the new 2nd generation effect pigments, irrespective of the number of aspecular angles. Appearance parameters such as sparkle, texture and diffuse coarseness are becoming more and more important. All these visible characteristics of effect pigments therefore evade capture by a "simple" measurement.

Colour measurement and colour management play a central role in quality control systems employed in many branches of industry and consumer goods production, and especially the paints industry. The reason is that paints characterised and approved on the basis of the supplier's instrumental colour data are becoming an ever greater cost factor. Furthermore, the sharing and transfer of data between suppliers and customers is a question not only of providing information, but also one of corporate culture, trust, and mutual respect. For this reason, it is essential that customer and supplier be in precise agreement about measuring angles and geometries when it comes to product analyses and specifications.

## 3.1 Optical principles behind special effect pigments

The optical principles used to differentiate and classify special effect pigments [3] (see Section X.4) are wholly different from those employed for white pigments, coloured pigments, black pigments, and metallic pigments.

**White pigments:** The pigment particles scatter all the incident light equally in all directions (diffuse reflection).

**Coloured pigments:** The pigment particles absorb certain wavelengths of visible light and scatter the rest (selective absorption).

**Black pigments:** The pigment particles absorb all wavelengths of visible light (complete absorption).

**Metallic pigments:** These consist of thin metal (mostly aluminium) platelets, are not transparent, and reflect all incident light in one direction. The pigment particles act like small mirrors, creating a reflecting metal lustre (metallic effect) if they are oriented in parallel during application.

**Interference pigments and transparent interference pigments:** These selectively reflect that portion of the light, which is directly incident on the smooth platelet surface. The other portion of the light enters the transparent particles and is again partly reflected either by interfaces within the pigments or by the underside of the platelets. The internally reflected light leaves the platelets, and can interact with adjoining pigment particles to create further (multiple) reflections. The partial reflection of the light and optical superposition create interference phenomena, which, together with multiple internal reflections, are responsible for the characteristic appearance of interference pigments.

**Multilayer and Fabry-Perot interference pigments:** These mainly utilise a metal-dielectric (a reflector impermeable to light) and the variable film thickness of the transparent dielectric layers.

**Diffractive pigments:** These reflect/transmit incident light on wedge-shaped grating structures and groove geometries (Echelette grating). Additional fringe spacing maximises diffraction of transmitted light in the direction of the incident light and diffraction of reflected light in the direction of the reflected light.

**Holographic pigments:** These pigments are based on highly oriented grooves (straight, parallel, equidistant alignment).

More details on the physical and mathematical calculations can be found in [4] and Section VII.2.

## 3.2 Measuring geometries for metallic pigments

In the early 1990s, metallic paints were already very popular in the automotive coatings industry when interference pigments emerged onto the scene and the best geometries for measuring the two effect pigments soon became a matter of debate. However, there was general agreement that the conventional measuring geometries 45°:0° and 0°:45° were inadequate. Metallic colours can be characterised effectively by measurements at three angles: the recommended DuPont Model, based on Alman's study, utilised 45° illumination under three aspecular angles, namely 15°, 45° and 110° [5]. This configuration was able to measure the light/dark flop typical of the metallic effect (see Figure II.13). However, because these DuPont geometries were restricted by patent, other instrument makers had to offer new aspecular angles. The development of the goniospectrophotometer, which was based on different concepts, was of little use for conducting comparative studies. The breakthrough came in 1974 with the "Trilac" concept proposed by Leres and Hofmeister. For more historical background, see [6, 7].

## 3.3 Measuring geometries for special effect pigments

A key characteristic of effect pigments is that the interference colour varies with the angle of illumination. According to the Fresnel equation for light reflection, as the illuminating

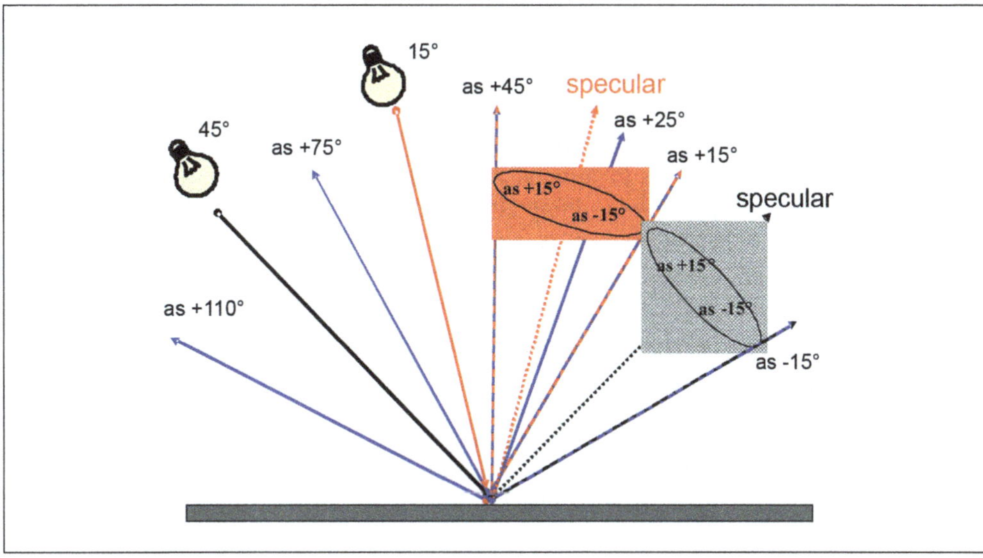


*Figure II.14: Measurement geometries ASTM E2539 "Practice for Multiangle Colour Measurement of Interference Pigments".*

angle increases, the reflection maximum shifts to shorter wavelengths, and is accompanied by an increase in the reflectance value. For a given layer thickness, the angle of incidence is crucial, because the interference colour depends on the phase difference of the reflected light rays. In theory, therefore, proper characterisation of an effect paint would require illuminating angles of 0° through to 90° and a constant viewing angle, e.g. as15°. However, the resultant mountain of data would be overwhelming for the practical user. Clearly, the amount of data needs to be reduced. Looking back, it is clear that it was the absence of standards, which led to the development of the goniospectrophotometer based on different concepts.

It was not until early 2000 that standards bodies such as DIN and ASTM published test procedures (DIN 6157-2 and ASTM E2194).

*Table II.5: CIE convention of illumination and observation for colour measurement geometries.*

| Method | Method 1 | Method 2 | Method 3 |
|---|---|---|---|
| **Discription** | Conventional | Aspecular | CIE |
| **Illumination** | Anormall | (Single) | Anormal |
| **Observation** | Anormal | Aspecular | Aspecular |
| **Standard** | E 2194 | E 2175 | |
| **Conventional metallic-pigments (ASTM E 2194, DIN 6175-2) 1st generation effectpigments** | 45°:-30°<br>45°:-20°<br>45°:0°<br>45°:30°<br>45°:65°* | 15°<br>25°<br>45°<br>75°<br>110° | 45°:as15°<br>45°:as25°<br>45°:as45°<br>45°:as75°<br>45°:as110 |
| **New (ASTM E2539) 2nd generation effectpigments** | 15°:0°<br>15°:-30°<br>45°:-60°<br>65°*:-50° | 15°<br>-15°<br>-15°<br>15° | 15°:as15°<br>15°:as-15°<br>45°:as-15°<br>65°:as15° |

** 65°:as15° Illumination has a conflict with observation position of conventional colormeasurement systems 45°:as110*

Comparison of instrument readings show that, up until then, only the visible behaviour of the 1st generation of effect pigments was being detected, but not that of the 2nd generation [8]. In March 2008, the members of ASTM E12 group "Colour and Appearance" voted to adopt ASTM E2539 "Practice for Multiangle Colour Measurement of Interference Pigments [9]." The agreed geometries are shown in Table II.5 and Figure II.14. A third type of illumination - measuring geometry 65°:as15° - is suitable only for bench top systems, e.g. ETA/STEAG, now known as the Datacolor-ETA System FX10. For design reasons, this illumination configuration conflicts with the measuring geometry 45°:as 110° used in portable systems. The German DIN standard will issue in the near future.

## *3.4 New measuring geometries - applications for special effect pigments*

Colour readings produced by varying the illuminating angle 15°, 45° and 65° and keeping the aspecular angle constant (as +15°, as -15°) can be portrayed two-dimensionally in the a*b* section of the L*a*b* colour chart by plotting the interference shift ("interference line"; see Figure II.15). The interference line characterises each interference pigment in the manner of a fingerprint. These geometries are, moreover, extremely helpful for identifying unknown paint formulations.

Figure II.16 shows the a* b* readings for a paint formulation containing different concentrations of blue pigment, alumina pigment and a blue interference pigment. The readings show that the "aspecular line", i.e. different aspecular angles at 45° illumination, is unsuitable for identifying or characterising the effect pigment in this paint formulation. However, the "interference line" in the a* b* diagram, measured at different illuminating angles but constant + and - aspecular angles, reveals the increasing concentration of the blue interference pigment. When the a* b* readings from the + and - aspecular angles are combined, a typical anchor shape emerges. The higher the concentration of the interference pigment, the more pronounced is the shape of the anchor. This investigation has been repeated for various

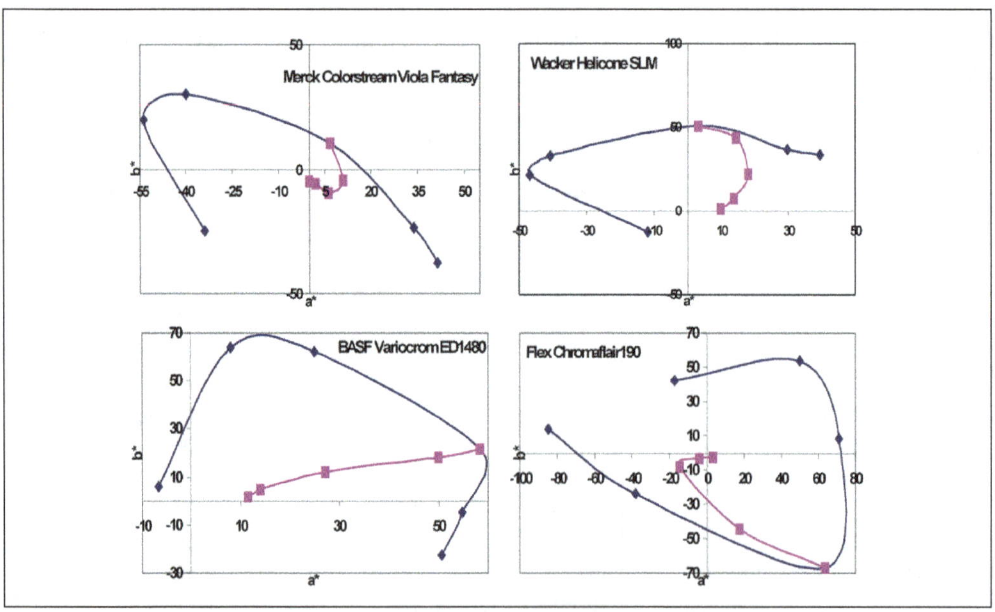


*Figure II.15: Interference and aspecular line of different special effect pigments (fullshade black background).*

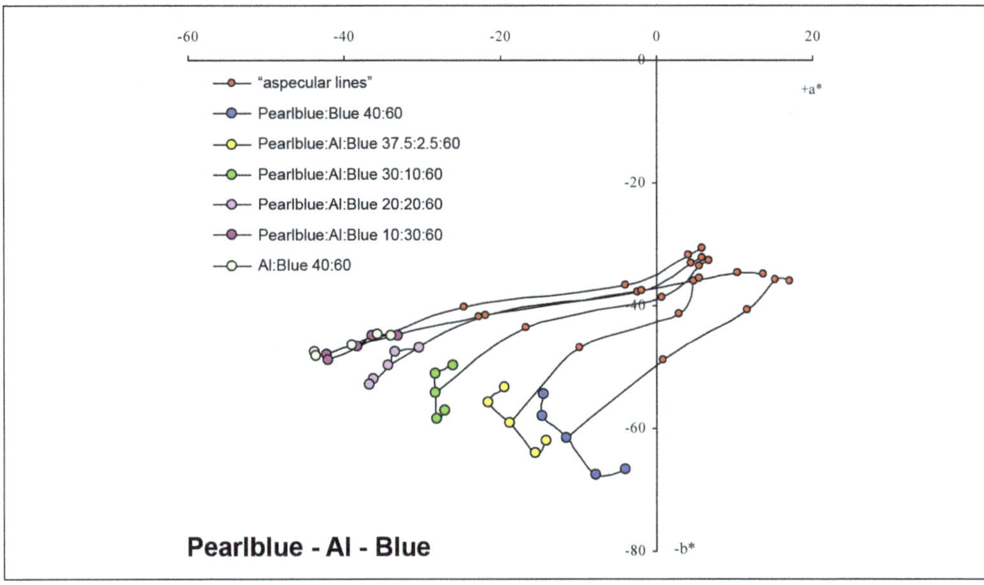


*Figure II.16: Mixture of special effect pigments like interference, aluminium and absorption pigments. The interference lines (anchor shape) are well observed with increasing concentration of the interference pigment in the mixture.*

interference and absorption pigments. The readings clearly demonstrate that interference/effect pigments can only be uniquely characterised by changing the illuminating angle.

### *3.4.1 New measuring geometries - new developments*

Many of the multiangle, bench-top colour-measurement systems, such as the Datacolor-ETA-System FX 10, Murakami GCMS-3B, Instrument Systems GON 360, and Zeiss GK311/M, meet the new requirements: variable illuminating angles, classic aspecular angle, + and - aspecular angles, and a reduced number of geometries. Nevertheless, a number of instrument manufacturers are working on the development of good handheld systems that are both efficient and easy to use.

At the 2008 American Coating Conference, *Nisper* [11] unveiled the X-Rite MA98 as a new tool for measuring an appearance property of a surface using a spatially under-sampled bidirectional reflectance distribution function (BRDF). This handheld instrument measures spectral reflectance curves for the five aspecular standard geometries at 15°, 25°, 45°, 75° and 110°. Aside from these five angles, its two 45° and 15° illuminators measure the respective as-15° aspecular angles. With the exception of

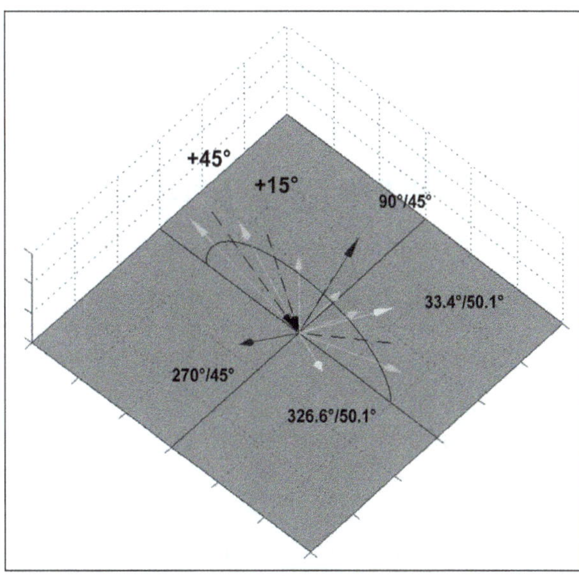


*Figure II.17: Measurement geometries of X-Rite MA98 (see Table II.5) in addition with "Out of Plane" geometries.*

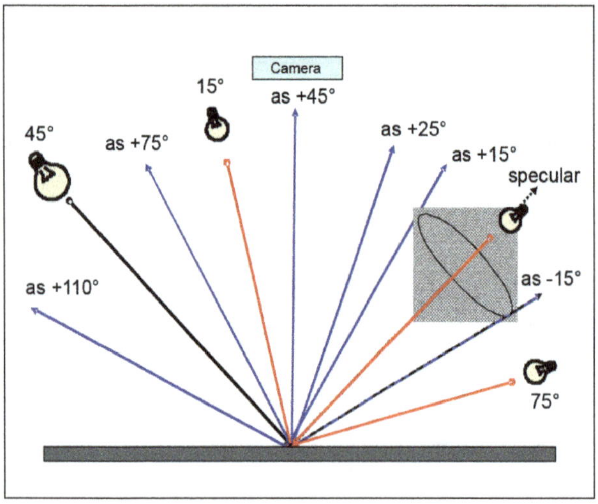


*Figure II.18: Measurement geometries of BYK-mac for colour measurement and sparkle characterization.*

measuring geometry 65°as15°, this system complies with ASTM E2539. That illumination configuration conflicts with the measuring geometry 45°:as 110° and was excluded from the portable version for design reasons.

All the angles of measurement above yield "in plane" readings. To measure appearance properties, the BRDF of a surface is measured out of plane with two sets of measuring geometries (see Figure II.17). X-Rite promoted a new software tool (a three-dimensional mathematical model). The company promises that by using and calculating the results from all these geometries the user can measure special effects, such as sparkle and coarseness, and also troubleshoot defects whether due to application or to pigmentation. The software is currently being widely tested.

Quality control, e.g., of automotive coatings, requires not only colour measurements but, increasingly, appearance-related readings as well. Drawing on research [12] by Akzo Nobel and Merck KGaA, BYK Gardner GmbH launched the BYK-mac. One of the reasons for this was the growing popularity of Xirallic effect pigments which exhibit living sparkle, i.e. numerous tiny yet highly intense (often white) spots of light that show up under an intense unidirectional light source, such as the sun.

This new instrument uses a digital camera and pixel intensities to calculate sparkle and coarseness. The sparkle is measured under direct illumination at -15°, 45° and 75°. The digital camera is located in the 0° position. Coarseness is measured under diffuse illumination. For the purposes of colour measurement, the instrument measures spectral reflectance curves under the five standard geometries as15°, as25°, as45°, as75° and as110° plus as-15° (see Figure II.18) in accordance with the new ASTM E2539.

# *4 Sample preparation*

*Gerhard Rösler, translated by Manfred Binder*

As described in ISO 15528, paint samples need to be representative of the paint in question. The right sampling device depends on the type of sample, the type of container, the hazard potential and the quantity of sample needed The sampling device should be:

- easy to handle
- easy to clean (smooth surfaces)
- commercially available
- durable and inert to the product

**Examples**: Ladle or scoop (for surface sample), plunging siphon, dip tank (samples taken from different depths), branch pipe (continuous sampling during pumping). When checking a batch, it is necessary to take samples from several containers:

| | |
|---|---|
| 2 to 8 containers in a batch: | 2 samples, |
| 9 to 25 containers: | 3 samples, |
| 26 to 100 containers: | 5 samples, |
| 101 to 500 containers: | 8 samples, |
| 501 to 1000 containers: | 13 samples, |
| More than 1000 containers: | (number of containers/2)$^{0.5}$ = number of samples |

Sample preparation is described in ISO 1513.

A pre-inspection to assess visual disturbance includes:

- skin formation
- change in consistency
- separation into layers
- visible contamination
- clarity and colour
- sediment

### *Sample panels complying with ISO 1514*

Before samples are prepared for colour measurement, it is important to properly define the substrates. One possibility is to use the substrate for which the paint is made. In that event, the pre-treatment regime is agreed on and is recorded in the laboratory report. For individual tests, special sample panels prepared according to detailed instructions can be used. Alternatively, different kinds of standardised sample panels can be used which must be prepared according to ISO 1514 before they are coated.

The procedure for coating the panel must be specified and the panel must always be prepared in the same way. The coating can be applied by spraying, brushing, with a coating knife or the panel might just be air-dried or stoved. The coating may also have to be post-treated in accordance with the manufacturer's specifications.

### *Notes on sample properties*

- If the sample is thermochromic, i.e. it changes colour reversibly when the temperature changes, maintain a specified temperature when measuring it. (For example, use a temperature-controlled sample holder or allow the sample enough time to adapt to the regulated room temperature).
- Non-thermochromic samples must be pretreated until the sample colour is stable.
- The colour of non-opaque samples is decisively influenced by the substrate and the layer thickness. These must therefore be determined by one of the methods described in ISO 2808.
- The sample must be free of scratches, dust and other contaminants (e.g. fingerprints).
- For process-control purposes, consider taking representative samples. The number of samples must be agreed by the parties involved. For accurate results, samples should be flat. Take several measurements at different locations on the sample; the precise number depends on the size and uniformity of the sample. There must be no outliers or drift in the individual readings. The measuring area should be as large as possible.
- The geometrical conditions for measuring flat samples do not apply to curved samples. The more curved the samples are, the greater will be the measuring uncertainty and the more the readings will differ from those yielded by flat samples. Samples should therefore be measured at the point of least curvature and the measurement spot should be as small as possible. This will minimise the geometrical error. For improved repeatability of measurements, use a holder specially designed for the sample.

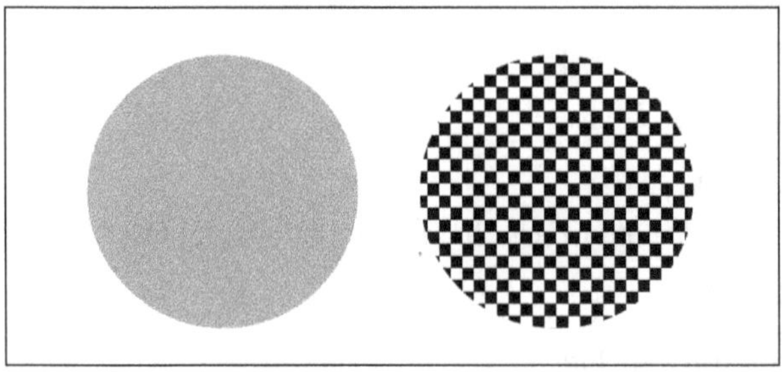

*Figure II.19: Sample with patterns.*

- A given measuring geometry may yield different readings for samples that differ in their surface properties, e.g. as regards grain, gloss. This is desirable in the quality assurance of finished products, as it matches the visual impression given. But it may be undesirable for paint formulation as the colour itself needs to be optimised first. The correct surface is generated in the second step of production (see Section II.6).
- Limitations: The type of colour measurement described here looks at a specific area on the surface and captures the average value. The human eye, however, sees an image and any patterns therein. Consider an extreme example: a measurement spot that is half-black and half-white will be interpreted by the instrument as grey. If this sample is compared with a real grey, the instrument will fail to detect a difference, whereas the human eye will see the pattern straight away. In industrial use, samples are often homogeneous; consequently, such patterns are of no significance if the measured surfaces are chosen carefully (see Figure II.19).

# 5 Recommended colourimetric conditions

*Gerhard Rösler, translated by Manfred Binder*

**Light source $D_{65}$** (daylight) should be used for measurements.

The metamerism index is akin to a colour difference, i.e. it is an overall evaluation only, without individual colourimetric components. Determination of metamerism entails mathematically calculating the chromaticity coordinates for several different light sources (this is only possible for instrument types B, C, D, and E). Apart from $D_{65}$ (daylight), types A (incandescent) and F11 (energy saving lamp) are often used. This gives rise to three metamerism indices: between $D_{65}$ and A, between $D_{65}$ and F11 and between A and F11.

An alternative to the metamerism index is to test that the tolerances for all required light sources are met. This means that the tolerances applicable to $D_{65}$ must also apply to light sources A and F11. The advantage of this is that any deviations revealed are quantified in $\Delta L^*\Delta a^*\Delta b^*$.
**Observer:** 10°
**Colour space:** CIELAB
**Colour differences:** It is important to understand that in CIELAB colour space small colour differences (<3 units) cannot be perceived as equidistant. CIELAB colour space is based on data obtained by the painter *Albert Munsell* in 1915 and was created in increments of 10 units. In industry, tolerances of less than 1 unit are common and so measures must be taken to arrive at results that correlate visually. The human eye is not as adept at discriminating between small colour deviations in highly saturated samples as it is between large deviations. It is well to bear this in mind in quality control, so that time is not spent trying to arrive at unnecessarily small numbers that are not visible anyway. This streamlines the procedure, without disadvantaging the customer. This situation led to the development of colour difference correction formulas, such as CMC, CIE94, CIEDE2000, and a completely new colour space: DIN 99 (DIN 6176 ), see Section V.

## 5.1 Calibration of the colour-measuring instrument

Proper calibration of colour-measuring instruments is one of the most important factors governing reproducible results. Incorrect or poor calibration can have a negative impact and generate false readings downstream. For maximum accuracy, it is essential to accurately observe both the basic conditions for calibration (temperature, positioning of the calibration standard) and the manufacturer's calibration instructions. The calculations based on the readings for the black and white standards are performed automatically by the measuring instruments or by the software. For this reason, only the underlying principle will be described below.

## 5.2 Ambient conditions

Equipment should be calibrated under the same ambient conditions as subsequent sample measurement.

## 5.3 Black calibration

A black calibration standard is a standard that has either a small but known reflectance or no reflectance at all (e.g. a light trap). The purpose of black calibration is to determine the instrument's zero point. It must be carried out on the black standard supplied by the manufacturer. Black calibration (zero calibration) is always performed before a white calibration. Some devices perform zero calibration internally, and do not require external black calibration.

## 5.4 White calibration

For white calibration, the instrument is adjusted with a known white standard (which can be traced back to a National Metrological Institute (NMI), e.g. PTB, NIST, Metas) until each wavelength reading corresponds to the set point reflectance of the white standard. To this end, a multiplication factor is calculated for each wavelength, which serves to convert the raw data from the sensor to the set point values of the white standard. The outcome is that the NMI scale is transferred to the colour-measurement system and the subsequent measurements are based on absolute white. Real white standards have the highest-possible reflectance factors (over 80%).

The white standard is made from a solid material, such as ceramic, glass, or enamel. The instrument maker supplies the standard needed for daily calibration. Its reflectance data can be traced back to the scale of a National Metrological authority and the values are stored inside the instrument. White calibration must be done before work is started and must be repeated after breaks. Where large series of measurements are carried out or if there is a change in the ambient conditions (e.g. in temperature), white calibration must be repeated regularly.

The reflectance data of the working standard are stored in the instrument, which is why it is important that the correct standard for each instrument be used for calibration.

## 5.5 Calibration function

The calibration function $k(\lambda)$ is calculated from the set point values for the white standard and the readings returned by the internal sensor during the calibration. It is then used for subsequent measurements on unknown samples.

The corresponding colourimetric data are calculated from the reflectance data R(λ).

$$k(\lambda) = \frac{\text{set point value for white calibration std. } (\lambda) - \text{set point value for black calibration std. } (\lambda)}{\text{measurement white calibration std. } (\lambda) - \text{measurement black calibration std. } (\lambda)}$$

R(λ) of unknown sample =
(measurement unknown sample (λ) – measurement black calibration standard (λ)) • k(λ)

## 5.6 Control measurement

At specific intervals, it is advisable to check the accuracy of the readings by measuring stable colour standards. The manufacturer's instructions in this regard must be observed.

→ **Note:**
The white standard used for calibration must not be used for the control measurements. Instead, a different sample must be measured, e.g. a green standard.

## 5.7 Storage of calibration standards

All calibration standards must be protected against light, mechanical stress and contamination. Scratched or damaged standards must be replaced. Dirt (e.g. fingerprints) will interfere with the accuracy of the calibration and with all subsequent readings. Consequently, cleanliness is essential for reliable readings. Should the standard be contaminated, it should be cleaned gently in accordance with the manufacturer's instructions. Under normal conditions, a standard has a lifetime of one year. After that, the manufacturer or a certified third-party inspector should check that it complies with the specifications.

## 5.8 Laboratory report

The laboratory report records the following information:

- sample name, and date and time of measurement
- instrument name with details of the measuring geometry
- measuring conditions: type of illuminant, observer, number of readings
- colour coordinates and colour differences according to the standard.

# 6 Influence of the surface

*Gerhard Rösler, translated by Manfred Binder*

The best measuring geometry is always a matter of debate. The different assessments yielded by two samples differing only in their texture illustrate just how great the colour differences are when readings are obtained under different geometries. This is especially true of dark and/or coloured samples. In the case of plastic surfaces, it is the texture (both in terms of its macroscopic and microscopic topography) which exerts a major influence. For paints, it is mostly differences in gloss level, caused by different surface roughness (microscopic topography).

As described in Section II.2.1, the spherical geometry "specular included" (di:8°) is virtually unaffected by differences in texture and gloss level. This is an advantage when there is a need to colour-match a substrate, e.g. plastic, paint, irrespective of the surface structure. This geometry is therefore the most popular for colour matching. Spherical measurement with "specular excluded" (de:8°) clearly shows that excluding the specular component can-

*Table II.6: Colour differences under different measuring geometries (see Figure II.20).*

| Colour difference | Single-geometry measurement<br>Spherical geometry | | | Multi-geometry measurement | | | |
|---|---|---|---|---|---|---|---|
| | di:8° | de:8° | 45°:0° | as20° | as45° | as75° | as110° |
| ΔL* = | -0.03 | 0.07 | 2.38 | -0.95 | 2.38 | 3.11 | 3.14 |
| Δa* = | 0 | 0 | 0.07 | -0.02 | 0.07 | 0.18 | 0.28 |
| Δb* = | -0.07 | 0 | 0.61 | 0.1 | 0.61 | 0.9 | 0.78 |

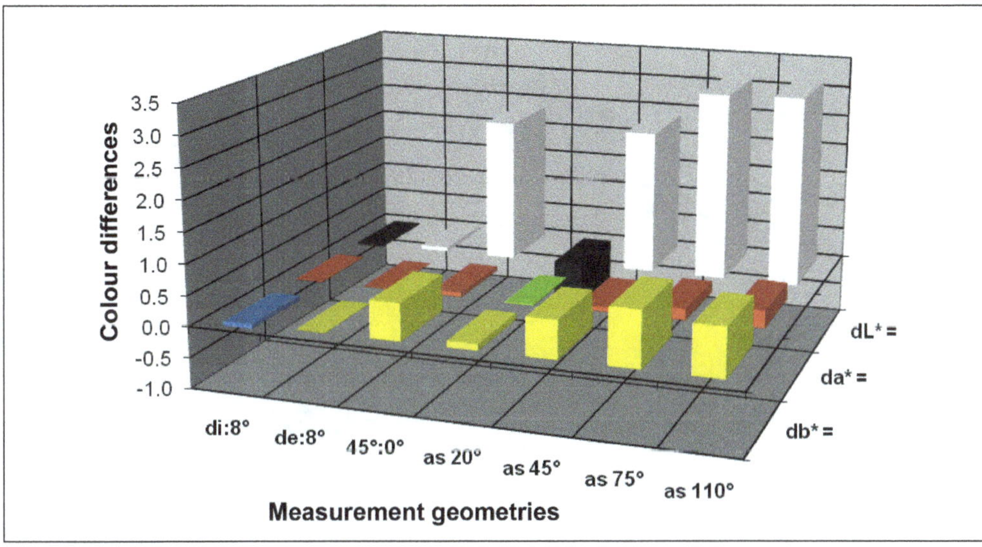


*Figure II.20: Dark blue plastic with different texture, measured under different geometries (see Table II.6).*

not work if the samples do not have high gloss. Under measurement at 45°:0°, a sample appears more than two units lighter and therefore less saturated (Figure II.20 shows the blue sample to be less blue, i.e. yellower).

Multi-geometry measurements at 45°:as20° up to as110° (i.e. from close to specular to far from specular) show the sample to be almost one unit darker at as20°, but lighter (by up to three units) under all other geometries. It is therefore clear that multi-geometry measurement is useful and necessary for samples with grained surfaces and not just for metallic finishes. Furthermore, grained samples also need to be assessed visually at different geometries if unpleasant surprises are to be avoided. In Figure II.20, the readings for the blue pair of samples under di:8° geometry show that the colouration is identical. This measurement fails to reveal any signs of different grain patterns, yet they may be clearly seen under, e.g., a medium viewing geometry (45°:0°).

The multi-geometry measurement or viewing clearly shows the different textures. Two samples are therefore only match if both the colouring and the texture are the same. Texture is angle-dependent, whereas the colouration mostly is not. The corollary is that differences in texture cannot be corrected by making changes to the colouration.

Nor is it possible to reproduce a high-gloss sample in the same colour in a matte finish, even though this is often desired by the industry. A colour match is only possible under a single viewing/measuring geometry – all other geometries will give rise to differences which will

*Table II.7: Colour differences at different measuring geometries (see Figure II.21).*

| Colour difference | Single-geometry measurement |  |  | Multi-geometry measurement |  |  |  |
|---|---|---|---|---|---|---|---|
|  | Spherical geometry |  |  |  |  |  |  |
|  | di:8° | de:8° | 45°:0° | as20° | as45° | as75° | as105° |
| ΔL* = | -1.76 | -1.73 | -1.5 | -3.66 | -1.5 | -0.27 | -0.09 |
| Δa* = | -0.03 | -0.01 | -0.04 | 0.1 | -0.04 | 0.02 | -0.04 |
| Δb* = | -0.02 | -0.01 | 0.11 | 1.03 | 0.11 | -0.68 | -0.63 |

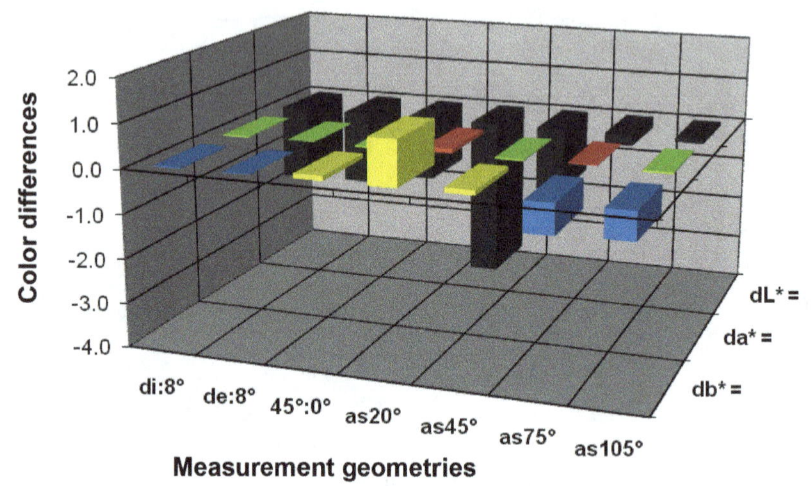


*Figure II.21: Beige leather and artificial leather with different lightness levels and different grain (see Table II.7).*

be evident both visually and colourimetrically. Where a colour match is specified, the only way to meet this requirement is to force the customer to specify the measuring geometry (i.e. the geometry for visual assessment) at which the match will be checked. If the colour match is then done properly, the two samples will look similar under this specific geometry, but different under other viewing conditions.

# 7 *Special case: optical brighteners and fluorescence*

*Gerhard Rösler, translated by Manfred Binder*

Samples that fluoresce are special cases. These pigments absorb short-wavelength energy, which is why the reflectance in that range drops. The energy is transformed into longer wavelengths and emitted at longer wavelengths. The various fluorescent colours that serve, for example, as "brightening agents" in white paper and textiles, or as high-visibility colours for warning signs are not easy to measure due to the mixture of reflectance colour and self-luminous radiation, which creates the fluorescence.

**Except** for these fluorescent colours, the type of illumination used for measurement is not important for any colour. For fluorescent colours, the spectral distribution of the illuminant must be the one specified, and must usually match that of daylight $D_{65}$. This match can only be approximate and requires additional, adjustable filters in the measuring instruments. Another method is to employ a special numerical calibration method, which determines the fluorescence mathematically from two measurements involving different UV-radiation content.

The ratio of the UV intensity to the intensity in the visible spectral range of blue fluorescent dyes can be determined by means of textile and plastic standards. These standards have defined whiteness values to which the instrument must be set. Sometimes, the spectral reflectance data exceed values of 100%, which the spectrophotometer must be capable of processing, without overload!

→ **Note:**
Fluorescent samples generate systematic measurement error in colorimeters with sphere geometry (see Section II.2.1) because the type of illuminant is changed due to absorption and emission by the sample.

→ **Note:**
However, instruments measuring under 45°:0° geometry (see Section II.2.2) are not affected, as the illuminant is not changed by the sample. The 45°:0° geometry is therefore recommended for optically brightened samples (white paper, white textiles) as well as for high-visibility warning colours for textiles.

# *8 Sources of error in colour measurements*

*Stephan Gauss*

As with any measurement, the practitioner must be aware of possible errors when making colour measurements. This is not problematic as long as he is cognizant of the various types of error that can occur, their approximate magnitude, and the ways in which they can be reduced or even eliminated. All these errors may cause the reading to be inaccurate (measuring accuracy) or lead to scattering when the measurements are repeated (measuring precision).

Training courses on how to analyse measurement systems for the purpose of quality assurance quite often invoke the example of darts thrown at a board. Apart from being distributed randomly about the board, the darts could all land close to each other, outside the bulls-eye. That is an example of high precision but low accuracy. If the darts are uniformly distributed about the bulls-eye, the throws have been highly accurate but not precise. These fluctuations from throw to throw or, in the case of colorimetry, from reading to reading, can be further split into specific and random errors. A further distinction is made between testing error and test results difference between two laboratories. In this book, we will not address error propagation or error evaluation.

The importance of calibrating the spectrophotometer has already been described in Section II.5.1. It will therefore be assumed in the following that the instrument has been serviced, calibrated and warmed up ready for operation.

There are three fundamentally different sources of error in colour measurements. On one hand, the sample may be flawed or be inhomogeneous and, on the other, the instrument may be faulty or operated wrongly. Colourimetric assessment of a colourant or paint is itself a process consisting of several steps, of which measurement merely constitutes the last.

## *8.1 Errors in sample preparation*

The colour properties of colourants (pigments, dyes) and coloured products (paints, coloured emulsions, varnishes, etc.) can for the most part only be assessed in the actual application. In other words, a pigment can be assessed only after it has first been converted into a paint, applied in the proper way (by draw-down, doctor blade, spraying, etc.) and then dried (in air, a furnace, UV light) – only then can the sample be measured with the aid of a spectrophotometer. This long chain of steps in the testing process provides plenty of scope for error. In industry, it is standard practice to measure both the sample and a reference in at least three different positions, and then to use the two mean values of each set of readings to calculate the colour difference. This serves to reduce the effect of sample inhomogeneity.

Any analyst conducting tests needs to know the difference between **repeatability** (repeat measurements by one analyst on the same sample) and **reproducibility** (the same sample measured by two or more analysts) ("gauge R&R"). These concern every step in the process and to the overall test. The scatter of the results must be in reasonable proportion to the specification tolerance, whether or not the specification is imposed in-house or externally by suppliers and customers. Numerous discussions in practice would suggest that this is not widely known, with the result that the scattering of readings obtained in colourimetric tests is often blamed on instrument error.

## *8.2 Instrument error*

From experience, instrument error is about one or two orders of magnitude smaller than error arising during sample preparation. Nonetheless, it needs to be discussed here because, instead of leading to scattering, it mostly leads to a shift in position (measuring accuracy). The term "position" is used in this context because colourimetric results are usually a position in 3-dimensional CIELAB space or the 3-dimensional distance (colour difference) between sample and reference. The scatter of the readings therefore takes the form of an error ellipsoid in CIELAB space. Most readers are familiar with error evaluations in one dimension or for one variable. They might not be so familiar with evaluations in three dimensions – or even four, if tinting strength is taken into account too. In such cases, it is common practice to talk about possible error sources in $dE^*_{ab}$. For a systematic description of assessments of three-dimensional error in colour measurements, see DIN 55600.

For spectrophotometers in a d:8° geometric arrangement, the cleanness of the white sphere must be checked regularly. Pigment dust or other extraneous matter must be kept at bay because, even when calibration is performed daily and a reference beam employed, these may still falsify the readings. Above all, it is essential to remember that the dirt is often similar in shade to the samples to be measured.

Whereas the monochromator is maintenance-free and aging resistant, the downstream electronics are not. Drift can occur in some or all of the channels, and in extreme cases the instrument might fail altogether. Routine calibration with the white and black standards will not reveal this problem. Instrument manufacturers usually enclose an additional standard tile whose spectrum rises and falls gently (i.e. green). Any drift in amplifying factors will lead to a change in the readings for that standard.

Optical instruments such as spectrophotomers have very few moving parts and so mechanical malfunctions often go undiscovered for long periods of time. Such parts might be the filter-wheel for UV calibration or the gloss trap. For transmission measurements, it is important to verify that the sample holder is correctly positioned inside the instrument.

In colour-measuring instruments, the wavelength information of the spectrum is associated with a specific position in the instrument. Any dislodging of the mechanics, e.g. due to impact, would lead to a change in the associated wavelength. In that event, the reflectance at the correct wavelength would be associated with a wrong wavelength. Routine servicing by trained personnel should identify and resolve such instrument problems.

## *8.3 Experimental error*

A large number of errors occur while the experiment is in progress. Not only that, but differences in how measurements are made at different stages can yield different results and give rise to discussions between the laboratories involved. These laboratories may be housed within the same company or be located at the customer and the supplier. These varying results are not measurement error as such but rather are procedural differences that produce different readings. At the very least, their causes must be understood.

***Causes of differences in readings***

Many paint tests are performed with **draw-down cards** and, as the coat of paint dries, the cards tend to bend a little. A certain amount of pressure therefore needs to be applied to the cards to ensure that they are planar at the port. Ideally, they should always be oriented in the same draw-down direction in front of the port. Multi-angle measurements, especially, are usually made by mounting the samples in a jig and applying a defined pressure at the port. The size of the aperture may also cause different readings. These measures are intended to prevent both stray light from entering the port and measuring light from getting lost, as that would affect the reading. Stray light is a minor source of error for spectrophotometers that employ a flash light.

However, **stray light** can influence the measurement in other ways. The Kubelka-Munk theory (see Section VII.4) assumes that the applied paint layer is opaque and scattering. This condition is not fulfilled in the case of mass tone applications. These are not opaque and the card substrate is visible, because that is the only way to assess the sample's transparency. As a result, the properties of the white card influence the reading. Not only should the whiteness (and the inevitable optical brightener) be sufficiently homogeneous, but its absolute value will also influence the reading. Thus, the use of different draw-down cards, e.g. by the customer and the supplier, can also yield different colour values.

A further cause in opaque applications can be the type of white used for **reductions**. This concerns not the whiteness of the substrate, but rather the titanium dioxide used in the paint. Understandably, the sample and reference must be based on identical white if meaningful colour differences are to be obtained. These considerations also apply when samples are being prepared for the calibration table (gauge samples). Otherwise the different shades of the $TiO_2$ powders would be superimposed on the pigment and paint readings.

For exact colour measurements of colourants, the intrinsic colour of the substrates used to prepare the samples is subtracted. In other words, allowance is made for the $TiO_2$ white of the reductions and, in the case of mass tone samples, the white of the card. All software programs for colour measurement offer this feature. However, the readings for the white employed in the samples must be used instead of the default setting for the theoretical ideal white. This requires that a corresponding **blank sample** (reference) be measured beforehand.

A further source of measuring differences is the **evaluation method** employed. Obviously, different results will be produced by different colour systems (CIELAB, CIE94, CMC,

CIEDE2000, DIN99, etc.). This applies to both colour location and colour differences. In addition, when colour data are stated, it is important to state whether or not the calculation was based on a **tinting strength adjustment**. Clearly, a sample with 105% tinting strength yields substantially different chromaticity coordinates from a sample with 100% tinting strength. There are several ways to calculate chromaticity coordinates after a tinting strength adjustment (see ISO 787-24 and ISO 18314). The most popular utilize K/S sum, weighted K/S sum and colour depth. As the evaluation software allows the user to choose the correction settings, different settings can be applied, but will yield different results.

These last few points do not constitute measuring errors in the classical sense but rather are a lack of correct information. However, they illustrate the point that thorough questioning is the way to establish the reason for differences in test results. Different measuring procedures do not always produce the same measurement differences. As a result, different adjustment procedures can remain undiscovered for a long time and may only become apparent when the reflectance curve has a specific shape.

# 9 Profiling of measuring instruments and colour management

*Gerhard Rösler, translated by Manfred Binder*

Ideally, all colour-measuring instruments of the same type will produce identical measurements and samples measured on any instruments will yield the same readings. In reality, this goal is only reached to a certain extent. It should be remembered that modern instruments can recognize over 10 million different colours. This means that the dynamics of a colorimeter are therefore 10,000,000 : 1. Normal measuring instruments, such as those used for determining length or voltage, have a dynamic range of 10,000 : 1, and only then if they are of good quality. Colour-measuring instruments must therefore be 1000 times as good as one-dimensional measuring devices. So, it is not a trivial matter to build a good colorimeter that can outperform the excellence of the human eye.

In the ideal scenario, perfect inter-instrument agreement is threatened by a number of causes. These include the accuracy of the absolute wavelength (an error of 1 nm can lead to $dE^* > 1$), the full width at half maximum, the calibration standard (e.g. contaminated standards lead to large errors), the aperture sizes, the illumination and observation beam paths, the linearity of the detectors at all wavelengths, stray light inside spectrophotometers, asymmetries and wavelengths, gravity differences of spectral filters and much more.

## 9.1 Commonly used colour standards for profiling colour-measurement instruments

Colour-measuring instruments are checked with the aid of colour standards. These are usually drawn from the 12-piece set of coloured ceramic tiles produced by Ceram (formerly BCRA) in England [13]. However, even these high-quality standards exhibit significant colour differences at different temperatures. Some yield a measuring difference of $\Delta E = 1$ when the tile temperature changes by 10 °C. As this property (thermochromism) is reversible, it would also be possible to use the coloured tiles with a colorimeter and to apply a corresponding correction to create a substitute thermometer. It should be noted that many other samples change colour with change in temperature (often yellow, orange, red and brown tones and mixtures thereof). Reliable measurements of such samples require that an agreed sample temperature be observed and maintained within maximum limits of 1 °C.

## *9.2 Set of colour standards for colour-measuring instruments*

After a colour-measuring instrument has been made, the manufacturer tests it against a set of colour standards to ensure that the instrument is operating within the specified tolerance. The most accurate instruments are specified to an average dE* <0.08 for the 12-piece set of samples. It can be useful to use colour systems that offer good visual correlation, such as CMC and DIN 99-(DIN 6176), instead of the CIELAB system because these systems offer better visual matching for small colour differences.

## *9.3 Instrument maker accuracy and profiling*

When a colorimeter is shipped, the manufacturer guarantees its accuracy. Operators should regularly verify the instrument's stability against a known colour standard (e.g. green). If this standard reveals discrepancies relative to previous measurements, the instrument will need to be serviced by the manufacturer, which will verify compliance with specifications by performing measurements with its own colour standard set.

The customer can further improve inter-instrument agreement by "profiling" the instrument. For this, the manufacturer delivers a certified set of standards, which the customer measures on a regular basis. The customer then uses special software to minimise any deviations from the set values. The software, for example, can allocate new set values for the white standard and can also correct wavelength assignment errors. Profiling typically reduces deviations to less than half the uncorrected value.

This approach therefore allows the operator to:

- check the instruments himself
- correct small deviations in between maintenance intervals

In this way, it is possible to ensure high measurement accuracy along the whole supply chain.

# *10 Non-contact colour measurement*

*Gerhard Rösler, translated by Manfred Binder*

Colour measurement is increasingly being introduced earlier and earlier into manufacturing processes for the purposes of corrective action or avoiding off-specification product. Non-contact, in-line colour measurements are the best approach. This applies particularly to the measurement of wet paint, where contact between the wet sample and the colorimeter would contaminate the instrument. Non-contact colour measurement can only employ directional measuring geometries because sphere geometries always have to touch the sample surface. Typical standardised measuring geometries are therefore 45°:0° and 0°:45°; however, non-contact multi-angle colour-measuring instruments can also be designed. It is important that ambient light does not influence the reading. This can be ensured by special methods, e.g. flash illumination, which has a much greater intensity than the ambient light during the brief measuring time. The ambient light can be measured separately and subtracted from the reading.

Non-contact measurement is helpful not only in the production process itself, but also in the laboratory, where powder samples, wet samples or adhesive surfaces can be measured. This prevents contamination of the high-precision colour-measuring instruments. Naturally, measurements of wet paint are no substitute for dry measurements conducted on the finished paint application. It is entirely possible to make timely, corrective interventions in wet paint, vis-à-vis previous batches. Furthermore, the wet paint measurement can be extrapolated onto the dry measurement.

There is a risk here, though, of not capturing further variables that influence the process between wet paint measurement and finished product. Tinting steps then take place at much shorter time cycles than would be possible if only the dry paint were to be measured. Colour measurement of intermediate products therefore is no substitute for measuring the finished product, but can often detect early process errors and eliminate further processing steps.

## 11 Literature


[1] DIN 6157-2, Tolerances for automotive paints - Part 2: Goniochromatic paints

[2] ASTM E2194-03: Standard Practice for Multiangle Colour Measurement of Metal-Flake Pigmented Materials, ASTM International, 2003

[3] Pfaff, G., Special Effect Pigments, Vincentz Verlag, Hanover, 2008

[4] Pedrotti, F. et al., Optik für Ingenieure, Springer Verlag, Berlin Heidelberg, 2008

[5] Rodrigues, A. B. J., Metallic flop and its measurement, JOCCA, 1992(4)

[6] Gabel, P. W., in [3]

[7] Gabel, P. W. in G. Pfaff, Encyclopedia of Applied Colour, Wiley VCH Verlag GmbH, 2010, gone to press

[8] Gabel, P. W., Goniochromatic colour measurement systems - the past 20 years, 9th Congress of the International Colour Association, Proceedings of SPIE Vol. 4421 (2002)

[9] ASTM E2539-08 Practice for Multiangle Colour Measurement of Interference Pigments

[10] Cramer, W. R., Gabel, P. W., Das gewisse Etwas, Dreiecksbeziehungen aus Bunt-Aluminium- und Interferenzpigmenten, farbe+lack 10 (2003) 1, p. 78

[11] Nisper, J., Major Advances in the Reliable Measurement of the Colour and Appearance of Special Effect Paints and Coatings, Charlotte, Proceedings (2008)

[12] Kirchner, E., De Haas, K., Njo, L., Roesler, M., Objektiv glitzern und funkeln, farbe+lack, Volume 114, 2/2008, p. 32

[13] Ceram, Queens Road, Penkhull, Stoke-on-Trent, Staffordshire ST4 7LQ, UK

# III Visual colour assessment

*Gerhard Rösler, translated by Manfred Binder*

## 1 Colour perception and colour deficiency

"Normal colour vision" is the term employed to describe the average colour perception of all humans. However, even people with normal colour vision perceive colours differently. "Colour deficiency" refers to colour perception that clearly deviates from what is termed normal, and can vary in severity. It is caused by total failure of or differences in the spectral sensitivity of one or more types of cone in the retina. Consequently, there are different types of colour deficiency. In the most extreme, all types of cone are defective and there is no colour vision of any kind – the person sees only grey shades. In that event, only the rods are active and so dark glasses must be worn because rods only work in low light levels.

A provisional diagnosis of colour deficiency can be obtained with the aid of pseudo-isochromatic samples (e.g. *Ishihara* Test; see Figure III.1). This test consists in looking at a number of pictures bearing a pattern of dots in different colours. People with normal colour vision recognize a symbol (typically a letter or a number) by virtue of the different colours in the pattern. People with a certain kind of colour deficiency cannot distinguish between these colours accurately. They either cannot see the symbol at all or they identify it wrongly. A more sophisticated but more complex method utilises the "anomaloscope" invented by *Willibald A. Nagel*.

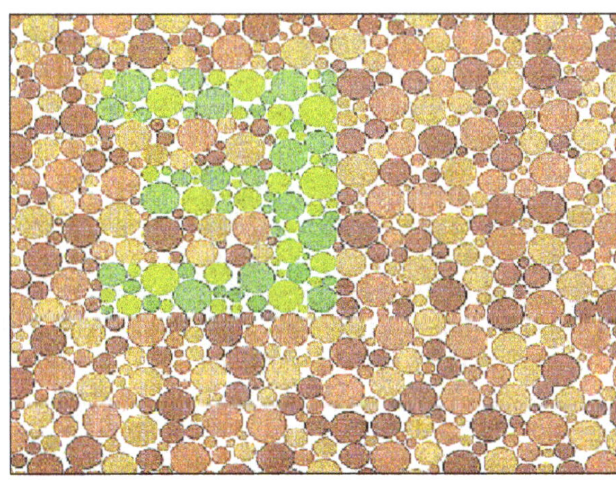

*Figure III.1: Pseudo-isochromatic panel.*

A much better way to assess colour vision and quantify different types of deficiencies is afforded by the Munsell Farnsworth 100 Hue Test (see Figure III.2). Many manufacturers use this test to check the colour vision of their colourists and their sub-contractors.

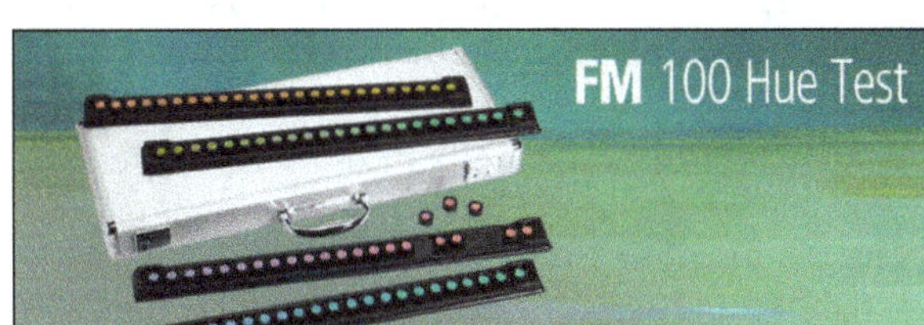


*Figure III.2: Farnsworth-Munsell 100 Hue Test.* Source: X-Rite

*Table III.1: Types and frequency of colour deficiency.*

| Type | Kind | Mail | Female |
|---|---|---|---|
| Anomalous trichromacy (one of the 3 cones is mutated) | Green deficiency | 4.9 % | 0.4 % |
| | Red deficiency | 1.0 % | 0.02 % |
| | Blue deficiency | very seldom | very seldom |
| Dichromacy (one of the 3 cones fails) | Green-red-blind (deuteranopia) | 1.1 % | 0.02 % |
| | Red-green-blind (protanopia) | 1.0 % | 0.02 % |
| | Blue-yellow-blind (tritanope) | 0.002 % | 0.001 % |
| Monochromacy (2 of the 3 cones fail or even all 3 fail In this case only the very light sensitive rods are active) | Only brightness sensitivity (black – gray – white) but all 4 types show different spectral sensitivity | 0.003 % | 0.003 % |

# 2 *Light booths*

A vast array of light sources, illumination directions, and viewing directions is available for performing visual colour assessments in everyday work. Therefore, prior to shipment, the specifications must be chosen carefully so that errors can be uncovered in good time. The colorist in quality control must be able to avail of high-quality test conditions for a colour assessment day and night. It is also important for the supplier and the customer to use similar light booths, as that helps to foster good communication between them. Criteria to take into account:

- The light source used in the application
  - **Daylight $D_{65}$** is the most important light source for many applications and is the international standard for average daylight. However, good artificial daylight is not easy to generate. Filtered halogen light has traditionally yielded the best approximation. There are also fluorescent lamps that simulate $D_{65}$. Caution is advised here, however, as only high-quality lamps are suitable for daily work. These are called "965" types. However, even they are inferior to filtered halogen lamps with a D65 light source.
  - **Incandescent lamp** (light source **A**) is easy to simulate with a tungsten lamp. This often the light source of choice when the finished product is used in rooms with incandescent lamps.
  - **Energy-saving lamps (F11)** are easy to simulate with fluorescent tubes. These are necessary for products displayed in warehouses or in rooms lit with energy-saving lamps.
  - **Office lamps** Especially in the USA, light source **F2** (CWF = cool white fluorescent) is used.
- The illumination direction
  - Diffuse or direct?
    The direction relative to the surface normal of the sample is important, especially for samples with effect surfaces.
- The viewing direction
  - **Direct** viewing is always used.
  - The direction relative to the surface normal of the sample is important, especially in the case of effect samples.

Most light booths (Figure III.3) have more than one light source. Typical combinations are D65, A and F11 or F2. It is important that there be no interruption when switching from

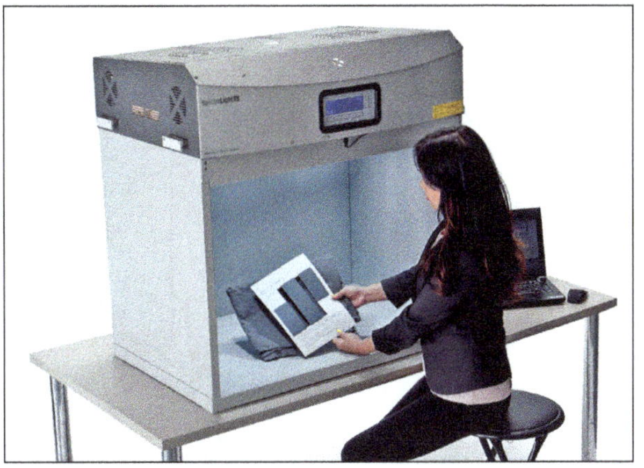

*Figure III.3: Light booth SPL QC.* Source: X-Rite

one to the other. For the assessment, a standard specimen is placed in the light booth for visual comparison with a sample taken from current production. Metameric samples are identified by alternating between light sources and consist of a pair of samples which look different to greater or lesser extents under different light sources. Typically, the standard specimen and sample will give a good match under one light source, but not under a different one. The job of the manufacturer is made much easier if the customer agrees to a colour match under only one light source. Often, though, a good match is required under different light sources. This greatly increases the effort involved in colour matching, as only a perfect spectral match will avoid metamerism.

The sample is very often placed on the floor of the light booth. This configuration causes it to be illuminated perpendicularly from the top (0°). The observer sits in front of the booth, looking at the sample from an angle (45°). For this reason, visual assessments provide very good correlation with the 45°:0° or 0°:45° measuring geometry (the results of both geometries are identical).

***Tips on visual colour assessment***

Take extreme care when assessing high-gloss specimens, because the reflection of the (normally grey) back wall of the booth appears on the sample's surface. This will cause dark or coloured samples to appear lighter and less saturated, and that reduces the ability to perceive differences. For good assessments, the back wall of the light booth should be black. The perception of colour differences in samples is also influenced by the surroundings. If dark samples are assessed in bright surroundings, the eye will adjust to the bright surroundings. The colour difference between a dark pair of samples will then be less apparent than in dark surroundings. For accurate assessments, the surroundings should match the colour of the samples.

A light booth is too small to assess items such as entire cars or large, mounted interior parts. In this case, harmony rooms must be set up, in which the upper parts of the light booths (luminaires) are mounted to the ceiling and the walls are painted in grey or black. The entire set of luminaires or groups of them are remote-controlled for the purpose of switching between the different light sources so that several colorists can see the colour simultaneously.

# 3 Visual colour assessment of samples with effect coatings

Samples with effect coatings, such as metallic paints and even textured surfaces like grained leather, impose special requirements. Here, it is important to view the samples from different angles. There are three different viewing conditions:

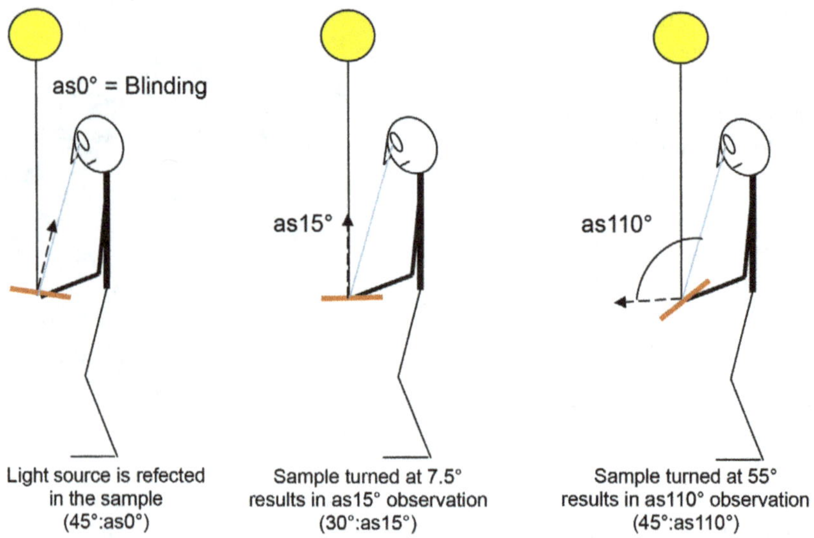


*Figure III.4: Multi-angle geometry colour assessment with sample modulation (light source and observer have fixed positions) - easy to realize, often used for quick assessments. Instrument: ER11 (1982).*

- sample modulation
- illuminant modulation
- viewer modulation

The following figures and descriptions show one implementation for each of the above.

## 3.1 Method: Sample modulation

A colorist holds a pair of samples in one hand, the light shines down from above and he moves the arm holding the samples (sample modulation). When the sample is moved into a certain position, the colorist can detect the specular angle at which the illumination is reflected in the sample. Figure III.4 shows this procedure. This is the reference position for further assessment and represents as0° in the definition of multi-angle geometries. In this position, no visual assessment is possible because the colorist is blinded by the illumination. If the sample is now rotated through 7.5° out of this position, the viewing angle is as15°, which is the first standardised viewing condition. The aspecular angle is thus twice the angle of rotation, relative to as0°. Therefore, to view the sample at as110°, the colorist must rotate it through 55°. It is important to note that, in sample modulation mode with fixed illumination and fixed observer, the aspecular angle is always twice the angle of sample rotation: 55° rotation = 110° aspecular angle. This is frequently overlooked.

The other geometries as25°, as45°, as75° lie between as15° and as110° and are not shown in the figure.

## 3.2 Method: Illumination modulation

Some commercial instruments have a module for modulating the illumination: different light sources are switched on separately to create different viewing conditions (see Figure III.5). The viewing angle and the angle of the sample remain constant. The advantage of

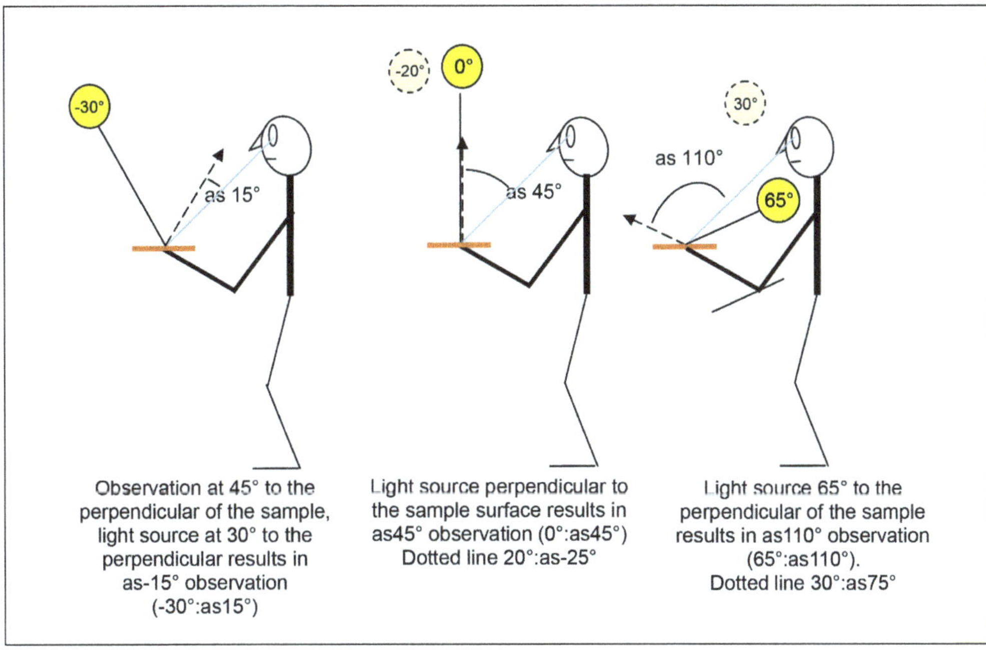


*Figure III.5: Multi-angle geometry assessment with illumination modulation (sample and observer are in fixed positions). Instrument: Macbeth AE 64x Series (1991), ER51 (1988).*

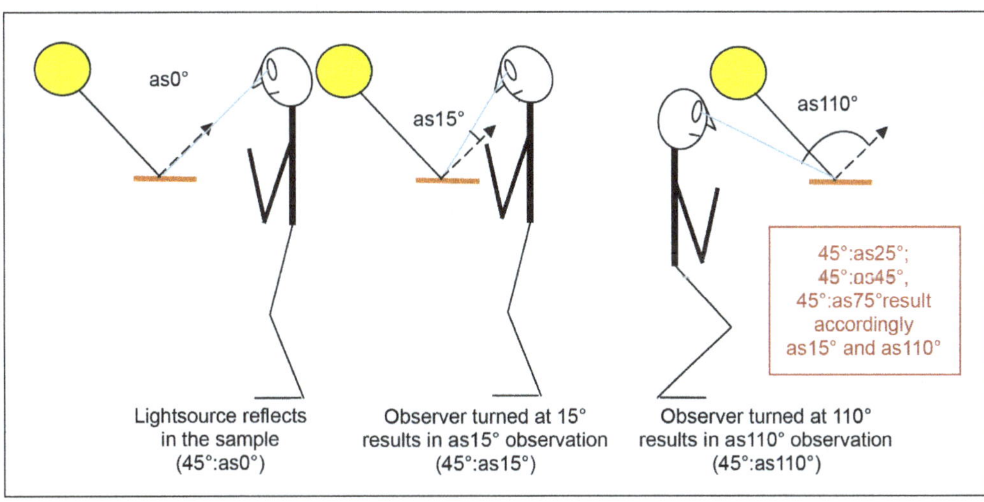


*Figure III.6: Multi-angle geometry assessment with observer modulation (sample and illumination are in fixed positions). Instruments: X-Rite MA6X Series (1991), MA9x Series (2008), BYKMac (2007).*

this procedure is that the viewing conditions can be changed quickly without the need to move sample or observer. The new viewing condition is achieved by simply switching lamps.

## 3.3 Method: Observer modulation

Many instruments operate on this principle (see Figure III.6). The illumination and the sample are in fixed positions, while the observer (measuring port) changes position. The results are the same as for illumination modulation.

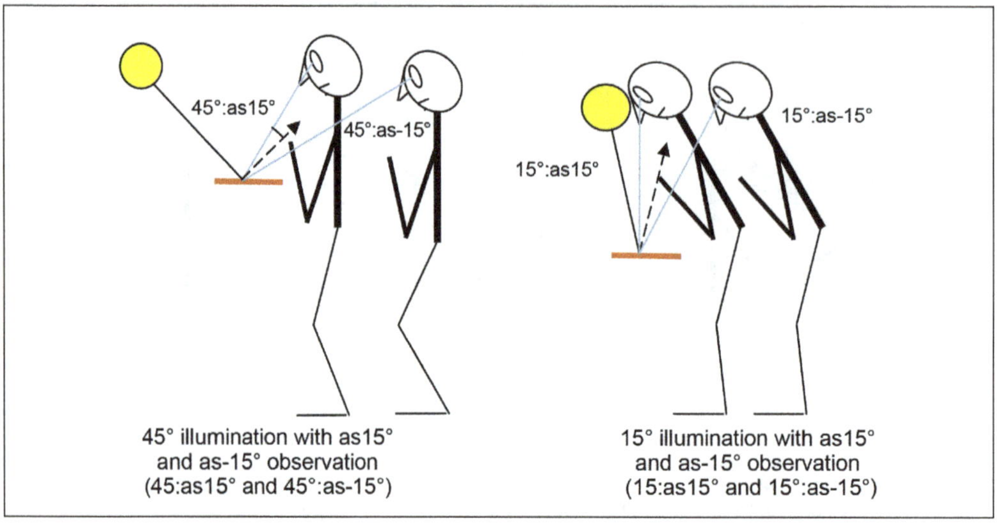


*Figure III.7: Multi-angle geometry assessment of interference pigments. Instrument: MA98.*

### 3.3.1 Assessment of special effect pigments by the method of observer modulation

As the colour flop of interference pigments occurs very close to the specular, it has proved necessary to standardise new measuring geometries (see Figure III.7). Besides illumination at 45° and viewing at as15° and as-15°, a second illumination was specified at 15° from the surface normal of the sample, with observation at as15° and as-15°. These additional geometries support good characterisation of interference pigments.

## 3.4 Summary of visual assessment of effect-coated samples

Multi-angle colour measurement has been state of the art in the automotive industry for many years. Unfortunately, people often become confused when doing visual assessments and end up comparing apples with oranges. Only if the geometric conditions of instrumental colour measurement are the same as for the visual assessment will the two results match. And only then will there be a good match between what is seen and what is measured. Often, the respective parties fail to properly specify the terms of visual assessments. This contrasts with the clearly defined geometries of the measuring instruments, which are helping to foster good communications between suppliers and manufacturers. Nowadays, discussions tend to be based on clearly defined, reproducible and communicable hard data.

# IV Colour-order systems

*Wilhelm Kettler*

## 1 Introduction and definition

According to *Günther Wyszecki* [1] a colour-order system is a rational method or plan of ordering and specifying a limited domain of colours or all object colours by means of a set of material standards to adequately represent the whole set of object colours under consideration. Many different colour-order systems can be conceived of as satisfying this definition, and indeed a number of colour-order systems have already been developed. These differ in their labelling and ordering principles, in their physical realisation and number of colour standards, and in the size of the colour domain, which they are intended to represent [2, 3].

Colour standards are an important element of each colour-order system. They support the main purpose of a colour-order system, namely, labelling of all colours by direct visual comparison.

When allowance is made for the various methods of colour reproduction and colour labelling, colour-order systems fall into three broad groups. The first is primarily based on the principles of additive colour mixing, the second creates systematically varying colour standards based on the principle of subtractive colour mixing, while the third consists of colour systems whose material standards are ordered systematically in line with the principles of perceptual uniformity and are labelled accordingly. Non-systematically structured collections of colour samples have acquired particular importance in industrial practice and should not go unmentioned in this context.

## 2 Psychometric scales

An important aspect of any systematic ordering of colours is the scaling method. Various principles have been devised for finding general rules to divide objects into categories and arrange them according to their individual characteristics.

Psychometric scales afford a way of assigning numbers to physical stimuli on the basis of the particular attributes which generate the stimuli. Of the many different types of psychometric scales, the following four in particular are relevant to **colour-order systems** [4-7]:

- **Nominal scales** merely specify whether things are equal or not. Numbers are assigned to stimuli in such a way that, where a pair of stimuli have the same numbers for a given attribute, they are considered equal. Fan decks issued by many paint manufacturers contain examples of nominal scales.
- **Ordinal scales** possess all the properties of nominal scales and additionally assign numbers in such a way that the numbers are ranked according to the ranking of the attributes being scaled. A given stimulus that has a higher value on the scale than another is perceived as having more of that attribute. All colour-order systems are based on ordinal scales.

- **Interval scales** possess all the properties of ordinal scales and the additional property that differences (intervals) between the numbers characterise the size of the corresponding perceived differences of the attributes. The majority of colour-order systems are based on interval scales.
- **Ratio scales** are interval scales with the additional property that they possess a natural origin. This means that the origin of the scale corresponds to the stimulus for which the attribute has zero magnitude. This additional property means that the numbers on the scale are proportional to the perceived magnitudes of the attributes being scaled. Many colour-order systems are based on scales that have this property.

This hierarchy of one-dimensional psychometric scales, proposed by *Stevens* [8], describes various levels of measurement, with the two highest (interval and ratio scales) being used in modern colour-order systems. Nominal and ordinal scales are non-measurement or categorical scales, while interval and ratio scales are measurement scales. The higher the scale level, the greater is the information content of the corresponding data and the more the data lend themselves to statistical operations. It is generally possible to transform data from a higher scale level to a lower scale level, with loss of information; the reverse approach is not possible.

# 3 Colour scales

The perceived colour of an object can always be described in terms of three independent partial attributes [9–12]. Thus, colour-order systems are always also based on three independent scales. The choice of the three attributes to be scaled varies considerably between different colour-order systems, but three classes of attributes for descriptive characterisation are common to all of them: **hue, chroma**, and **lightness**. All major colour-order systems implemented in systematic colour swatches are based on these three elements.

- To change the **lightness** of a colour, it is only necessary to raise or lower its spectral reflectance curve proportionally. This effect can be achieved by additive mixing with ideal black or ideal white.
- The **hue** is the property of a colour by which the chromatic colours (red, green, blue, yellow, etc.) differ from the achromatic (white, grey, black) ones.
- The term **chroma** denotes the property of a colour of arbitrary hue to appear more or less chromatic at the same lightness level. A mere desaturation of a given colour shade can be achieved by blending with an additive mixing of equally bright grey. Where both components have different lightness levels, chroma and lightness will change simultaneously.

# 4 Colour notation systems

## 4.1 CIELAB, CIELUV, and DIN99

The most widespread colour systems are the **CIELAB76** and **CIELUV76** models recommended in 1976 by the **CIE**[1], which meet the demand for the definition of perceptually uniform colour spaces [13]. Both models are purely mathematical in nature and do not exist as physical colour atlases. However, because of their colorimetric origin, they can be utilised

[1] *Commission Internationale de l'Éclairage*

for output-device-independent colour coding. While **CIELAB76** is mainly serves the colorimetric characterisation of object colours, **CIELUV76** finds application for coloured light sources (monitors, TVs) because additive colour mixtures are easy to visualise. In practical application, the two colour systems have turned out to be not perfectly uniform in terms of their chroma scale.

In the newer colour space **DIN99** [14, 15] the three axes of the **CIELAB76** colour space have been rescaled, and this has substantially reduced deviations from uniformity.

## *4.2 Munsell colour system*

The *Munsell* system was developed on the basis of perceptual uniformity and is one the most widespread colour systems [16, 17]. It has been repeatedly modified over the decades. It groups all object colours about an **achromatic axis** formed by a perceptually uniform distribution of the various grey planes between black and white. As in the tristimulus system, this axis is oriented perpendicular to planes through the colour gamut spanned by colours having the same luminosity. Within these planes, colour saturation gradually increases with increase in distance from the achromatic axis, with the interspaces also scaled so as to be perceptually uniform.

The **hue circle** is divided into five principal hues (R - red, Y - yellow, G - green, B - blue, P - purple) and five intermediate hues (YR - yellow-red, GY - yellow-green, BG - blue-green, PB - purple-blue, RP - red-purple) labelled by different letter symbols (Figure IV.1). In order to label the 40 different hues existing in the Munsell system, the letter symbols are prefixed by numbers (0 to 10), e.g. 2.5PB, 5P, and 7.5GY. In addition, a number between 0 and 100 may be assigned to each hue.

Hues of equal **saturation** (Munsell Chroma C) are located on circles about the achromatic axis, where C = 0. Moreover, these hues are visually equidistant. About the centre of the coordinate system, 16 equidistant concentric circles are defined. The denominator of a fraction, /2, /4 etc., designates the chroma (saturation) plane. Thus, in each plane, any chromaticity is denoted by the polar coordinates **Munsell hue H** and **Munsell chroma C**. By adding further planes of equal chromaticity, but different lightness (**Munsell value V**), the colour system is completed to yield an irregularly shaped colour solid. For pigment reasons, not all planes of constant luminance can be filled homogeneously. Only in red (R) and yellow-red (YR) regions can chroma values of 16 be reached.

The vertical central axis forms the 10-part grey scale (V = 10 for ideal white and V = 0 for ideal black) and the respective lightness graduation for all hues of a plane. The quantity $V$ (Munsell value) is related to the luminance $A \equiv Y$ of the CIE chromaticity system by a 5$^{th}$ order polynomial:

*Equation IV.1* $$Y = 1{,}1913V - 0{,}22533V^2 + 0{,}23352V^3 - 0{,}020484V^4 + 0{,}0008194V^5$$

A plot of this relationship is shown in Figure IV.1.

The notation for colours in the Munsell system takes the form *HV*/C. For instance, 2.5YR5/10 describes a yellowish-red hue (*H*) with a lightness value of 5 (*V*) and a chroma level of 10 (C). However, it should be stated that, although the Munsell colour-order system is based on visually equidistant scales for hue (*H*), lightness (*L*), and chroma (*C*), the sizes of the steps on the scales are different[18]. Munsell lightness and chroma are ratio scales, while Munsell hue is an interval scale.

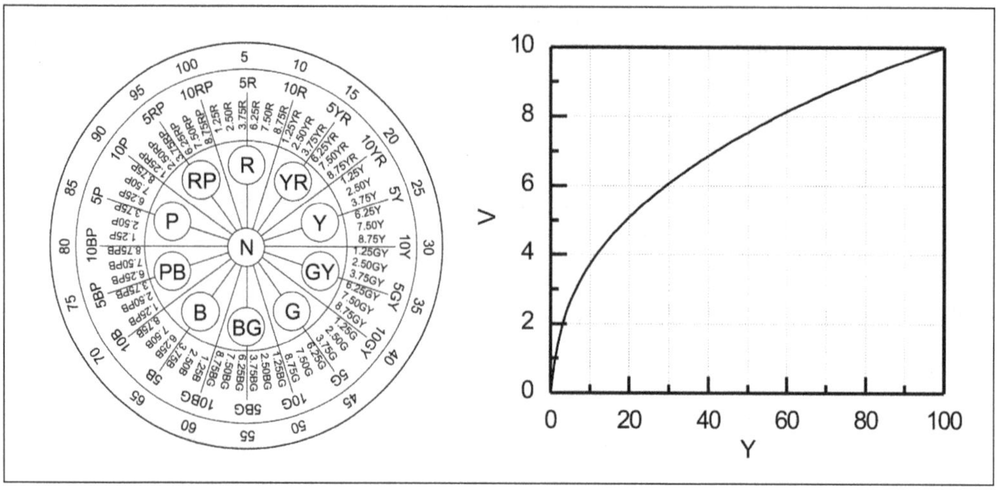


*Figure IV.1: The left diagram shows the division of a value plane of the Munsell system in radial directions of equal hue and concentric circles of constant chroma. Colour counting proceeds clockwise. The right diagram shows the relationship between Munsell value V and luminance Y.*

The Munsell colour system is available in two variants: a collection of matte (about 1325 colour swatches) and a collection of glossy standards (about 1600). The colour swatches are offered in the two-volume *Munsell Book of Colour* or as a collection of loose colour sheets.

## 4.3 DIN colour system

The DIN colour system was developed by *M. Richter* at the suggestion of the German Institute for Standardisation (DIN) [9, 19]. Colours are labelled in terms of three parameters **hue** (*T*, DIN Farbton), **saturation** (*S*, DIN Sättigungsstufe), and **darkness** (*D*, DIN Dunkelstufe). Hue *T* and saturation *S* characterise the chromaticity and are uniquely valence-metrically defined by the colour coordinates *x* and *y* in the chromaticity diagram. The system is based on a colour circle of identical saturation and divided into 24 perceptually equal steps, starting at yellow (1) and proceeding clockwise via orange, red, purple, violet, blue, and green, to close again at yellow (Figure IV.2). The angles between the various hue directions are not constant in the chromaticity diagram; the different areas of the chromaticity diagram are non-uniformly filled by the curves of constant saturation. The loss of uniformity in the chromaticity diagram can be traced back to the fact that the **CIE** chromaticity space is not based on equally perceived quantities.

In the DIN colour system, DIN hues are numbered 1 to 24. To characterise DIN saturation, a 16-part uniform scale was introduced, in which the achromatic point is assigned a saturation $S = 0$, this figure gradually increasing with increase in distance from the achromatic point.

DIN darkness $D$ is related to the luminance $Y$ of the colour viewed and to the analogous quantity $Y_0$ of the optimal colour of same chromaticity belonging to this colour. DIN darkness D is calculated from the given relative lightness $h = Y/Y_0$ as follows:

*Equation IV.2*

$$D = 10{,}0 - 6{,}1723 \cdot \log\left(40{,}7\,\frac{Y}{Y_0} + 1\right)$$

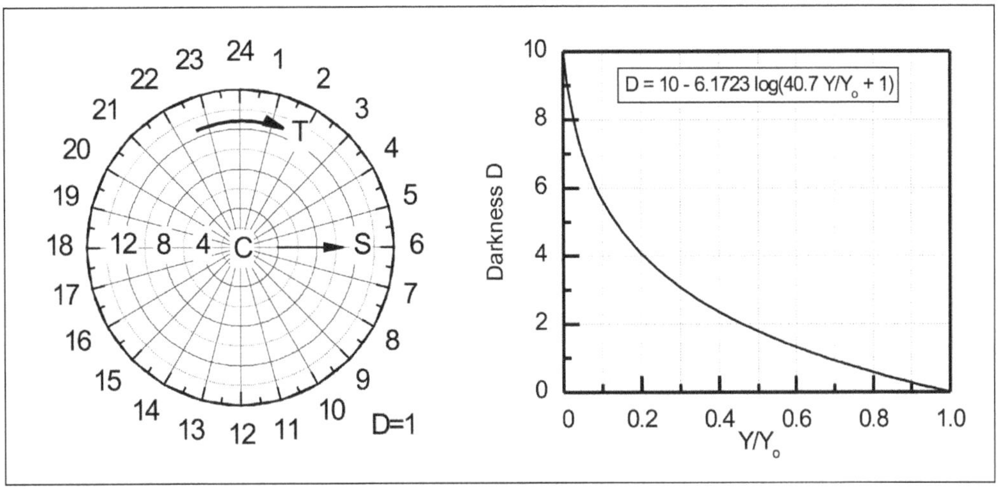


*Figure IV.2: The left diagram shows the 24-part uniform colour circle of the DIN 6164 colour system. The concentric circles form 16 uniform steps, which cannot be achieved for all hues. The centre represents the achromatic point. The right diagram serves to determine the degree of darkness D of the DIN 6164 colour system.*

In this equation, DIN darkness $D=0$ is assigned to ideal white and $D=10$ is assigned to ideal black. Figure IV.2 shows the variation in $D$ between these two extreme values. The three elements DIN hue, DIN saturation, and DIN darkness form a three-part colour code, namely *T*:*S*:*D*.

Equation IV.2 shows that, for DIN darkness, there is no simple correlation with the perceived lightness. As in the Munsell system, the hue is an interval scale in the DIN system, while DIN saturation is a ratio scale but one, which is different for each plane.

The colours in the DIN colour system 6164 are arranged in the form of a spherical coordinate system. The resultant colour solid is a rotation cone, in which areas of constant *T* value are depicted as half-planes with a common vertical achromatic axis, whose black point ($D=10$) coincides with the tip of the spherical sector.

## 4.4 NCS colour system

The *NCS* (**N**atural **Co**lour **S**ystem) was developed in Sweden and standardised in 1981 [20-22]. It comprises more than 16,000 colours, which are characterised and tabulated by the NCS colour code. Like the opponent colour theory of *Hering*, the NCS is based on the six elementary colours white (W), black (S), yellow (Y), red (R), blue (B), and green (G). The four chromatic colours, which lack any similarity to white and black, are equidistantly arranged on a colour circle and have a theoretical maximum chroma of $c = 100$. The axis of achromatic colours is perpendicular to this colour circle and intersects its centre so that the *NCS* colour solid takes the shape of a double cone (see Figure IV.3). **Blackness** ($s$), **chromaticness** ($c$) and **hue** ($\phi$) serve as the three independent elementary parameters. Whiteness is not shown separately, as it results directly from the relation $F=s+w+c=100$.

The colour circle of the mid-plane is divided into 400 parts, i.e., 100 hues denoted by corresponding numerical values are located between two elementary colours of each quadrant. For instance, the notation Y30R, means that this hue is composed of 70% yellow and

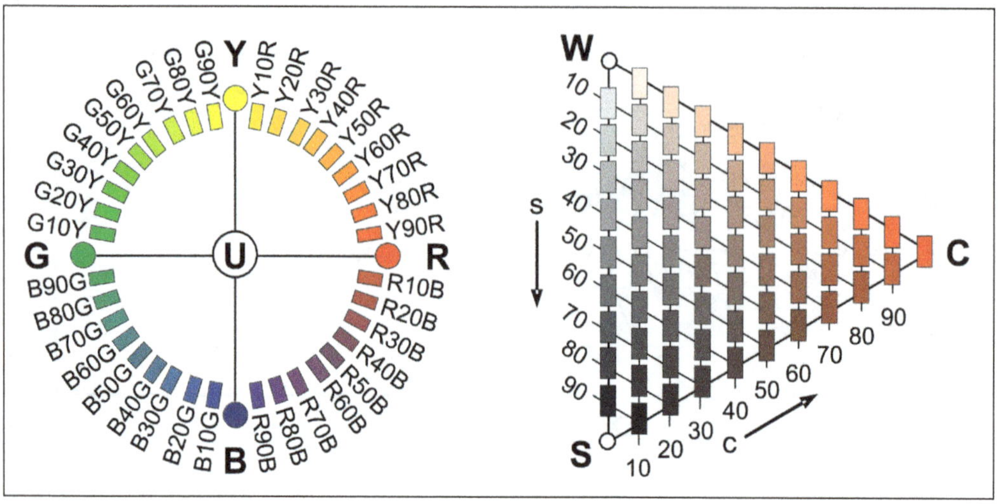


*Figure IV.3: The left diagram shows the NCS colour circle with the four elementary colours red, blue, yellow, and green. The right diagram shows an NCS colour triangle, which is oriented perpendicularly to the colour circle. The vertical $\overline{WS}$ represents the grey axis, where only the blackness is marked. At the apex is the elementary colour C. The line $\overline{SC}$ specifies the chromaticness c, while line $\overline{WC}$ indicates the whiteness, which is not explicitly labelled.*

30% red. The percentage refers exclusively to colour perception and does not permit any conclusions to be drawn about colour mixtures of an optical or material nature. Vertical sections through the NCS solid colour yield triangles (see Figure IV.3), whose vertices are white (*W*), black (*S*), and full colour (C). The triangles of equal hue enable the chromaticness (*c*), the blackness (*s*), and the whiteness (*w*) to be determined. The full notation for a colour in NCS takes the form 4020-Y30R. The first two digits quantify the blackness ($s = 40\,\%$) and the next two the chromaticness ($c = 20\,\%$) (it follows that whiteness $w = 40\,\%$). This block of digits determines the shade. This combination of characters and digits yields a hue $\phi$ that is yellow (*Y*) with 30% red (i.e. yellowness of 70%).

The scales for whiteness, blackness, yellowness, redness, blueness, chromaticness, and each of the four hue scales are all ratio scales for similarity. However, the question remains as to whether they are also ratio scales for the perceived residual total colour difference.

## 4.5 OSA-UCS colour system

In the years from 1947 to 1974 the OSA colour system [23, 24] was developed by the "Committee on Uniform Colour Scales" of the "Optical Society of America". The goal of the project was the production of 500 colour charts to illustrate the **best uniform colour space** by means of a regular rhombohedral lattice with the proviso of smallest uniform distances in three dimensions. In 1961, first evidence of additivity errors provided experimental proof of a negative curvature of this colour space and the consequence that its Euclidicity was impossible. Six years later, the committee was convinced that a perfectly regular rhombohedral sample collection of the colour space for a fixed background does not exist and it relaxed the goal to produce the best approximation for such a lattice against a neutral Munsell 6/background. The production of the **OSA-UCS** colour collection consisting of 558 colour samples was not completed until 1976. For 424 of the 558 colour samples, there are 12 nearest neighbours with visually equal colour differences from the reference point. From this subset, 398 uniform colour scales with different numbers of colour samples and different increments can be selected [24]. Only about 50 of these scales contain neutral grey and complementary colours.

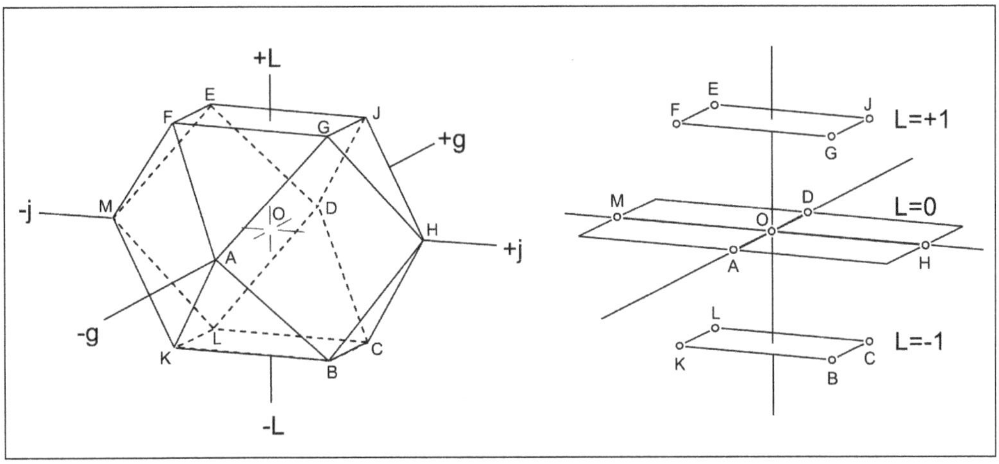


*Figure IV.4: The left diagram shows the OSA cuboctahedron with 12 equidistant positions of nearest neighbours about a central reference position. As the right diagram shows, they are located on quadratic lattices in horizontal planes.*

As in case of the Munsell system, the over-riding goal of the OSA committee was the attainment of visually equal distances between colours. The choice of a rhombohedral lattice with the geometry described below made it possible to realise this equidistance in three dimensions.

The main focus was on the global and local uniformity, with the ultimately achieved accuracy limited by the strongly non-Euclidean properties of the colour space.

The coordinate system ultimately devised by the committee is based on axes with opponent colour character, with identifiers ***L*** for **lightness,** ***j*** (derived from the French word *jaune* for yellow) for the **yellow-blue axis**, and ***g*** for the **red-green axis**. Hence the OSA code for a colour consists of a combination of three digits, e.g. *L:j:g = 3:1:5*. These coordinates span a 3-dimensional Euclidean space. For *L* = 0, the colour in question has the same lightness as a medium grey with a reflectance level of 30 %. However, the reflectance level of chromatic colours with L = 0 is usually less than 30 %, since the chromaticity also contributes to the lightness. *L* is negative (positive) for colours which are darker (lighter) than the background. Values for *L* vary between -7 and +5. The second digit of the *OSA* notation denoted by *j* represents the yellowness of a colour; *j* is positive for yellowish and negative for bluish colours, with values ranging roughly between -6 and +11. The third digit labelled with *g* specifies the greenness of a colour; *g* is positive for greenish colours and negative for reddish colours over a range of about -7 to +5. For all achromatic colours, both coordinates *j* and *g* are equal to zero.

However, the designations redness, greenness, yellowness, and blueness associated with the axes must not be taken too literally. Although positive *j*-values represent yellow shades of high lightness, *j* must not be interpreted as a yellow-blue axis, since negative *j*-values separate blue and purplish colour regions. Similarly, a positive *g*-value does not specify a green, but almost perfectly separates blue and green colours, with pink colours (not red) located at the negative end of the *g*-axis. Therefore, *j* and *g* should be interpreted as abstract symbols that are not related to colour names.

The basic geometric shape needed for describing the rhombohedral lattice of the OSA-UCS system is the cuboctahedron that remains after the eight edges have been removed from a cube (see Figure IV.4). A sample in the centre has 12 nearest neighbours located at the vertices of the geometric shape. The entire colour solid is developed by extending the unit cuboctahedron in all space directions where this is technically possible. Due to the nature of the lattice and the orientation of the axes, the unit-length of the *L*-axis is smaller by a factor of $1/\sqrt{2}$ than the unit lengths of the *j*- and *g*-axes. This has to be accounted for when computing total colour differences:

*Equation IV.3*
$$\Delta E = \sqrt{2(\Delta L)^2 + (\Delta j)^2 + (\Delta g)^2}$$

Strictly speaking, the derived formulas for *L*, *j* and *g* may not be applied to the computation of large colour distances, as they only represent committee results for paired comparisons of large hexagonal plates against a grey background. Furthermore, the assessment results have been squeezed into a Euclidean space, although clear evidence of significant deviations from Euclidicity did exist. For instance, the sum of visually assessed differences between the six colours of equal lightness located at the perimeter of a hexagon, all of which are at the same distance from the central grey in a *j-g* diagram, would be significantly higher than the six-fold value of their assessed common differences from the grey.

## *4.6 RAL design system*

The RAL design system was published in 1993 by the RAL institute and is a colour atlas of about 1700 swatches for visualising the colour gamut of a special assortment of 16 solid colourants [25]. The ordering principle of the atlas follows the standardised **CIELAB76 system** and uses the coordinates **hue-angle** ($h_{ab}$), **lightness** ($L^*$), and **chroma** ($C^*$). Each of its 36 pages (plus three special pages in the yellow colour region) represents one hue-angle on the colour circle divided into 360 degrees, i.e. the hue-angle difference from one page to the next is $\Delta h_{ab} = 10°$. The design of each page corresponds to a matrix, in which planes of equal lightness are ordered in rows; the vertical columns represent the chroma, which increases from the inside towards the outside. The distances between the rows and between the columns each represent 10 units of the respective base parameter (lightness, chroma). An extra page, not belonging to the colour circle, represents the grey axis and its immediate vicinity.

The notation in the **RAL Design system** is composed of a triplet of numbers, in which the first denotes the hue-angle and thus the page number within the atlas, the second is the lightness and the third is the chroma. The coordinates for lightness and chroma uniquely define the position of a colour sample within a page. For instance, an atlas colour with the code 270 30 20 corresponds to a dark blue with a hue-angle of $h_{ab} = 270°$, a lightness of $L^* = 30$ and a chroma of $C^* = 20$. In order to prevent any confusion, all hue-angle values must be specified in three digits (e.g. 010 instead of 10).

All colour samples in the RAL design system are semi-glossy in appearance. This represents a compromise between the glossy surface coatings in automotive applications and the matte surface texture of architectural paints. The average gloss level *G* of the colour samples, as determined in accordance with DIN 67530 and ASTM D 523, is about $G(20°) = 22$, $G(60°) = 65$, and $G(85°) = 80$. Colours with higher gloss levels can yield higher values for chroma, relative to the colour samples of the RAL design system – a shift of $\Delta C^* = +5$ is well within the realms of possibility. Less glossy surface textures give rise to paler colours and can certainly reduce the chroma by the same amount.

# 5 Colour-naming systems and colour-card collections

## 5.1 RAL system

The RAL[1], organisation, founded in 1925, is the recognised body in Germany for the general management of collaborative projects undertaken by industrial regulatory bodies and public authorities aimed at promoting quality and defining delivery terms in respect of quality [26]. RAL is sponsored by the central organisations of industry, trade, and consumers as well as federal ministries. A special task of RAL consists in collaborating with the relevant business sectors to publish a colour register to act as a basis for industrial colouring.

The colour cards in the collection of colour samples have practically normative character for public institutions and many economic organisations, such as the railways, post office, police, road traffic authorities, and fire services. Moreover, RAL colours are used almost exclusively in the field of commercial vehicle coatings (trucks, agricultural machines). The only purpose of the four-digit numbers for coding RAL colours is to create order within the system. A four-digit code cannot be used to further characterise colours with respect to lightness, chroma, and hue. To organise those RAL names which are additionally used in practice, all RAL colours have been given supplemental names, such as "RAL 6002 Leaf Green". Use of the supplemental names is recommended for shipping because it avoids confusion that may arise from transposed digits.

The **principal register** RAL 840-HR is a collection of base colours used by German industry. It consists of about 190 individual colours. The sample surfaces are semi-glossy and have reflectometer values of about 23 (65) at an angle of 60° (85°). The register is subdivided into nine groups of colours, with the first digit of each code indicating the group; all further digits number the colours within that group. Vacant numbers or gaps are the result of deletions. These colour samples are binding for assessments of matte surface coatings.

Glossy colours are covered by the gloss register RAL 841-GL which exists in parallel to RAL 840-HR. It takes account of the fact that coloured surfaces that differ substantially in gloss cannot be reliably compared with each other, either visually or instrumentally. In addition to the roughly 190 solid colours, both registers contain 15 colour samples, which have been prepared with interference pigments and so satisfy the modern trend for creative colour designs.

The two principal RAL registers described above are supplemented by a series of smaller registers for different fields of application, e.g. retro-reflecting sheets (RAL F-7), safety colours (RAL F-14), camouflage paints (RAL F-9), and planning tools for designers (RAL F-12), with some 560 colours in total.

## 5.2 British Standards Institution (BSI)

In the United Kingdom, the British Colour Council began its efforts to standardise colours back in the 1930s. Its activities were always divided up according to the areas in which the colours were used. Some colour collections are now only of historical or very specific importance, while others are still consulted occasionally. However, technical standards

*[1] Original abbreviation for „Reichsausschuß für Lieferbedingungen"; the updated current designation is „RAL Deutsches Institut für Gütesicherung und Kennzeichnung e.V."*

BS 4800 – Colours for Construction Purposes (the successor to BS 2660) and BS 381 C – Colours for Special Applications – have gained great importance and are also frequently used internationally to order colours.

BS 4800 represents a collection of 88 colours, which were not gathered successively, but rather were defined in collaboration with industry. They are visualised on eight charts, which order the colours on a grey scale from one to eight. The colour label, e.g. C 37 for blue lilac relates to chromaticity and lightness. Of particular interest is that the *Munsell* notation for each colour is specified; naturally, this only indicates the nearest colour and as such only serves as a rough classification guide. As the eight charts and the matte paint finishes are quite small, individual charts measuring 100 x 150 mm$^2$ are available to facilitate exact colour matching. In the first edition of BS 4800, BSI published the CIE chromaticity coordinates and the luminosity of all colours.

BS 381 C currently consists of about 100 glossy colour charts which have been integrated in one register and labelled with a three-digit code. The theoretical colour coordinates are available; however, they are not given on the colour charts. Again, the register is a collection of colour samples and not a colour-order system.

### 5.3 Pantone colour system

The Pantone system is widely employed for specifying and reproducing **ink colours** [27]. However, in practice, Pantone colours are also matched with mixing systems used in automotive and industrial application areas. Within the system, colours are named on the basis of Pantone numbers. Since no transformation exists between this number system and colorimetric-based colour systems, communication between customer and supplier is only possible with the aid of physical colour samples. As the colourants used for the **colour samples** do not possess long-term stability, the colour collections have to be renewed on a regular basis. Isomeric matches are not always possible with pigment assortments used in the automotive industry.

The main component of the Pantone System, the Pantone Colour Formula Guide, is not a colour-order system, as it is not founded on continuous scales, as there are no actual physical embodiments, and as assessment conditions are not standardised. It is rather a **colour-naming system** consisting of 1012 colour samples. Each of these colour samples is assigned a numeric identifier that can be used to communicate the desired colour to a printer. The printer mixes the colour sample from a given Pantone formula. Generally the mixed colour is an acceptable approximation of a given colour sample.

A second collection of colour samples, the Pantone Process Colour Imaging Guide, consists of 942 colour swatches that can be simulated reasonably accurately with a four-colour printing process (CMYK).

## 6 Link between colour-order systems

The large number of colour-order systems contrasts sharply with the underlying intention of making it easy to specify and communicate colour information. Due to the fact that colour-order systems are defined by means of discrete colour samples, exact conversion between the systems is generally not possible. Direct comparison reveals hardly any equivalent colour positions and so only geometrically nearest neighbours in colour space can be identified.

This problem can only be resolved by recourse to transformation models that permit sufficiently precise conversion of colour information between the various colour-order systems, including the necessary interpolation. However, to avoid the need for implementation of $N^2$ bijective transformations between N colour-order systems, it would make sense for a central system to be devised as a way of simplifying the problem. This role could probably be played by the CIE XYZ system, because it serves as the starting system for all uniform colorimetric models. For the purpose of specifying the colour stimulus of all eligible colour-order systems, it would be necessary to measure them all with a spectrophotometer under the appropriate measurement geometry (45°/0°, d/8°). The resulting reflectance spectra would then enable deviating assessment conditions to be simulated. However, this approach has two major limitations. Bijective mappings between physical colour samples and the CIE XYZ system are not available for all eligible colour-order systems. Furthermore, the readings of the various colour-order systems have all been generated under different conditions. These differences stem from the use of instruments with different geometric properties, spectral resolution and different illuminants for processing the reflectance data. For instance, the *Munsell* system was developed for assessment under illuminant C, whereas colour comparison with samples from the NCS system is predicated on the use of illuminant $D_{65}$. A generally satisfactory solution still needs to be developed.

# 7 Decisions

All of the colour-order systems discussed above have advantages and disadvantages in practical use. Probably the most important aspect in this context is the required level of accuracy. *Kelly* and *Judd* [28] have described six levels of accuracy, which may both assist with choosing a colour-order system and meet the requirements of the respective problem. This multi-stage system ranges from a system purely based on colour naming (e.g. red, yellow, blue) at the lowest level to generic colour measurement at the highest level.

**Colour-order systems** have an advantage that should not be underestimated, namely that they can easily be understood due to the availability of material standards for visual assessment. Thus, colour comparisons are possible without the need to use colour-measurement equipment. The number and distance between the colour samples can often be adapted to the application at hand. Furthermore, the samples can be arranged in different ways to suit different purposes. Many colour-order systems are moreover calibrated in CIE tristimulus values which can be generated for colours, if required.

On the other hand, colour-order systems have some disadvantages that restrict their usefulness in industrial practice. Many are in use today, yet accurate conversion from one to the other is not straightforward. Furthermore, some important gaps exist between the colour samples which, for colour-matching purposes, can only be filled by means of interpolation. Visual comparison between colour standards and colour trials is strictly only valid if the same assessment conditions (spectral distribution of the light source, luminance, geometric arrangement, etc.) can be ensured that were originally used to assess and calibrate the system. In addition, different observers may end up with different assessment results (observer metamerism), even when dyschromatopsic observers are excluded; this effect is amplified with increasing difference between the spectral properties of the standard and the sample. Manufacturing tolerances, aging effects, and signs of wear can likewise lead to discrepancies between the current collection and the original standards. When colour standards are being matched in practice, it might happen that they are located outside the miscible colour volume of a given colourant assortment and thus can no longer be matched isomerically. Furthermore, the majority of colour-order systems cannot be used for primary light sources unless special auxiliary equipment is employed.

On account of all these practical limitations and manufacturing costs, the number of colour samples in an atlas is limited. A computer equivalent does not suffer from these restrictions because of its ability to interpolate. Additionally, an **electronic atlas** offers an opportunity to display out-of-gamut colours in an acceptable approximation on output devices (monitors, printers). Likewise, the influence of textured surfaces can be simulated electronically with satisfactory accuracy. A large number of programs are already available on the market that can implement individual colour-order systems and convert electronically from one colour-notation system to another.

In addition to the colour-order systems described above, many more are utilised in practice but could not be discussed here due to space limitations. Readers can find further information on this topic in the books of *Kühni* [2] and *Kühni and Schwarz* [3].

## 8 Literature

[1] Wyszecki, G, *Farbsysteme, Musterschmidt-Verlag, Göttingen, Germany, 1960*

[2] Kuehni, R. G., *Colour Space and its Divisions. Colour Order from Antiquity to the Present,* John Wiley & Sons, Inc., Hoboken, NJ, USA, 2003

[3] Kuehni, R. G., Schwarz, A., *Colour Ordered: A Survey of Colour Systems from Antiquity to the Present,* Oxford University Press, Cambridge, UK, 2008

[4] Stevens, S. S., *Psychophysics,* John Wiley & Sons, LTD, New York, 1975

[5] Bartleson, C. J., Grum, F., eds., *Optical Radiation Measurements. Visual Measurements,* Vol. 5, Academic Press, New York, USA, 1984

[6] Gescheider, G. A., *Psychophysics. The Fundamentals,* Lawrence Erlbaum Associates, Publishers, Mahwah, NJ, USA, 3rd edition, 1997

[7] Norton, T. R., Corliss, D. A., Bailey, J. E., *The Psychophysical Measurement of Visual Function,* Butterworth, Heinemann, New York, USA, 2002

[8] Stevens, S. S., "On the theory of scales of measurement", Science 103, p. 677, 1946

[9] Richter, M., *Einführung in die Farbmetrik, de Gruyter, Berlin, Germany, 1981*

[10] Wyszecki, G., Stiles, W. S., *Colour Science: Concepts and Methods, Quantitative Data and Formulae,* John Wiley & Sons, Inc., New York, USA, 1982

[11] Rhodes, S. A., "Colour notation systems", in Colour Engineering: Achieving Device Independent Colour, Green P., MacDonald, L., Publisher., Wiley-SID Series in Display Technology, John Wiley & Sons, LTD, Chichester, UK, 2003

[12] Robertson, A. R., "Colour-order systems: an introductory review", Color Res. Appl. 9, p. 234, 1984

[13] Schanda, J., *Colorimetry: Understanding the CIE System,* John Wiley & Sons, LTD, New York, USA, 2007

[14] "Farbmetrische Bestimmung von Farbabständen bei Körperfarben nach der DIN99-Formel", Deutsches Institut für Normung, DIN 6176, Beuth-Verlag GmbH, Berlin, Germany, March 2001

[15] Witt, K., "Vektorielle Farbabstandsbewertung mit der DIN99-Formel", Farbe + Lack 111, p. 86, 2005

[16] Munsell, A. H., *A Colour Notation,* Ellis, Boston, MA, USA, 1905-16

[17] Newhall, S. M., Nickerson, D., Judd, D. B., "Final report of the O.S.A. subcommittee on the spacing of the Munsell colours", J. Opt. Soc. Am. 33, p. 385, 1943

[18] Judd, D. B., Wyszecki, G., *Colour in Business, Science and Industry,* John Wiley & Sons, Inc., New York, USA, 1975

[19] DIN Colour Chart, "Teil 1 (1980): System der DIN-Farbenkarte für den 2° Normalbeobachter. Teil 2 (1980): Festlegungen der Farbmuster. Teil 3 (1981): System der DIN-Farbenkarte für den 10° Normalbeobachter. Beiblatt 50 (1981): Farbmaßzahlen für Normlichtart C", Deutsches Institut für Normung, DIN 6164, Beuth-Verlag GmbH, Berlin, Germany, 1981

[20] "Swedish standard for colour notation", Swedish Standards Institution (SIS), SS 01 91 00, 1972-73

[21] "Swedish standard colour atlas", Swedish Standards Institution (SIS), SS 01 91 02, 1979

[22] Hard, A. Sivik, L., Tonquist, G., "NCS natual colour system from concept to research and applications", Colour Res. Appl. 21, p. 129, 1996

[23] MacAdam, D. L., "Uniform colour scales", J. Opt. Soc. Amer. 64, p. 1691, 1974

[24] MacAdam, D. L., "System of OSA committee on uniform colour scales", in Colour 77 (Third congress of the international colour association, Troy, N. Y.), Billmeyer, F. W., Wyszecki, G., publisher, p. 300, Adam Hilger, Bristol, UK, 1978

[25] Gall, L., "Das RAL-Design-System - ein neuer Farbenatlas nach CIELab", Farbe + Lack 98, p. 863, 1992

[26] "Fakten zu RAL-Farben", RAL Deutsches Institut für Gütesicherung und Kennzeichnung e. V., St. Augustin/Bonn, 1985

[27] Eiseman, L., *Colour: Messages & Meanings. A PANTONE Colour Resource*, Hand Books Press, Gloucester, MA, USA, 2006

[28] "Colour: Universal language and dictionary of names", NBS Special Publication 440, U. S. Government Printing Office, Washington, D.C., September 1976

# V Instrumental colour difference assessment

*Wilhelm Kettler*

## 1 Introduction

Colour difference equations are important tools in the production of coloured goods. Production efficiency and customer satisfaction are heavily reliant on extensive colour checks on such materials. Industrial experience shows that individual colourists vary substantially in their ability to assess colours and colour differences. The implementation of objective control methods is therefore highly desirable. Based on a limited number of visual assessment data sets or different formal approaches concerning colour difference perception, over the years various metrics have been developed, which have found their way to industrial applications with more or less success[1-3]. The most important modern colour and colour difference systems are presented and compared below. The chapter concludes with an outline of extensions necessary for assessing goniochromatic colour shades – a metric needs to be developed here that unifies the appearance arising from hue and the texture of effect pigments.

## 2 Geometric structure of colour difference models

The geometric structure of existing colour difference models can be reduced to the following three basic types[1]:

*Equation V.1 to V.3*

1. $$\Delta E = \frac{1}{k_E}\sqrt{(\Delta L^*)^2 + (\Delta a^*)^2 + (\Delta b^*)^2} = \frac{1}{k_E}\sqrt{(\Delta L^*)^2 + (\Delta C^*)^2 + (\Delta H^*)^2}$$

2. $$\Delta E = \frac{1}{k_E}\sqrt{\left(\frac{\Delta L^*}{k_L S_L}\right)^2 + \left(\frac{\Delta C^*}{k_C S_C}\right)^2 + \left(\frac{\Delta H^*}{k_H S_H}\right)^2}$$

3. $$\Delta E = \frac{1}{k_E}\sqrt{\left(\frac{\Delta L^*}{k_L S_L}\right)^2 + \left(\frac{\Delta C^*}{k_C S_C}\right)^2 + \left(\frac{\Delta H^*}{k_H S_H}\right)^2 + R_T\left(\frac{\Delta C^*}{k_C S_C}\frac{\Delta H^*}{k_H S_H}\right)}$$

where $\Delta L^*=L^*_s - L^*_r$, $\Delta C^*=C^*_s - C^*_r$, and $\Delta H^*=C^*\cdot\Delta h_{ab}=C^*\cdot(h_{ab,s} - h_{ab,r})$. The reference colour position (standard) is given by the coordinate triple $\mathbf{F}_r=(L^*_r,a^*_r,b^*_r)^T$, while the geometric position of a comparison colour shade (sample) is given by $\mathbf{F}_s=(L^*_s,a^*_s,b^*_s)^T$.

[1] *Note that the symbols utilised in the colour difference equations not always correspond to the notation adopted in the respective technical standards. For consistency reasons a uniform notation has been preferred at this point.*

Quantities $S_L$, $S_C$, and $S_H$ represent weighting functions and generally depend on $a^*$, $b^*$, and $L^*$, while $k_L$, $k_C$, and $k_H$ are scaling parameters which serve to account for deviations from reference conditions during a visual assessment. Parameter $k_E$ can be used to scale the order of magnitude of the total residual colour difference and to adjust it to non-standardised assessment conditions or special tolerance thresholds.

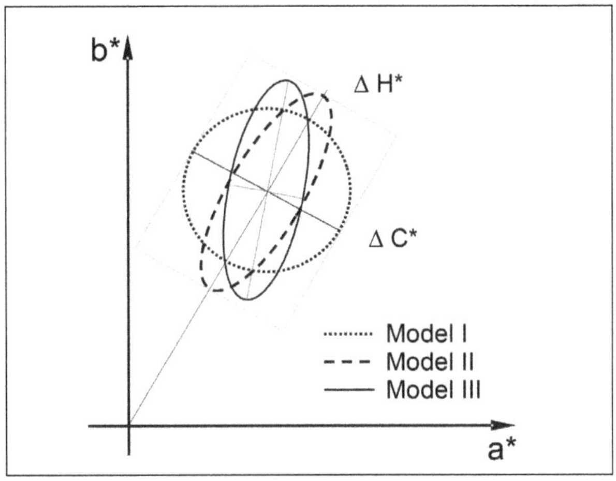


*Figure V.1: The colour positions of tolerance solids of various colour difference models as projected onto the a*-b* plane of colour space CIELAB76.*

The first model (type I) is a **Euclidean colour space**, in which the distance between two positions $\mathbf{F}_r$ and $\mathbf{F}_s$ is given by their direct vector connection. A three-dimensional space, all colours having the same colour difference to a reference position $\mathbf{F}_r$ define the surface of a sphere. In a two-dimensional projection of the tolerance solid onto the $a^*$-$b^*$ plane of the CIELAB76 colour space leads to a circle (see Figure V.1). Colour spaces **CIELAB76** and **DIN99** belong to this type.

Model II is a generalisation of the Euclidean colour space of model I and defines the line element of a non-Euclidean or **special Riemannian colour space**[2]. The representation of this metric in CIELAB76 colour space describes the locus of ellipsoids, with the semi-axes $a = k_C \cdot S_C$, $b = k_H \cdot S_H$, and $c = k_L \cdot S_L$, where the $c$ axis is parallel to the $L^*$ axis, and the $a$ axis in the lightness plane $L^*_B$ of the reference always points towards the achromatic axis. The third semi-axis $b$ is perpendicular to the two other semi-axes. Projection of the tolerance solid onto the $a^*$-$b^*$ plane of CIELAB76 colour space yields an ellipse, whose major semi-axis points towards the origin of the coordinate system (see Figure V.1).

The third model (type III) defines a further **special Riemannian colour space**, whose line element, compared to model II, includes a mixed **rotation term** scaled with $R_T$, which influences the orientation of semi-axis $a$ in the geometric representation in CIELAB76 colour space. This semi-axis does not necessarily point towards the achromatic axis and so the ellipsoids of model II undergo a rotation about the $c$ axis parallel with the lightness axis. Projection of the tolerance solid onto the $a^*$-$b^*$ plane of the CIELAB76 colour space yields an ellipse, whose major semi-axis does not necessarily point towards the origin of the coordinate system (see Figure V.1)

# 3 Colour difference model CMC($k_L$:$k_C$)

In 1984, the Colour Measurement Committee (CMC) of the Society of Dyers and Colourists in the UK took the challenge to reduce known weaknesses in established colour difference models (CIELAB76, JPC79, etc.) for instrumental colour testing. The structure of the published colour difference model is oriented on the generalised CIELAB76 colour

*[2] In Euclidean spaces the line element is defined by $ds^2 = \Sigma_i dx_i^2$. This is a special case of a more general form of the line element es $ds^2 = \Sigma_{i,j} g_{ij}\, dx_i dx_j$ investigated by Riemann, where the metric fundamental quantities $g_{ij}$ again are functions of the coordinates $x_k$. For Euclidean spaces $g_{ij}=0$ for $i \neq j$ and $g_{ij}=1$ for $i=j$ is fulfilled.*

difference equation and includes local correction functions that give a weighting to the specific CIELAB76 colour difference values $\Delta L^*$, $\Delta C^*$, and $\Delta H^*$ [4]:

*Equation V.4*
$$\Delta E^*_{CMC}(k_L : k_C) = \sqrt{\left(\frac{\Delta L^*}{k_L\, S_L}\right)^2 + \left(\frac{\Delta C^*}{k_C\, S_C}\right)^2 + \left(\frac{\Delta H^*}{k_H\, S_H}\right)^2}$$

where weighting functions
$$S_L = \begin{cases} 0{,}511 & \text{für } L^* < 16 \\ \dfrac{0{,}040975 \cdot L^*}{1 + 0{,}01765 \cdot L^*} & \text{für } L^* \geq 16 \end{cases}$$

$$S_C = 0.638 + \frac{0.0638C^*}{1 + 0.0131C^*}, \quad S_H = S_C\,(T \cdot F + 1 - F)$$ and the auxiliary variables

$$F = \sqrt{\frac{(C^*)^4}{(C^*)^4 + 1900}}$$

$$T = \begin{cases} 0.56 + |0.2 \cos(h_{ab} + 168°)| & \text{for } 164° \leq h_{ab} \leq 345° \\ 0.36 + |0.4 \cos(h_{ab} + 35°)| & \text{elsewhere} \end{cases}$$

The two parameters $k_L$ and $k_C$ serve to provide further task-specific and sector-specific scaling possibilities. In the **CMC model**, the analogous parameter for scaling the hue contribution is set to $k_H$=1, while the textile sector generally uses $k_L$=2 and $k_C$=1. Colour difference results determined in this way are then referred to as **CMC(2 : 1).** In this weighting function constellation, the lightness difference $\Delta L^*$ is rated two times less than the chroma difference $\Delta C^*$. This reflects the fact that, as a result of shadows in drapery, lightness differences are expected in textiles in any event and therefore less attention is paid to them. In other sectors, e.g. the paint industry, $k_L$=1.3 has proved to be more effective; here, too, the parameter of the chroma term is usually set to its standard value of $k_C = 1$.

As can be seen from the equations given above for $S_L$ und $S_C$, both functions initially increase rapidly with increase in lightness $L^*$ or chroma $C^*$, converging to a limiting value at high values. The function for $S_L$ varies from 0.511 (for $L^* = 0$) to 1.48 (for $L^* = 100$), i.e. at high absolute lightness level the **CMC($k_L$:$k_C$) equation** gives a lower weighting to lightness differences $\Delta L^*$ than to low absolute lightness values. On the other hand, $S_C$ varies from 0.638 to 2.76 for $0 \leq C^* \leq 100$, so that hue differences of achromatic colours are weighted more heavily than brilliant variants. This accounts for the fact that the CIELAB76 equation is known to over-estimate chroma differences of brilliant colours by up to a factor of five.

The behaviour of correction function $S_H$ is not so easy to understand, as it exists in a multiplicative relationship with $S_C$. Likewise, $\Delta H^*$ is progressively weighted less with increase in chroma $C^*$. At the periphery, the hue rays of the CIELAB76 system are spaced too far apart from each other, and this needs to be corrected accordingly. This is achieved by means of a transformation which distorts the originally spherical tolerance solid into ellipsoids, whose major axis is oriented along the direction of the hue rays. In this way, hue differences are given a stronger weighting than chroma differences. With increase in brilliance, tolerance ellipsoids in respect of $\Delta C^*$ become increasingly elongated. The CMC($k_L$:$k_C$) equation gives higher emphasis to hue differences $\Delta H^*$ in the case of more achromatic colours, and less emphasis in the case of more brilliant colours.

The use of the CMC($k_L$:$k_C$) model to compute small colour differences for textiles is recommended in British (BS 6923) and American (AATCC 173) technical standards and in 1996 was integrated by ISO into the technical standard for the assessment of light fastness of textiles (ISO 105-J03).

# 4 Colour difference model BFD($k_L$:$k_C$)

In 1986, *Luo* und *Rigg*[5,6] published a detailed overview of their efforts to consolidate various data sets from visual assessments of different materials. The authors collated results from 13 experiments carried out on glossy paint materials, printing inks, matte paint materials, and textiles in various preparations. Analysis of these data sets led to the identification of 132 credible ellipses in the CIE chromaticity diagram (see Section 1.7.2). However, a substantial difference in the size and orientation of the ellipses becomes apparent in each region of this chromaticity diagram. This can be attributed to differing experimental conditions, observer variability, and differences in the quality criteria. Extensive statistical analysis yielded a combined data set which formed the basis for the definition of a new **colour difference equation BFD($k_L$:$k_C$)**. In this definition of the colour difference model, the generalised colour difference model proposed by the CIE for the first time was extended by a rotation term:

*Equation V.5*
$$\Delta E = \sqrt{\left(\frac{\Delta L^*}{k_L\, S_L}\right)^2 + \left(\frac{\Delta C^*}{k_C\, S_C}\right)^2 + \left(\frac{\Delta H^*}{k_H\, S_H}\right)^2 + R_T\left(\frac{\Delta C^*}{k_C\, S_C}\,\frac{\Delta H^*}{k_H\, S_H}\right)}$$

The new equation includes an additional mixing term, which was needed so that the underlying data sets could be described. The weighting functions and parameters in the above equation are computed as follows:

$$S_L = 1; \quad S_C = 0.521 + \frac{0.035\bar{C}^*}{1 + 0.00365\bar{C}^*}; \quad S_H = S_C\,(G \cdot T' + 1 - G) \quad \text{where}$$

$$G = \sqrt{\frac{(\bar{C}^*)^4}{(\bar{C}^*)^4 + 14000}} \quad \text{and}$$

$$T' = 0.627 + 0.055\cos(\bar{h}_{ab} - 254°) - 0.040\cos(2\bar{h}_{ab} - 136°) + 0.070\cos(3\bar{h}_{ab} - 32°) + 0.049\cos(4\bar{h}_{ab} + 114°) - 0.015\cos(5\bar{h}_{ab} - 103°)$$

Computation of the interaction term $R_T = R_H \cdot R_C$ requires the following auxiliary quantities to be evaluated:

$$R_H = -0.260\cos(\bar{h}_{ab} - 308°) - 0.379\cos(2\bar{h}_{ab} - 160°) - 0.636\cos(3\bar{h}_{ab} + 254°) + 0.226\cos(4\bar{h}_{ab} + 140°) - 0.194\cos(5\bar{h}_{ab} + 280°)$$

$$R_C = \sqrt{\frac{(\bar{C}^*)^6}{(\bar{C}^*)^6 + 7 \cdot 10^7}}$$

The lightness scale of the *BFD* equation is defined as follows:

$$L_{BFD} = 54.6\log_{10}(Y + 1.5) - 9.6$$

All quantities with overbars denote mean values (*s* = sample, *r* = reference):

$$\bar{C}^* = \frac{1}{2}(C^*_r + C^*_s)$$

$$\bar{h}_{ab} = \frac{1}{2}\begin{cases}(h_{ab,r} + h_{ab,s}) & \text{for } \bar{h}_{ab} \le 180° \\ (h_{ab,r} + h_{ab,s} - 360°) & \text{for } \bar{h}_{ab} > 180°\end{cases}$$

For reasons unknown, the *BFD* colour difference model has been virtually ignored in industry.

# 5 Colour difference model CIE94

Proceeding from the generalised CIE colour difference equation

*Equation V.6*

$$\Delta E^*_{94} = \sqrt{\left(\frac{\Delta L^*}{k_L S_L}\right)^2 + \left(\frac{\Delta C^*}{k_C S_C}\right)^2 + \left(\frac{\Delta H^*}{k_H S_H}\right)^2}$$

corresponding to model II, the two technical CIE committees TC 1-28 and TC 1-29 attempted to define improved position-dependent weighting functions $S_L$, $S_C$, and $S_H$ with a view to reducing deviations of the CIELAB76 model from uniformity to an acceptable level. For this purpose, all the available data sets of object colours were screened for statistical significance and for properly documented experimental conditions. Only the three data sets of Witt [7, 8], *Luo-Rigg* [9], and *RIT-DuPont* [10] satisfied the criteria and were used to optimise the weighting functions. The outcome of the analyses are the comparatively simple relationships[11]

$$S_L = 1; \quad S_C = 1 + 0.045C^*; \quad S_H = 1 + 0.015C^* \quad \text{where} \quad k_L = k_C = k_H = 1$$

under reference conditions. Correction of the lightness difference scale does not seem necessary, and the chroma and hue difference terms are weighted by linear functions whose independent variable is the chroma of the standard. For the chroma $C^*$ in the weighting functions, the recommendation is to use

$$C^* = C^*_r \quad \text{oder} \quad \frac{1}{2}(C^*_r + C^*_s) \quad \text{oder} \quad \sqrt{C^*_r \cdot C^*_s}$$

where the first option is not invariant with respect to permutation of sample (index *s*) and standard (index *r*).

# 6 Colour difference model CIEDE2000

The various efforts to define a colour difference model that conforms with visual perception culminate in the **CIEDE2000 colour difference equation**[12,13]

*Equation V.8*

$$\Delta E^*_{00} = \sqrt{\left(\frac{\Delta L'}{k_L S_L}\right)^2 + \left(\frac{\Delta C'}{k_C S_C}\right)^2 + \left(\frac{\Delta H'}{k_H S_H}\right)^2 + R_T\left(\frac{\Delta C'}{k_C S_C}\frac{\Delta H'}{k_H S_H}\right)}$$

developed by the CIE with the new auxiliary coordinates

$$L' = L^*; \quad a' = (1 + G)a^*; \quad b' = b^*; \quad C' = \sqrt{(a')^2 + (b')^2}; \quad h' = \frac{180°}{\pi}\arctan\left(\frac{b'}{a'}\right)$$

$$\text{where} \quad G = \frac{1}{2}\left(1 - \sqrt{\frac{(\bar{C}^*)^7}{(\bar{C}^*)^7 + 25^7}}\right)$$

For the difference values, the following relations have to used:

$$\Delta L' = L'_s - L'_r; \quad \Delta C' = C'_s - C'_r; \quad \Delta H' = 2\sqrt{C'_s \cdot C'_r} \cdot \sin\left(\frac{\Delta h'}{2}\right)$$

$$\Delta h' = \begin{cases} 0 & \text{for } C'_s \cdot C'_r = 0 \\ h'_s - h'_r & \text{for } C'_s \cdot C'_r \neq 0 \text{ and } |h'_s - h'_r| \leq 180° \\ h'_s - h'_r - 360° & \text{for } C'_s \cdot C'_r \neq 0 \text{ and } (h'_s - h'_r) > 180° \\ h'_s - h'_r + 360° & \text{for } C'_s \cdot C'_r \neq 0 \text{ and } (h'_s - h'_r) < -180° \end{cases}$$

The functional relationship for the weighting functions has been defined as follows:

$$S_L = 1 + \frac{0.015\,(\bar{L}' - 50)^2}{\sqrt{20 + (\bar{L}' - 50)^2}}; \quad S_C = 1 + 0.045\,\bar{C}'; \quad S_H = 1 + 0.015\,\bar{C}'T \quad \text{where}$$

$$T = 1 - 0.17\cos(\bar{h}' - 30°) + 0.24\cos(2\bar{h}') + 0.32\cos(3\bar{h}' + 60°) - 0.20\cos(4\bar{h}' - 63°)$$

For the purpose of evaluating the interaction term $R_T = -R_C \cdot \sin(2\Delta\theta)$ between hue and chroma, the two auxiliary quantities

$$\Delta\theta = 30e^{-\left(\frac{\bar{h}' - 275°}{25°}\right)^2} \quad \text{and} \quad R_C = 2\sqrt{\frac{(\bar{C}')^7}{(\bar{C}')^7 + 25^7}}$$

need to be evaluated. The quantities with overbars denote arithmetic mean values of the respective colorimetric auxiliary coordinates, i.e.,

$$\bar{L}' = \frac{1}{2}(L'_s + L'_r); \quad \bar{C}' = \frac{1}{2}(C'_s + C'_r);$$

$$\bar{h}' = \begin{cases} \frac{1}{2}(h'_s + h'_r) & \text{for } |h'_s - h'_r| \leq 180° \text{ and } C'_s \cdot C'_r \neq 0 \\ \frac{1}{2}(h'_s + h'_r + 360°) & \text{for } |h'_s - h'_r| > 180°; \; (h'_s + h'_r) < 360° \text{ and } C'_s \cdot C'_r \neq 0 \\ \frac{1}{2}(h'_s + h'_r - 360°) & \text{for } |h'_s - h'_r| > 180°; \; (h'_s + h'_r) \geq 360° \text{ and } C'_s \cdot C'_r \neq 0 \\ h'_s + h'_r & \text{for } C'_s \cdot C'_r = 0 \end{cases}$$

For $\bar{h}'$ >180°, the mean value of the hue angle has to be reduced by 180°.

An improvement in the computation of total residual colour differences in industrial colour difference assessment is achieved by correcting lightness, chroma, and hue difference contributions and by accounting for a chroma-hue interaction in colour perception. It can be assumed that the recommended CIEDE2000 model yields reliable results under the implied reference conditions of $k_L = k_C = k_H = 1$. If there are considerable deviations between assessment and reference conditions, parameters $k_L$, $k_C$, and $k_H$ can be invoked for correction purposes.

The full designation of the colour difference model for industrial colour difference assessment is CIE 2000($\Delta L'$, $\Delta C'$, $\Delta H'$) colour difference model, denoted by the symbol $\Delta E_{00}$ and in abbreviated form CIEDE2000.

However, caution must be exercised when computing hue-angle differences if both hue angles are located in different quadrants of the colour space. If, for example, a colour difference pair has hue angles of 30° (reference) and 300° (sample), the simple difference of 270° leads to a wrong arithmetic sign in the equation for hue-angle difference. To correctly determine the hue-angle difference, the absolute difference of both hue angles must first be computed. If this absolute difference exceeds 180°, a value of 360° has to be added to the smaller hue angle before subtraction is carried out. In the given example, the correct hue-angle difference would be a value of −90°.

Similarly, mean hue angles must be computed carefully if the hue angles of a colour difference pair are located in different quadrants of the colour space. For the example above of two hue angles of 30° (reference) and 300° (sample), the mean value of 165° is not correct. In this case, too, before the mean value is computed, an angle of 360° must be added to the smaller hue angle if the absolute hue-angle difference exceeds 180°. The correct mean value of the given sample pair is therefore 345°.

This colour difference model is intended to fully replace predecessor models, such as CMC(kL:kC) and CIE94. Unfortunately, the CIELAB76 model will continue to live on, since the colour difference model CIEDE2000 is not based on an independent Euclidean colour space with vectorial distance measure that ensures the path independence of differences between two colour positions.

# 7 Colour space DIN99

Colour space DIN99 [14-16] is a Euclidean space spanned by three orthogonal axes $L_{99}$, $a_{99}$, and $b_{99}$. These three new coordinates replace the $L^*$, $a^*$, and $b^*$ coordinates of the CIELAB76 system and are related to them by non-linear transformations. Due to the Euclidicity of the DIN99 coordinate system, the total residual colour difference is given by the equation

*Equation V.9*
$$\Delta E_{99} = \frac{1}{k_E}\sqrt{(\Delta L_{99})^2 + (\Delta a_{99})^2 + (\Delta b_{99})^2} = \frac{1}{k_E}\sqrt{(\Delta L_{99})^2 + (\Delta C_{99})^2 + (\Delta H_{99})^2}$$

For the new lightness coordinate, $L_{99}$, the following formal relation has been defined:

*Equation V.10*
$$L_{99} = \frac{p_1}{k_E}\ln(1 + p_2 L^*)$$

Before the coordinates of the missing two opponent chromatic axes $a_{99}$ and $b_{99}$ can be calculated, some auxiliary quantities

$$e = a^* \cos(\phi_1) + b^* \sin(\phi_1), \quad f = -p_3 a^* \sin(\phi_1) + p_3 b^* \cos(\phi_1),$$

$$G = \sqrt{e^2 + f^2}, \quad h_{ef} = \arctan\left(\frac{f}{e}\right)$$

must first be assigned. The two quantities $e$ and $f$ result from clockwise rotation of the coordinate system through the angle $\phi_1$ about the $L^*$ axis, whereby, because $p_3 < 1$, the $a^*$ axis is additionally compressed. These auxiliary quantities can be used to derive the new chromaticity coordinates

*Equation V.11a to d*
$$h_{99} = \frac{180°}{\pi} h_{ef} + \phi_2$$

$$C_{99} = \frac{\ln(1 + p_5 G)}{p_4 k_{CH} k_E}$$

$$a_{990} = C_{990} \cos(h_{ef})$$
$$b_{990} = C_{990} \sin(h_{ef})$$

where the corresponding quantities $C_{99}$ and $h_{99}$ are stated in the cylindrical coordinate system.

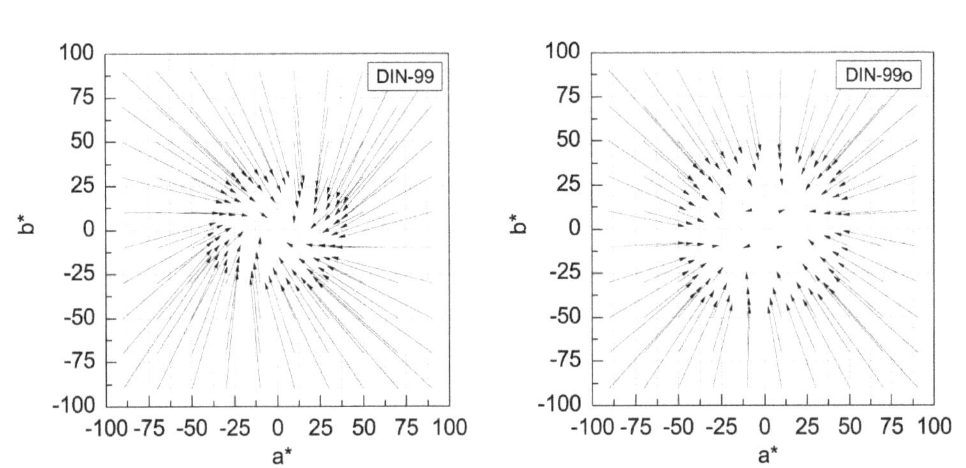


*Figure V.2: Projection of a quadratic grid of the CIELAB76 system into the DIN system (left diagram) and the DIN99o system (right diagram) when displayed in the $a^*$-$b^*$ plane in the form of an arrow diagram for a lightness level of $L^*$=50. The arrow base represents the colour position in CIELAB76 reference space and the arrow head the position in target space.*

All values for the parameter set are shown in Table V.1, in which the first set belongs to the DIN99 model[14]. This parameter set has been optimised under the constraint that the DIN99 model in the prediction of total residual colour differences closely resembles that of the CIE94 model. The second parameter set for the optimised DIN99o metric was derived by adapting the model to the CIEDE2000 equation[16].

The consequences of the DIN99 transformation on the colour positions of the CIELAB76 system can be studied by applying the transformation instructions given above to a regular grid of CIELAB76 coordinates of constant mesh size and analysing the results in graphical form. Since $L_{99}$ only depends on $L^*$ and not on $a^*$ and $b^*$, the projection onto the principal plane $a^*$-$b^*$ is independent of $L^*$. Figure V.2 shows these results in the form of an arrow diagram, in which the arrow base corresponds to the reference position in the CIELAB76 system and the arrow head to the transformed new position in colour space DIN99. A close scrutiny of Figure V.2 reveals that the quadratic starting area of edge length 180 is considerably compressed in size. The shape of the transformed area resembles a barrel-shaped distorted rectangle, whose edges are additionally rotated relative to the reference square.

The projection of colour positions onto the $L^*$-$a^*$ ($L^*$-$b^*$) principal plane depends on the size of the $b^*$ ($a^*$) coordinate (see transformation equations). For this reason, several cross-sections need to be analysed if the changes in the transition from the CIELAB76 to the DIN99 system are to be understood.

*Table V.1: Parameter sets for the two colour space models DIN99[14] and DIN99o[16].*

| Parameter | DIN99 | DIN99o |
|---|---|---|
| $p_1$ | 105.51 | 303.67 |
| $p_2$ | 0.0158 | 0.0039 |
| $p_3$ | 0.7 | 0.83 |
| $p_4$ | 0.045 | 0.0435 |
| $p_5$ | 0.045 | 0.075 |
| $\phi_1$ | 16° | 26° |
| $\phi_2$ | 0° | 26° |

Furthermore, distortions of CIELAB76 colour space in the

region of brilliant red colours, which generally have comparatively high $b^*$ values of 20 units or more, are reduced considerably; the new positions of these colours are located closer to the positive $a_{99}$ axis, thereby indicating a reduction of the $b_{99}$ value compared to $b^*$.

# 8 Parametric effects

The colour difference equation originally generalised by CIE[11] is of type II and comprises constant factors $k_E$, $k_L$, $k_C$, $k_H$, and weighting functions $S_L$, $S_C$, and $S_H$, for correcting the lack of uniformity of the CIELAB76 equation. The model requires that the reference conditions for the colour difference assessment be defined[3]. A viable comparative judgement of colour differences is only possible under standardised **viewing conditions**. If the colour difference stimuli satisfy the reference conditions, the parametric factors have to be set to their standard values, i.e., $k_E = k_L = k_C = k_H = 1$. Deviations from the reference conditions have to be accounted for by modified parametric factors. The reference conditions (see Table V.2) presume a non-textured sample pair in direct edge contact with each other against a neutral background of lightness $L^* = 50$, which is illuminated by a $D_{65}$ light source with an illuminance of 1000 *lux*. The assessment distance needs to be adjusted so that the viewing angle of the sample pair exceeds a value of 4°.

Not all of the colour difference models discussed above have used the same **reference conditions** for the underlying visual assessments. The impact of parametric effects on visual assessment results has been compiled by the CIE in a technical report[18]. Parameter $k_E$ can be used to vary the sensitivity of the respective colour difference equation. Relative to a given residual colour difference, sensitivity is reduced for $k_E > 1$ and increased for $k_E < 1$. Adjustment of $k_E$ is needed if the sample pair has no direct edge contact in a visual assessment. With increase in distance between the colour samples, the perceived visual colour difference progressively decreases. At a distance of about 0.5° between sample pairs, colour discrimination requires that a sensitivity adjustment of $k_E \cong 1.5 - 2.6$, relative to the reference conditions, be made to the difference metric, with the exact figure depending on the lightness level[18].

A change of **ambient conditions** (lightness, hue) induces a different adaptation on the part of the observer, which can only be reproduced by means of an appropriate appearance

*Table V.2: Reference conditions for visual assessments of object colours as defined by CIE[17].*

| Reference condition | Specification |
|---|---|
| Illumination | $D_{65}$ simulation |
| Illuminance | 1000 lx |
| Surroundings | Uniform, neutral grey with $L^* = 50$ |
| Viewing mode | Object |
| Sample size | Greater than 4° |
| Sample separation | Minimum distance, in edge contact |
| Magnitude of colour difference | 0 – 5 $\Delta E^*_{76}$ |
| Sample texture | Homogeneous, without visual non-uniformities |

*[3] By publishing the colour difference model CIEDE2000 also the CIE has abandoned the path of virtue and did join the general trend to improve models by introducing further parameters instead of establishing a reliable database as an outcome of basic and unified experiments and to build upon it an improved (and possibly) Euclidean metric.*

metric (see, e.g. CIECAM97s [19-21] or CIECAM02 [22,23]). When changes occur in background lightness, it is necessary to check whether making an adjustment to the $k_L$ factor would be sufficient or whether the $S_L$ function needs to be redefined.

As can be seen from the difference between the colour-matching functions of the 2° and 10° standard observers, a change in sample size may also give rise to a changed perception of colour stimuli of a sample pair.

With increasing **gap** between two specimens, the lightness contrast decreases and an observer becomes more tolerant to colour differences. In colorimetry, this effect can be accounted for by raising the $k_E$ factor.

Likewise, the **texture** of a sample pair to be assessed has an impact on colour perception and can be accounted for by adjusting the scaling factors. In applications in the textile industry, it has been found that provision can be made for the influence of shadowing effects by choosing a relative weighting of $k_L : k_C = 2 : 1$ in the CMC equation to improve the correlation between visual and instrumental assessment results. On the other hand, glossy colour pairs have been successfully assessed in practice with a parametric ratio of $k_L : k_C = 1.3 : 1$.

Changing the **illuminant** of a light source might reveal metameric problems which are caused by a metameric pigmentation of a sample pair or which might be a consequence of observer metamerism. While a change in illuminant also shifts the chromaticity coordinates, they stay constant under the other conditions. The lightness of the surround is a very influential factor. It is therefore best to assess carbon blacks against a black surrounding. For samples with sharp edge-to-edge contact, colour differences are enhanced by the phenomenon of simultaneous contrast.

# 9 Comparative analyses of the performance of modern colour difference models

The existence and use of different colour difference models and the fact that none of the metrics developed so far has gained general acceptance indicate that none of the models is perfect and that there is a permanent need for improvement. Only the CIELAB76 and DIN99 models span Euclidean colour spaces; all other metrics discussed are special manifestations of Riemannian spaces. Attempts to "Euclidise" them have only been partially successful [24-27]. Owing to the presence of a rotation term, model III has defied complete analytical Euclidisation, but was more or less amenable to numerical treatment [28].

Practical application of the CIELAB76 system in the textiles industry [29, 30] soon revealed that the contours of tolerance limits in CIELAB76 colour space are not spheres of constant radius, but rather ellipsoids of variable size. In particular, the perceived lightness differences are underestimated by the difference measure $\Delta L^*$ and the perceived hue differences are overestimated by the difference measure $\Delta C^*$ [31]. This asymmetry needs to be eliminated in a more uniform model.

The mixed rotation term appearing in the BFD and CIEDE2000 colour difference equations had been introduced into the formalism to correct for the slanted tolerance ellipsoids in the blue colour region (vicinity of negative $b^*$ axis). Experiments have revealed that this correction is necessary to ensure compliance of the model with visual assessment results in this colour region.

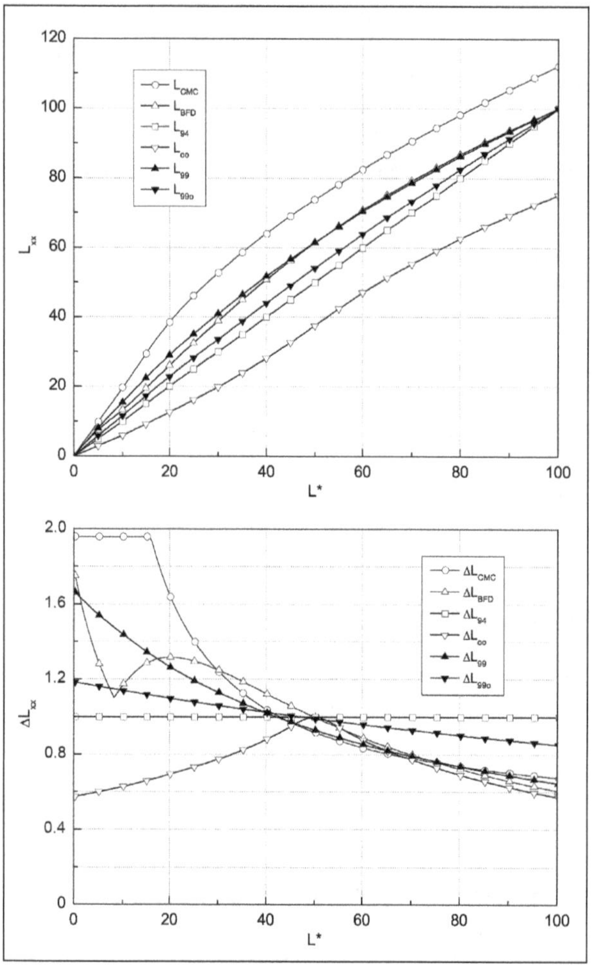

*Figure V.3: Comparison of absolute lightness scales of various colour spaces (upper diagram) and their sensitivities (lower diagram) with colour space CIELAB76 as reference, and $\Delta L$ =1.0. In the case of the non-Euclidean colour difference spaces, the absolute scales have been computed by integrating the line elements. All the colour-difference functions shown are valid for the parameter set $k_L = k_C = k_H = 1$.*

## 9.1 Normalisation

*Völz*[31] has pointed out two problems, which arise when different colour difference models are compared. On one hand, not all metrics are invariant concerning permutation of reference and sample and, on the other, they are not consistently normalised to the smallest perceptible colour difference and hence are not commensurable. According to a hypothesis by *Schrödinger*[32] the distance between two colour stimuli having positions $P_1$ and $P_2$ in a non-Euclidean colour space is obtained by counting the number of just perceptible steps ($ds_o$=const.) between $P_1$ and $P_2$ for an arbitrarily chosen path. For a given path, this number of just perceptible steps assumes its smallest value. After normalisation to the smallest perceptible step size $ds_o$, this minimum value can be interpreted as the colour difference between the two colour stimuli. Only after this normalisation step colour difference equations are commensurable. In the case of the CMC equation, this deficiency can be remedied by dividing all $S_x$ weighting functions by the value 0.638, which results at the origin of the coordinate system ($C^* = \Delta L^* = \Delta C^* = 0$) for $\Delta H^* = 1$.

## 9.2 Lightness scale

The weighting functions $S_L$, $S_C$, and $S_H$ are used to adjust the semi-axes of the ellipsoids which define the tolerance volume and the perceptibility threshold in colour space CIELAB76. In all colour-difference models, the lightness term can be considered separately and integrated analytically, thereby permitting straightforward quantitative comparison of the lightness scales. The upper diagram of Figure V.3 shows the variation in the lightness $L^*_x$ of all discussed colour-difference models having lightness $L^*$ in CIELAB76 colour space. For the non-Euclidean colour-difference models (CMC, BFD, CIEDE2000), the lightness functions shown were determined by analytic integration of the respective line element. Clearly, the lightness scales of the CMC and CIEDE2000 colour-difference models are normalised differently compared to the four other models (BFD, CIE94, DIN99, DIN99o), all of which assume the same function value $L^*_{xx} = 100$ for $L^* = 100$. Since the lightness functions of CIELAB76 and CIE94 are identical, $L^*_{94}$ divides the first quadrant of the coordinate system into domains of equal size. In the lightness interval shown, $L^*_{BFD}$, $L^*_{99}$, $L^*_{99o} > L^*_{94}$, with maximum deviations of about 10 units at medium lightness values.

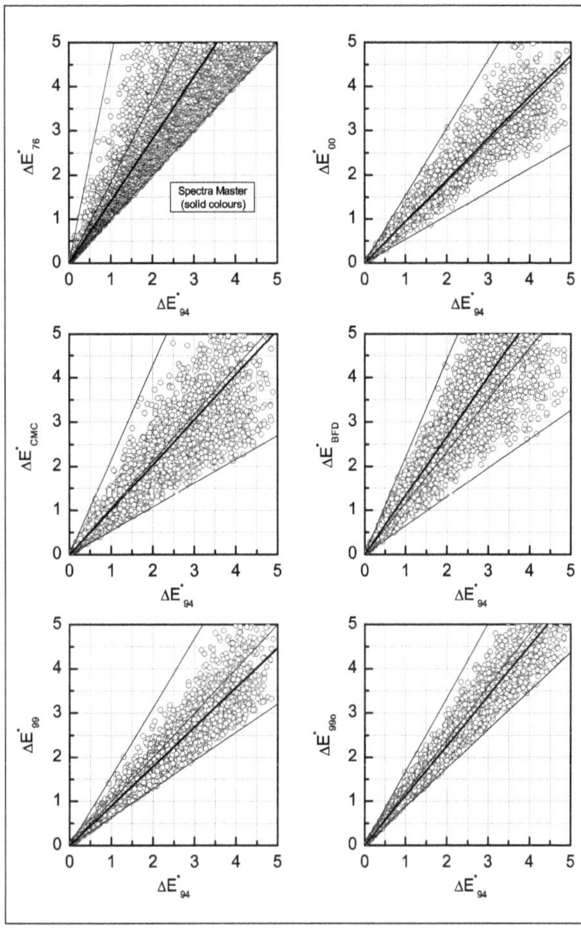


*Figure V.4: Correlation between the determined residual colour difference values of various metrics. The reference data set consists of about 3300 solid colours, whose colour positions are homogeneously distributed over the colour gamut of a refinish mixing system. The respective comparison partner was generated artificially by means of a random generator under the boundary condition $\Delta E^*_{76} \leq 5$. The thick solid lines represent the average correlation between the respective colour-difference spaces, while the thin solid lines enclose or bisect the discrete distributions.*

Particularly interesting is the plot of lightness differences as a function of $L^*$ for a reference difference of $\Delta L^* = 1$ (see Figure V.3) for the various colour-difference models. With the exception of the CIEDE2000 colour-difference equation, all the metrics shown follow the same trend. Relative to the CIELAB76 model, they yield higher values for $L^* < 50$ and lower values for $L^* > 50$. While the DIN99 and DIN99o models exhibit monotonically decreasing curves with increasing lightness, anomalies can be observed at low lightness values for the CMC(1.3:1) and BDF metrics. In case of the CMC model, this is a consequence of the division of the lightness scale into two domains having formally different $S_L$ weighting functions. The anomaly of the BFD model is a direct result of the division of the CIELAB76 lightness function into two domains with $L^* \gtrless 8$. There is no colorimetric basis for this anomalous behaviour of the BFD model. It is merely a mathematical artefact which escaped the notice of the model's developers.

As already mentioned, colour-difference model CIEDE2000 behaves completely differently. Due to the mathematical structure of the lightness weighting function, $S_L$, symmetry with respect to a lightness level of $L^* = 50$ is enforced. At the maximum of $\Delta L^*_{00}$ for $L^* = 50$, both CIEDE2000 and CIELAB76 models are the same when it comes to assessment of lightness differences. For $L^* \gtrless 50$, the sensitivity in lightness difference assessment decreases monotonically down to a value of about 0.6. While this behaviour is in conformity with other colour-difference models for $L^* > 50$, CIEDE2000 assesses lightness differences for $L^* < 50$ as being markedly less than those of the CIELAB76 equation, a fact which contrasts sharply with the CMC, BFD, and DIN models treated above and with practical experience in colour

*Table V.3: Specific parameters for characterising the correlation between the metrics of various colour-difference spaces relative to the CIE94 model. Quantity $\Delta\varphi = \varphi_{max} - \varphi_{min}$ denotes the acute angle of the respective discrete triangular distribution.*

| Ratio | $\tan\varphi_{max}$ | $\tan\varphi_{min\,x}$ | $\langle\tan\varphi\rangle$ | $\varphi_{max}$ | $\varphi_{min\,x}$ | $\langle\varphi\rangle$ | $\Delta\varphi$ | $\bar{\varphi}$ |
|---|---|---|---|---|---|---|---|---|
| $\Delta E^*_{76} / \Delta E^*_{94}$ | 4.6318 | 1.0000 | 1.4043 | 77.82 | 45.00 | 54.54 | 32.82 | 61.41 |
| $\Delta E^*_{CMC} / \Delta E^*_{94}$ | 2.1439 | 0.5395 | 1.0193 | 64.99 | 28.35 | 45.55 | 36.65 | 46.67 |
| $\Delta E^*_{BFD} / \Delta E^*_{94}$ | 2.2079 | 0.6535 | 1.3383 | 65.63 | 33.17 | 53.23 | 32.47 | 49.40 |
| $\Delta E^*_{99} / \Delta E^*_{94}$ | 1.5602 | 0.6417 | 0.8928 | 57.34 | 32.69 | 41.76 | 24.66 | 45.02 |
| $\Delta E^*_{99_O} / \Delta E^*_{94}$ | 1.6718 | 0.8742 | 1.1278 | 59.11 | 41.16 | 48.44 | 17.95 | 50.14 |
| $\Delta E^*_{00} / \Delta E^*_{94}$ | 1.5298 | 0.5364 | 0.9431 | 56.83 | 28.21 | 43.32 | 28.62 | 42.52 |

laboratories [31]. The formal mathematical structure of the weighting function, $S_L$, for lightness differences can be considered an attempt to account for the finding that, when sample pairs are being visually assessed, maximum discrimination is achieved if the background colour closely resembles the colour of the two specimens to be assessed ("crispening" effect [3, 33]).

## 9.3 Chroma and hue scales

The chroma and hue scales of the various colour-difference models cannot easily be compared with each other, as they differ not only in their weighting functions, but also in the calculation of the unweighted specific difference values. An additional complexity arises from the fact that weighting functions for the hue difference partly depend on both the chroma and the hue angle. Lack of space precludes a detailed discussion of these relationships.

## 9.4 Total colour difference

The situation is simpler when it comes to comparing the total residual colour difference of all models. For a general comparison of the various colour spaces based on physical object colours, consider the Spectra-Master colours from DuPont. This selection comprises about 3300 colour samples which, to an approximation, homogeneously fill out the colour gamut of a typical refinish mixing system. If the colour positions $(L^*,a^*,b^*)$ of these colour samples are taken as colour standards in colour space CIELAB76 and test colours are produced by means of a random generator in accordance with $\Delta E^*_{76} \leq 5$ (see foot note [4]), the correlation between the metrics of the various colour spaces shown in Figure V.4 is obtained. In all cases, the CIE94 model serves as the reference on the abscissa of each panel.

The simple structure of the weighting functions of the CIE94 model makes it an easy matter to interpret the simulated comparison of the CIELAB76 and CIE94 colour-difference models. Since, for the weighting functions of the special Riemannian CIE94 colour-difference space $S_L = 1$, $S_C \geq 1$ and $S_H \geq 1$ apply to all colour positions, the inequality $\Delta E^*_{94} \leq \Delta E^*_{76}$ must always be fulfilled for all values of the total residual colour difference. This statement is clearly confirmed by Figure V.4. As may also be seen from the data set, deviations between the CIELAB76 and CIE94 metrics can be up to a factor of 4.6. Apart from that, the discrete simulation data cover a broad range of the triangle above the bisectrix. This reflects the fact that the CIE94 model substantially modifies the CIELAB76 metric. Extensive application of

[4] *$\Delta E^*_{76} \leq 5$ is the recommended range of application for the CIELAB76 equation.*

the CIE94 model in industrial colouristic practice shows that deviations of the CIELAB76 model from uniformity have been significantly reduced by the introduction of appropriate weighting functions and that the number of wrongly assessed colour difference pairs can be restricted to such an extent that globally valid tolerances are definable and colour-region-dependent local tolerances largely become obsolete. The transition from the CIELAB76 to the CIE94 metric represents a paradigm shift.

Due to the complex mathematical structure of the weighting functions of all other colour-difference models and the additionally modified specific colour-difference values, simple general comparisons cannot be derived in these cases. The computed discrete distribution functions all fill the shape of acute angles that point towards the origin of the coordinate system. They can be characterised by specifying the angle enclosed by the two sides of the triangular-shaped distribution, namely the angle

*Equation V.12* $$\langle\varphi\rangle = \frac{1}{2}\left(\varphi_{max} + \varphi_{min}\right)$$ of the bisectrix, and the angle

*Equation V.13* $$\langle\varphi\rangle = \frac{180°}{\pi}\arctan\left(\frac{1}{N}\sum_{i=1}^{N}\frac{\Delta E^{*}_{xx}(i)}{\Delta E^{*}_{94}(i)}\right)$$

of the centre of gravity straight line (see Table V.3). Only in the case of colour difference models DIN99 and CIEDE2000 the centre of gravity straight lines are below the bisectrix of the first quadrant; in all other models studied, they are located above. Besides the CIELAB76 metric, the BFD equation exhibits the greatest deviation from $\tan\varphi = 1$. Only in case of the *CMC* equation the matched straight line comes near the bisectrix of the first quadrant of the coordinate system. As in the discussion of the lightness scales of all models, the difference in normalisation becomes apparent here, too. Table V.3 shows that the size $\Delta\varphi = \varphi_{max} - \varphi_{min}$ of the acute angle varies between 18° for DIN99o and 36.7° for CMC. The latter observation is in contradiction to the frequently reported equivalence of the CMC and CIE94 metrics.

The smaller the difference between the angles $\bar{\varphi}$ and $\langle\varphi\rangle$, the greater the symmetry of the respective data distribution about the bisectrix. As expected, the difference between these two angles is highest for the CIELAB76 model at $|\Delta\varphi| \approx 6.9°$, while for all other metrics $|\Delta\varphi| < 4°$.

Comparison of the deviations of the metrics of all colour-difference spaces to the CIE94 model allows two groups to be identified. The first includes the CMC and BFD metrics, with maximum deviations of factor 2.14 and 2.21, respectively. Consequently, the cone angles of the discrete distribution functions of these two models are substantially larger than for all other metrics if the CIELAB76 system is excluded here. In the second group comprising the DIN99, DIN99o, and CIEDE2000 metrics, the extreme differences are markedly smaller, with factors 1.56, 1.56, and 1.53. Not only upward deviations can be observed, but also downward deviations. However, with the exception of the CIEDE2000 model, these factors are smaller than for the upward deviations.

The simulation results discussed above indicate that the metrics of all modern colour-difference spaces (CIE94, DIN99, DIN99o, CIEDE2000) are only marginally different and hence the observed differences in the assessment criterion proposed in the literature [34] are probably not significant.

# 10 *Appraisal of the current state of colour difference metric*

The publication of the CIELAB76 colour space by the CIE can certainly be considered a milestone in the development of colorimetry. This model has quickly gained global acceptance and, despite the shortcomings already known at the time of publication, has replaced almost all competing models in applications. The motto expressed in technical committees *"If something goes wrong in each colour-difference equation, at least let it be done consistently in the future!"* has thus proved somewhat prophetic. Unfortunately users have shown strong inertia with regard to all successor models that offer significantly improved conformity with visual perception, as a result of which none of these models has been able to gain a comparable level of acceptance, and the CIELAB76 metric is still very popular.

*Kühni*[35] has correctly pointed out that the various colour-difference models came about through matching with empirical data sets from psychophysical experiments, whose boundary conditions in some cases differed substantially as the experiments were being performed. Consequently, colour-difference models may well describe a subset of these empirical data while, in the case of other independent data sets, the mean error may be substantially higher. Consequently, as has been suggested by Kühni[35], there should be a concerted action to sample the colour space anew under rigidly controlled conditions in order that a new (where possible, Euclidean) metric may be developed. This is the only way to put an end to the uncontrolled growth of metrics and to achieve the desired progress for practical applications.

Furthermore, there is no commonly accepted method of quantifying the performance level of a colour-difference model in relation to a given set of assessment data. In the literature, resort is frequently made to an average of three different parameters (*PF/3*; **p**erformance **f**actor), which assumes zero for perfect predictions and, with increasing failure rate, continuously approaches a value of 100 for complete failure[34]. Based on the sets of assessment data used for the development of the CIEDE2000 equation, the predictions of the CIELAB76 metric score a value of *PF/3* ≈ 56, i.e., only about 50 % of the predictions are in conformity with visual perception. The CMC equation lies in the range *PF/3* ≈ 38, while the CIEDE2000 model stands at *PF/3* ≈ 33. Tests carried out in various industries show that the CIEDE2000 equation does not perform significantly better on a statistical basis than the CMC equation. This necessarily raises the question as to the significance of the observed *PF/3* difference between both models. At this stage, a fundamental shortcoming of the assessment model becomes apparent, namely that it eludes the application of standard statistical tests for estimating the significance of different performance characteristics.

All modern colour difference systems still employ the coordinates of the CIELAB76 model. In practice, this has turned out to be a disadvantage since, with the exception of the Euclidean DIN99 system, the non-Euclidean metrics always intermingle two different colour spaces. Specific colour differences are still interpreted in the CIELAB76 model, while total colour differences are characterised in one of the successor models. Only a direct transformation of XYZ chromaticity coordinates to a uniform new Lab colour space without a detour via the CIELAB76 model will make it possible to break old habits and to lay down a uniformly employed system in technical standards and to firmly establish in practice.

# 11 Model extensions for goniochromatic colours

Experience shows that the cube roots in the equations for computing CIELAB76 coordinates of bright or highly chromatic object colours tend to overestimate colour differences. Furthermore, in the case of effect colour shades deviations from uniformity in the CIELAB76 equation prove to be angle-dependent. The various colour difference measures can be improved by introducing specific angle-dependent weighting functions. The likewise angle-dependent total colour difference $\Delta E^*(\varepsilon)$ between two colour positions in an $N_\varepsilon$ dimensional space can, e.g., be reduced to a single coordinate by means of a weighted average

*Equation V.14*

$$(\Delta E^*)^2 = \sum_{i=1}^{N\varepsilon} w_i \left\{ \left( \frac{\Delta L^*(\varepsilon_i)}{k_L S_L(\varepsilon_i)} \right)^2 + \left( \frac{\Delta C^*(\varepsilon_i)}{k_C S_C(\varepsilon_i)} \right)^2 + \left( \frac{\Delta H^*(\varepsilon_i)}{k_H S_H(\varepsilon_i)} \right)^2 \right\} = \sum_{i=1}^{N\varepsilon} w_i \, (\Delta E^*(\varepsilon_i))^2$$

where ε denotes the aspecular angle, and the weights are normalised in accordance with

$$\sum_{i=1}^{N\varepsilon} w_i - 1$$

to ensure that the total colour difference is of a comparable magnitude to single-angle assessments. Formally, this expression describes a weighted metric colour difference between two reflectance surfaces in a $3 \times N_\varepsilon$ dimensional space with a suitably large number of weighting functions. In the simplest case, all weighting functions are constant and have to be chosen in a way that the metric distance correlates with visual perception. It has to be taken into account that, in the case of gonioapparent sample pairs, lightness differences are perceived more strongly at high aspecular angles than close to the specular angle, whereas hue differences are generally perceived more strongly here than lightness differences. The DIN6175 P2 model discussed in the context of colour tolerances has been structured accordingly.

All the colour-difference models presented above were developed for solid colours. When they are applied to effect colours, it is found that, with progressively increasing lightness level $L^*$ increasing lightness differences are tolerated, so that at least the weighting of the lightness scale needs to be adjusted accordingly. By analogy, the CMC model, which evolved from assessments of textile samples, already accounted for the impact of texture on the rating of lightness differences of bright colours by introducing an adequate weighting function $S_L$. In psychophysical experiments involving blends of an aluminium pigment and carbon black with different lightness levels, *Rodrigues* and *Locke*[36] demonstrated that the observed deviation from uniformity for metallic surface coatings can be substantially mitigated by introducing the weighting function

*Equation V.15*

$$S_L = \begin{cases} 1 & \text{for } L^* \leq 29.4 \\ 0.034\, L^* & \text{for } L^* > 29.4 \end{cases}$$

into the CIE94 colour-difference model. In gonioapparent colour shades, platelet-like effect pigments give rise to a visually perceptible texture or coarseness and to sparkle. This visual texture seems to be partly responsible for the fact that in visual assessments observers tolerate higher lightness differences. From its formal structure, this weighting function closely resembles the lightness weighting function of the CMC equation when the lightness parameter is set to $k_L = 2$.

For very dark colours, this study has failed to provide conclusive results, as this lightness range is only based on assessments of metallic panels at far-aspecular geometries, where the visual texture of samples is barely perceptible. However, there are no clear indications of higher sensitivity at low $L^*$ levels.

From experience, the aforementioned extensions of the **metric for effect colours** lead to a useful instrument for colour assessment. However, in practice, there are always cases where, despite an instrumental colour match within the realised range of viewing angles, visually perceptible differences between sample pairs give rise to complaints. Generally these differences can be traced back to deviations in the visual texture of both colour samples which gets lost as colorimetric information due to integration over the realised measurement spot. This could be considered a kind of **conditional colour match**, as the reduction in dimension causes the loss of a degree of freedom which cannot be clearly reproduced in a reverse calculation.

Hence, an instrumental comparison of a pair of gonioapparent colour shades requires that the appearance, consisting of colour and texture components, be assessed holistically. For this goal to be achieved, a substantial extension to the colour metrics is required which, on the one hand, entails the establishment of a metric to assess texture differences and, on the other, unifies colour and texture difference metrics into an overarching metric of the total appearance. The first steps in the defining of a **texture metric** have seen some basic developments incorporated into novel commercial measurement systems [37, 38, 39]. At the moment the performance of this texture metric needs to be validated in practical application before harmonization of colour and texture metrics can be considered. For this reason, only a qualitative overview of the problem of texture assessment can be given in the following.

The **visual texture** of an effect colour is the two-dimensional non-uniformity of colour and lightness which occurs across the surface of an object due to small, platelet-like pigment particles embedded in a surface coating and which depends on their size and topology. The inhomogeneities are related not only to spatial lightness variations, but also to spatially dependent changes in chromaticity coordinates. These variations manifest themselves in irregular patterns on arbitrary length scales. Insofar, the concept of visual texture accounts for the nature of variations and their length scales. Of the various features of visual texture, two in particular emerge from the practical colouristic characterisation of effect colours, namely the

- coarseness[5] and the
- sparkle[6].

**Coarseness** is the perceived contrast of the irregular light/dark pattern of effect colours under diffuse illumination conditions. The perceived magnitude of coarseness depends on the lightness differences between the bright and dark regions and the length scale of the specimen.

**Sparkle** is the global phenomenon of several or multiple point light sources, whose brightness causes them to drastically stand out from their surround. Furthermore, they may differ substantially in colour from their surround. The sparkle becomes immediately apparent to an observer when the point light sources are switched on and off by changing the geometry defined by the illumination and viewing angles. The sparkle phenomenon can only be perceived under intense uni-directional illumination conditions. The effect is so strong that all other aspects of the visual texture are of minor importance. With diminution in directional character of the illumination conditions, the sparkle progressively loses intensity and disappears completely in the limiting case of perfectly diffuse illumination. In mixed

[5] *An alternate and frequently used designation of the described appearance feature is „graininess".*

[6] *An alternate and frequently used designation of the described appearance feature is „glint impression".*

illumination conditions with diffuse and directional components, coarseness is likewise visible, which, as a further feature of visual texture, is completely different from sparkle. It is less intense and more pattern-like.

Diffuse and directional illumination conditions arise outdoors when the sky is overcast and on sunny days under clear blue sky, respectively. However, these extreme conditions of perfectly diffuse or ideal directional illumination frequently do not occur in outdoor applications, and both light distributions interfere with each other, thereby creating mixed illumination conditions. In particular, this applies to artificial viewing conditions in light booths when visually assessing sample pairs under controlled conditions. When visual texture is being assessed in a light booth, the relative contributions of coarseness and sparkle have to be determined. Separation of both contributions can only be achieved through prior quantification of the visual texture features under the extreme conditions of purely diffuse or directional illumination, which afterwards is also transferable to other light booths.

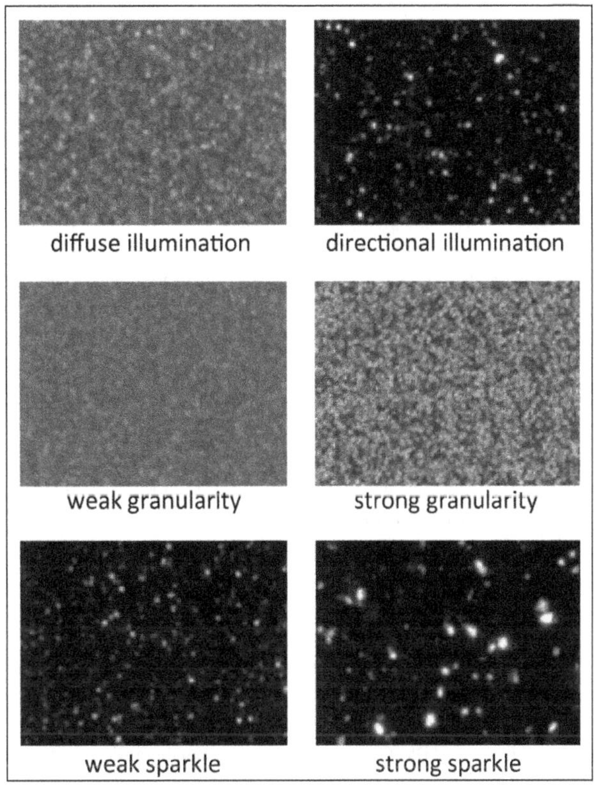


*Figure V.5: Images of various platelet-like pigments embedded in surface coatings that lead to different visual texture. The upper pair of samples shows the two features "coarseness" (left image) and "sparkle" (right image). The second row of images compares two samples of different coarseness levels, while the third row of images depicts a pair of samples exhibiting distinctly different degree of sparkle.*

## 12 Literature

[1] Kuehni, R. G., Colour Space and its Divisions. Colour Order from Antiquity to the Present, John Wiley & Sons, Inc., Hoboken, NJ, USA, 2003

[2] Kuehni, R. G., Schwarz, A., *Colour Ordered: A Survey of Colour Systems from Antiquity to the Present*, Oxford University Press, Cambridge, UK, 2008

[3] Schanda, J., *Colorimetry: Understanding the CIE System*, John Wiley & Sons, LTD, New York, USA, 2007

[4] Clarke, F. J. J., McDonald, R., Rigg, B., "Modification to the JPC79 colour difference formula", *J. Soc. Dyers and Colourists* **100**, p. 128 and 281, 1984

[5] Luo, M. R., Rigg, B., "BFD(l:c) colour-difference formula Part I - Development of the formula", *J. Soc. Dyers and Colourists*} {\bf **103**}, p. 86, 1987

[6] Luo, M. R., Rigg, B., "BFD(l:c) colour-difference formula Part II - Performance of the formula", *J. Soc. Dyers and Colourists* **103**, p. 126, 1987

[7] Witt, K., "Three-dimensional threshold of colour-difference perceptibility in painted samples: variability of observers in four CIE colour regions", *Colour Res. Appl.* **12**, p. 128, 1987

[8] Witt, K., "Parametric effects on surface colour-difference evaluation at threshold", *Colour Res. Appl.* **15**, p. 189, 1990

[9] Luo, M. R., Rigg, B., "Chromaticity-discrimination ellipses for surface colours", *Colour Res. Appl.* **11**, p. 25, 1986

[10] Berns, R. S., Alman, D. H., Reniff, L., Snyder, G. D., Balonen-Rosen, M. R., "Visual determination of suprathreshold colour-difference tolerances using probit analysis", *Colour Res. Appl.* **16**, p. 297, 1991

[11] Industrial colour-difference evaluation, CIE Publication No. 116 (Vienna, Austria), CIE, 1995

[12] Luo, M. R., Cui, G., Rigg, B., "Derivation of a rotation function for the new CIE colour difference formula", *Proc. Conf. Colour and Visual Scales* (National Physical Laboratory, Teddington, UK), 2000

[13] Luo, M. R., Cui, G., Rigg, B., "The development of the CIE 2000 colour difference formula: CIEDE2000", *Colour Res. Appl.* **26**, p. 340, 2001

[14] *Farbmetrische Bestimmung von Farbabständen bei Körperfarben nach der DIN99-Formel*, Deutsches Institut für Normung, DIN 6176, Beuth-Verlag GmbH, Berlin, Germany, March 2001

[15] Cui, G., Luo, M. R., Rigg, B., Roesler, G., Witt, K., "Uniform colour spaces based on the DIN99 colour-difference formula", *Colour Res. Appl.* **26**, p. 340, 2002

[16] Witt, K., "Vektorielle Farbabstandsbewertung mit der DIN99-Formel", *Farbe + Lack* **111**, p. 86, 2005

[17] *Colorimetry, CIE Publication No. 15.3*, 3.Edn., (Vienna, Austria), CIE, 2004

[18] Parametric effects in colour-difference evaluation, CIE Publication No. 101, (Wien, Austria), CIE, 1993

[19] Luo, M. R., Hunt, R. W. G., "The structure of the CIE 1997 colour appearance model (CIECAM97s), *Colour Res. Appl.* **23**, p. 138, 1998

[20] Luo, M. R., Hunt, R. W. G., "A chromatic adaptation transform and a colour inconstancy index", *Colour Res. Appl.* **23**, p. 154, 1998

[21] Luo, M. R., Hunt, R. W. G., "A revision of the CIECAM97s model", *Colour Res. Appl.* **25**, p. 260, 2000

[22] Moroney, N., "The CIECAM02 colour appearance model," in Proceedings of the Tenth Colour Imaging Conference: Colour, Science, Systems, and Applications, p. 23, 2002

[23] A colour appearance model for colour management systems: CIECAM02, CIE Publication No. 159, (Vienna, Austria), CIE, 2004

[24] Völz, H. G., "Die Berechnung großer Farbabstände in nichteuklidischen Farbräumen", *DIE FARBE* **44**, p. 1, 1998

[25] Völz, H. G., "Transformation der CIE94-Formel in einen euklidischen Farbraum", *DIE FARBE* **44**, p. 97, 1998

[26] Völz, H. G., "Die Euklidisierung des CMC-Raumes zur Berechnung großer Farbabstände", *DIE FARBE* **45**, p. 1, 1999/2000

[27] Völz, H. G., "Euclidisation of the first quadrant of the CIEDE2000 colour difference system for the calculation of large colour difference", *Colour Res. Appl.* **31**, p. 5, 2006

[28] Urban, S., Rosen, M. R., Berns, R. S., Schleicher, D., "Embedding non-Euclidean colour spaces into Euclidean colour spaces with minimal isometric disagreement", *J. Opt. Soc. Am. A* **24**, p. 1516, 2007

[29] Kuehni, R. G., "Colour-tolerance data and the tentative CIE 1976 $L^*a^*b^*$ formula", *J. Opt. Soc. Am.* **66**, p. 497, 1976

[30] McLaren, K., "CIELAB hue-angle anomalies at low tristimulus ratios", *Colour Res. Appl.* **5**, p. 139, 1980

[31] Völz, H. G., *Industrial Colour Testing. Fundamentals and Techniques* (2. Edn.), Wiley-VCH, New York, USA, 2001

[32] Schrödinger, E., "Grundlinien einer Theorie der Farbenmetrik im Tagessehen", *Ann. Phys.* **63**, p. 397, 427, and 481, 1920

[33] Wyszecki G., Stiles, W. S., *Colour Science: Concepts and Methods: Quantitative Data and Formulae*, John Wiley & Sons, Inc., New York, USA, 1982

[34] Guan S. S., Luo,M. R., "Investigation of parametric effects using small colour difference pairs", Colour Res. Appl. **24**, p. 331, 1999

[35] Kuehni, R. G., "Colour difference formulas: An unsatisfactory state of Affairs", *Colour Res. Appl.* **33**, p. 324, 2008

[36] Rodrigues, A. B., Locke, J. S., "Weighting function for the measurement of lightness differences in gonioapparent and dark colours", *9th Congress of the International Colour Association* (2001), published in Proceedings of SPIE} **4421**, p. 658, AIC, (Rochester, NY, USA), 2002

[37] "Beurteilung von Effektlackierungen: Den Gesamteindruck objektiv messen", JOT **1.2009**, p. 50, Byk-Gardner GmbH, Geretsried, Germany, 2009

[38] Feld, R., "Finding numbers for sparkle and shine", European Coatings Journal **02**, p. 18, 2009

[39] E. Kirchner, E., van den Kieboom, G.-J., Njo, L., Supèr, R., R. Gottenbos, R., Observation of visual texture of metallic and pearlescent materials, Col. Res. Appl. 32, p. 256, 2007

# VI Definition and application of colour tolerances

*Wilhelm Kettler*

## 1 Tolerance and acceptance

The tolerance of colour differences and their definition is a multi-faceted issue for which there is no general solution [1-3]. In *MacAdam*'s experiment [4] (see Section I.8.4), tolerances are considered equivalent to the perceptibility thresholds for colour differences. Delivery specifications in the automotive industry commonly contain "sub-threshold" tolerance agreements, which give rise to considerable manufacturing outlay that cannot be justified from a colorimetric point of view. The situation is different in the textiles industry, which tolerates specimens with "supra-threshold" colour differences to the colour standard. In this case, such tolerance thresholds can be considered acceptance thresholds, whose specific values can be defined with sufficient statistical significance in appropriately designed visual assessment experiments. The printing industry, too, employs acceptance thresholds, which, in some cases, are a multiple of MacAdam's thresholds for the perceptibility of colour differences and can thus be classified more or less as colour similarity.

Thus, only the **perceptibility threshold** for colour differences can be uniquely defined for observers with normal colour vision across multiple applications; the **acceptability threshold** generally cannot be specified. Both the variance in the ratings of different observers and the variance among individual observers in repeated assessments yield much higher values for acceptance as a measure of quality than for perceptibility, and in such decisions a higher degree of subjectivity generally has to be tolerated. The application in question or the field of application determines this threshold, which customers and suppliers then must adopt and observe as a binding agreement.

### 1.1 Euclidean colour space

In Euclidean colour systems, e.g. **CIELAB76** and **DIN99** (see Sections I.8 and V.7), the colour positions of all colour samples, whose residual colour differences $\Delta E^*$ with respect to a reference colour position are smaller than or equal to an agreed colour tolerance, are located within a sphere of radius r, which is identical with the $\Delta E^*$ tolerance. Since the metric is assumed to be uniform, it is equally tolerant in all spatial directions. However, as is widely known, the CIELAB76 equation deviates substantially from uniformity and so is less suitable for agreeing on global colour tolerances. On the contrary, the fact that this metric deviates from uniformity needs to be addressed in the agreement on locally valid tolerances. This will require that colour samples be collected and assessed at the respective colour centres in order that local tolerances may be defined. The more uniformly these colour samples are distributed about a reference position, the more robust the derived tolerance metric will be.

The introduction of an alternative uniform metric would be less error prone and above all less labour intensive in practice. Hitherto, only a single Euclidean colour space (after CIELAB76) has been published in technical standards with which deviations from uniformity in CIELAB76 colour space can be considerably reduced, namely DIN99. Consequently, all analyses and interpretations relating to CIELAB76 are transferable to this alternative metric which, as regards the assessment of total colour difference, is very similar to colour-difference space **CIE94**.

### *1.2 Non-Euclidean colour space*

Colour spaces such as **CIE94**, **CIEDE2000**, **CMC** and **BFD** (see Section V) are non-Euclidean and ought to be rendered Euclidean (if at all possible) in order that a corresponding, three-dimensional, new coordinate system may be defined. Tolerances of these colour spaces are no longer direction-independent in CIELAB76 colour space. The special Riemannian structure of these colour difference spaces in $L^*a^*b^*$ coordinates yields tolerance solids in the shape of ellipsoids whose surface separates regions of accepted and rejected colour matches.

## *2 Psychophysical measurements*

By means of psychophysical experiments the sensations of human colour perception are directly assessed. The choice of stimulus configuration and the definition of the question to be addressed to the proband offer plenty of scope for an investigation into the properties of **colour perception**. In order that general and quantitative statements may be derived from this subjective method, the question needs to be phrased in such a way that it can be answered by the proband as accurately and reproducibly as possible.

In the **method of equal stimuli**, the observer is required to match a presented variable stimulus $S_{var}$ to a constant reference stimulus $S_{ref}$. The colour-matching functions, for example, were determined in this way.

In one variant of a related method for determining **thresholds** (i.e. barely perceptible differences), the reference stimulus $S_{ref}$ and variable stimulus $S_{var}$ are set up so as to match at the beginning of an assessment. The observer then has to incrementally change $S_{var}$ until a perceptible difference between $S_{ref}$ and $S_{var}$ occurs. This was the basic experimental method which *MacAdam* [4] used to determine his threshold ellipsoids.

In the **method of paired comparisons**, pairs of stimuli are presented to an observer. Their task is to identify that stimulus of the pair which is either higher or lower than the other (i.e. $S_1 \gtrless S_2$) with respect to a predefined perception attribute (e.g. lightness, total colour difference).

A highly reliable method is that of **constant stimuli**, in which a constant stimulus is repeatedly presented to an observer for assessment. When absolute thresholds are measured, the stimulus will not always be given the same rating. The method can be generalised, for example, by mixing $N > 1$ levels of stimuli which are then assessed in a high number of presentations, with each level being repeated several times in random order. This approach ensures that the probability of detection is a function of stimulus level (psychometric function). This method can be combined with the method of paired comparison, in which case a constant stimulus is paired with many others and is repeatedly assessed.

A further method assigns an absolute rating to a selection of stimuli. In a variant of this **method of judgement scale**, a **reference stimulus** with an associated perception rating (e.g. 100) is presented to the observer. Their task now is to assess other presented stimuli in relation to the reference stimulus in such a way that ratios between the stimuli are preserved.

The interpretation of results obtained in psychophysical experiments is not always a trivial matter, as they are subject to systematic and statistical errors. Statistical methods are indispensable for the experimental design and analysis. Systematic errors must be largely ruled out through carefully choosing and controlling experimental conditions.

# 3 *Visual colour assessment*

In any type of colour reproduction (painting, printing, etc.), there is usually quite a large residual colour difference between the batch and the standard. Allowance is made for this by, e.g., an agreement between manufacturer and customer on tolerable colour differences. Tolerance agreements presuppose the use of a practicable **colour difference equation**, which in turn is based on functioning colorimetry. The model is based on a preceding colour measurement which must be designed according to the laws of colour stimulus specification. The sensible use of a colour difference equation requires an understanding of its origins and above all of its limitations.

***Interactions arising during colour assessment***

Discrimination between two specimens depends on many factors, which moreover may interact with each other:

- **Viewing conditions:** The appearance of a colour is the result of a complex interaction between the colour stimuli of the entire visual field that enter the eye and the visual system itself. Viewing conditions, e.g. the absolute brightness of the ambience, the surroundings, the illumination, and the size of the specimens can influence the sensation of colour in many ways. In particular, the presentation of the specimens has a decisive effect on the perception threshold. Where samples are seamlessly contiguous, even very small colour differences can be perceived, whereas, as the gap grows between a pair of specimens, increasingly higher colour differences are tolerated.
- **Observer:** Every observer has their own colour perception which can deviate to a greater or lesser extent from the "artificial" colorimetric standard observer. In such cases, the spectral sensitivity functions of the receptors (here in particular the cone apparatus) manifest themselves to different extents. In this sense, visual assessment does not take precedence over instrumental evaluation, as long as averaging of a sizeable number of observers is not carried out. Instrumental assessments are more objective than the visual kind, even when the underlying colour space is physiologically non-uniform.
- **Colour:** Where the adopted colour space deviates from uniformity, visually equal colour differences are assigned to different numerical values. According to the CIELAB76 formula, colour difference thresholds can vary from about $\Delta E^*_{76} = 0.2$ in the achromatic region to $\Delta E^*_{76} = 0.7$ in the region of brilliant red and yellow colours.

# 4 *Statistical threshold determination*

A method of threshold measurement will be outlined below to illustrate how tolerances are typically defined. This method is based on the quantal-response measurement, where the proband has to decide whether a single test stimulus (sample, standard) is higher or lower than a previously defined comparison stimulus (anchor pair). The rating is either "accepted" or "rejected". If test stimuli of different magnitudes are repeatedly presented to the proband at random, a range of stimuli intensities will be covered in which the proband will not always give the same response to stimuli of equal intensity. When a high number $n_i$ of constant stimuli $x_i$ are presented to the proband, a ratio $p_i = m_i/n_i$ of positive answers $m_i$ to the total number $n_i$ will occur. If the stimulus intensity is varied, the relative frequency $p_i$ of responses will change. Through interpolation, the cumulative function *P(x)* of relative frequencies of positive responses as a function of stimulus intensity will be sigmoidal. The threshold is defined as the stimulus intensity at which the relative frequency of positive responses reaches a value of 0.5 and coincides with the number of negative responses. It is assumed that this frequency distribution involving an increasing number of presentations approaches a probability distribution *P(x)* which in this case is also called a **psychometric function**. If the comparison stimulus or the decision criterion is chosen carefully, the behaviour will also prove optimal in extensive observations.

Several standard models exist for psychometric functions. In specific cases, such a standard model is adapted to the experimental data by choosing the model parameters appropriately. The following models are preferably applied.

***Psychometric model functions***

Owing to the central limit theorem, according to which the probability distribution of a sum of independent probability variables with increasing number of summands approximates a **cumulative Gaussian function**

*Equation VI.1*

$$P(x) = \frac{1}{\sqrt{2\pi}\, b} \int_{-\infty}^{x} dt\, e^{-\frac{(t-a)^2}{2b^2}}$$

this function recommends itself as a model for a psychometric function. As usual, parameters *a* and *b* denote the mean and the standard deviation of a random sample. By means of the probit transformation Equation VI.1 can be linearised. After substituting for $y = (t-a)/b$, the **cumulative Gaussian function** is transformed to

*Equation VI.2*

$$P(x) = \frac{1}{\sqrt{2\pi}} \int_{-\infty}^{(x-a)/b} dy\, e^{-\frac{y^2}{2}}$$

The relationship between *P(x)* and the upper integration limit is the subject of the ***probit transformation***

*Equation VI.3*

$$\text{probit } P(x) = \frac{x-a}{b}$$

For practical applications, the ***logistic function***

*Equation VI.4*

$$P(x) = \frac{1}{1 + e^{-(x-a)/b}}$$ or its extension

*Equation VI.5*

$$P^*(x) = g + (1-g)\, P(x)$$

is of particular interest, since it is a good approximation of the cumulative Gaussian function, but is much easier to handle analytically. The logistic function is linearized by means of the so-called **logit transformation**

*Equation VI.6*

$$\text{logit } P(x) = \ln\left(\frac{P(x)}{1-P(x)}\right) = \frac{x-a}{b}$$

which can be utilised with advantage to arrive at a simple estimation of the model parameters *a* and *b*.

# 5 An experiment to define colour tolerances

By way of a practical example of colour tolerance definition in an industrial application[5, 6], consider the production of a brilliant yellow colour shade with colour position **F** = $(L^*=69.8, a^*=10.6, b^*=89.0)^T$. When this colour is frequently produced, a large set of historical data will be available after some time has elapsed. If this database contains the results of instrumental and visual assessments, there is a good chance that data analysis will lead to the definition of a meaningful instrumental tolerance, without the need to systematically produce additional samples of different colour difference values to the standard. All the experimental results of the instrumental assessment and respective visual rating are summarised in Table VI.1.

Needless to say, the process of data collection can be shortened by preparing colour samples which are distributed in a systematic fashion about the colour standard and which can be divided into "accepted" and "rejected" categories. Given careful distribution of all test colours with respect of lightness, chroma, and hue differences, statistical analysis of the data set should support the definition of a practical tolerance or acceptance measure. The psychophysical experiment is performed by presenting the series of stimuli (32 in this case) to a representative panel of observers (22 here) for assessment. The reference stimulus is an achromatic anchor pair having the two colour positions $\mathbf{F_1} = (L^*_1=34.9, a^*_1=-3.4, b^*_1=-13.3)^T$ and $\mathbf{F_2} = (L^*_2=37.6, a^*_2=3.0, b^*_2=-13.8)^T$, whose residual colour difference of $\Delta E^{(ref)}_{94} = 2.84$ determines the quality metric. In the present example, the observers assess the reference and test stimuli (paired comparison) and decide in which cases the perceived intensity of the test stimulus **exceeds** (category: fail) or **falls below** (category: pass) the reference stimulus.

### *Colour differences in CIELAB76 colour space*

In an initial analysis, the corresponding projected colour difference values between respective test sample and standard are plotted for each pair of axes in **CIELAB76** colour space. The calculation of the chromaticity coordinates must factor in the assessment conditions agreed between the supplier and the customer. In the present case, the size of the samples to be assessed corresponds to the visual field of the 10° normal observer, who has to judge pairs of samples under daylight conditions (corresponding to illuminant $D_{65}$), with the specular excluded. Figure VI.1 shows all collected measurement data, with membership of the two categories pass and fail being indicated by different symbols. Surprisingly, the data

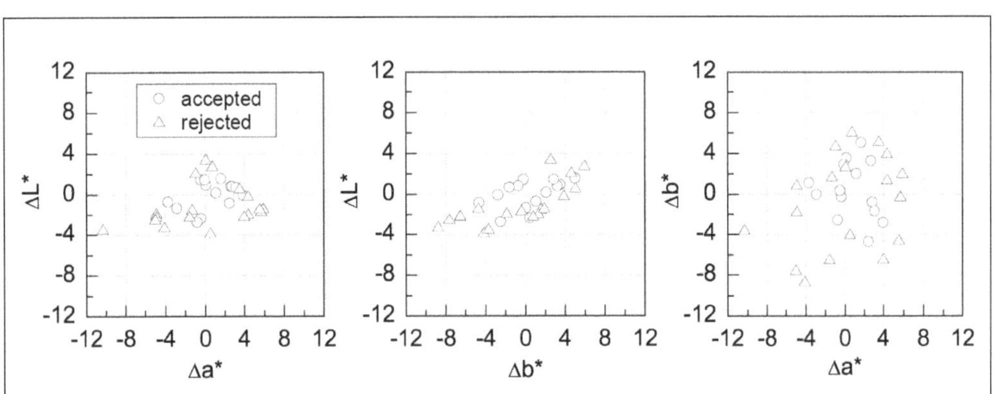


*Figure VI.1: Colour positions of 32 test samples prepared for the visual assessment experiment for defining an acceptance threshold. In each diagram the colour position of the standard coincides with the origin of the coordinate system.*

set is not consistent and cannot be divided into two distinct groups. Likewise, there is no indication for the rotational symmetry that would be expected on the basis of the CIELAB76

*Table VI.1: Results (instrumental and visual) of an assessment experiment*[5, 6] *for a brilliant yellow colour shade designed to define an acceptance level, in which a grey anchor pair with a total residual colour difference of* $\Delta E^*_{76} = 2.74$ *was used for the visual assessment.*

| | L* | a* | b* | C* | $h_{ab}$ | | | | | | | | |
|---|---|---|---|---|---|---|---|---|---|---|---|---|---|
| Standard | 69.80 | 10.60 | 89.00 | 89.63 | 83.21 | | | | | | | | |
| Sample | L* | a* | b* | C* | $h_{ab}$ | ΔL* | Δa* | Δb* | ΔC* | $\Delta h_{ab}$ | ΔH* | $\Delta E_{94}$ | visual judgement |
| 1 | 67.07 | 9.80 | 86.47 | 87.02 | 83.53 | -2.73 | -0.80 | -2.53 | -2.61 | 0.33 | 0.50 | 2.79 | Pass |
| 2 | 70.62 | 13.38 | 88.27 | 89.28 | 81.38 | 0.82 | 2.78 | -0.73 | -0.35 | -1.83 | -2.85 | 1.47 | Pass |
| 3 | 65.89 | 11.12 | 84.88 | 85.61 | 82.54 | -3.91 | 0.52 | -4.12 | -4.02 | -0.67 | -1.03 | 4.02 | Fail |
| 4 | 68.06 | 16.33 | 88.63 | 90.12 | 79.56 | -1.74 | 5.73 | -0.37 | 0.49 | -3.65 | -5.72 | 3.00 | Fail |
| 5 | 66.22 | 0.29 | 85.30 | 85.30 | 89.81 | -3.58 | -10.31 | -3.70 | -4.33 | 6.60 | 10.06 | 5.70 | Fail |
| 6 | 70.56 | 13.66 | 87.35 | 88.41 | 81.11 | 0.76 | 3.06 | -1.65 | -1.22 | -2.10 | -3.26 | 1.61 | Pass |
| 7 | 68.22 | 9.31 | 90.62 | 91.10 | 84.13 | -1.58 | -1.29 | 1.62 | 1.47 | 0.93 | 1.46 | 1.72 | Fail |
| 8 | 67.48 | 5.68 | 89.78 | 89.96 | 86.38 | -2.32 | -4.92 | 0.78 | 0.33 | 3.17 | 4.97 | 3.14 | Fail |
| 9 | 70.29 | 14.12 | 94.07 | 95.12 | 81.46 | 0.49 | 3.52 | 5.07 | 5.49 | -1.74 | -2.81 | 1.66 | Fail |
| 10 | 71.18 | 10.17 | 88.77 | 89.35 | 83.46 | 1.38 | -0.43 | -0.23 | -0.28 | 0.26 | 0.40 | 1.39 | Pass |
| 11 | 73.11 | 10.69 | 91.55 | 92.17 | 83.34 | 3.31 | 0.09 | 2.55 | 2.54 | 0.13 | 0.21 | 3.35 | Fail |
| 12 | 68.96 | 13.01 | 84.32 | 85.32 | 81.23 | -0.84 | 2.41 | -4.68 | -4.31 | -1.98 | -3.02 | 1.78 | Pass |
| 13 | 69.73 | 14.47 | 86.18 | 87.39 | 80.47 | -0.07 | 3.87 | -2.82 | -2.24 | -2.74 | -4.23 | 1.87 | Pass |
| 14 | 67.51 | 9.04 | 82.43 | 82.92 | 83.74 | -2.29 | -1.56 | -6.57 | -6.70 | 0.53 | 0.80 | 2.69 | Fail |
| 15 | 70.61 | 13.23 | 92.23 | 93.17 | 81.84 | 0.81 | 2.63 | 3.23 | 3.55 | -1.37 | -2.19 | 1.41 | Pass |
| 16 | 71.42 | 12.18 | 94.02 | 94.81 | 82.62 | 1.62 | 1.58 | 5.02 | 5.18 | -0.59 | -0.95 | 1.95 | Pass |
| 17 | 70.71 | 10.69 | 92.48 | 93.10 | 83.41 | 0.91 | 0.09 | 3.48 | 3.47 | 0.20 | 0.32 | 1.14 | Pass |
| 18 | 68.44 | 7.73 | 89.05 | 89.38 | 85.04 | -1.36 | -2.87 | 0.05 | -0.24 | 1.83 | 2.86 | 1.83 | Pass |
| 19 | 69.94 | 11.70 | 91.03 | 91.78 | 82.68 | 0.14 | 1.10 | 2.03 | 2.15 | -0.53 | -0.84 | 0.57 | Pass |
| 20 | 67.19 | 5.62 | 81.34 | 81.53 | 86.05 | -2.61 | -4.98 | -7.66 | -8.10 | 2.84 | 4.24 | 3.61 | Fail |
| 21 | 67.75 | 14.97 | 90.30 | 91.53 | 80.59 | -2.05 | 4.37 | 1.30 | 1.90 | -2.62 | -4.14 | 2.73 | Fail |
| 22 | 71.17 | 10.53 | 91.88 | 92.48 | 83.46 | 1.37 | -0.07 | 2.88 | 2.85 | 0.25 | 0.40 | 1.49 | Pass |
| 23 | 66.42 | 6.50 | 80.25 | 80.51 | 85.37 | -3.38 | -4.10 | -8.75 | -9.12 | 2.16 | 3.20 | 4.12 | Fail |
| 24 | 67.52 | 14.56 | 82.50 | 83.77 | 79.99 | -2.28 | 3.96 | -6.50 | -5.85 | -3.22 | -4.86 | 3.33 | Fail |
| 25 | 67.40 | 10.11 | 89.38 | 89.95 | 83.55 | -2.40 | -0.49 | 0.38 | 0.32 | 0.34 | 0.53 | 2.41 | Pass |
| 26 | 69.54 | 14.99 | 92.90 | 94.10 | 80.83 | -0.26 | 4.39 | 3.90 | 4.47 | -2.37 | -3.81 | 1.84 | Fail |
| 27 | 68.32 | 16.52 | 90.88 | 92.37 | 79.70 | -1.48 | 5.92 | 1.88 | 2.74 | -3.51 | -5.57 | 2.83 | Fail |
| 28 | 69.12 | 6.85 | 90.09 | 90.35 | 85.65 | -0.68 | -3.75 | 1.09 | 0.72 | 2.44 | 3.84 | 1.77 | Pass |
| 29 | 71.80 | 9.68 | 93.63 | 94.13 | 84.10 | 2.00 | -0.92 | 4.63 | 4.50 | 0.89 | 1.43 | 2.26 | Fail |
| 30 | 68.31 | 16.18 | 84.31 | 85.85 | 79.14 | -1.49 | 5.58 | -4.69 | -3.78 | -4.07 | -6.23 | 3.17 | Fail |
| 31 | 67.79 | 5.72 | 87.17 | 87.36 | 86.25 | -2.01 | -4.88 | -1.83 | -2.27 | 3.04 | 4.69 | 2.88 | Fail |
| 32 | 72.48 | 11.36 | 94.95 | 95.63 | 83.18 | 2.68 | 0.76 | 5.95 | 6.00 | -0.03 | -0.05 | 2.92 | Fail |

metric. Some samples with high colour difference values have been accepted (pass), while others with small difference values have been rejected (fail).

*Table VI.2: Sorted "pass" (accepted) and "fail" (rejected) decisions of the visual assessment experiment in Table VI.1 and the resulting cumulative frequencies.*

| Sample | ΔL* | Δa* | Δb* | $\Delta E_{94}$ | Visual judgement | Cumulative frequency [%] |
|---|---|---|---|---|---|---|
| 19 | 0.14 | 1.1 | 2.03 | 0.57 | Pass | 92.86 |
| 17 | 0.91 | 0.09 | 3.48 | 1.14 | Pass | 85.71 |
| 10 | 1.38 | -0.43 | -0.23 | 1.39 | Pass | 78.57 |
| 15 | 0.81 | 2.63 | 3.23 | 1.41 | Pass | 71.43 |
| 2 | 0.82 | 2.78 | -0.73 | 1.47 | Pass | 64.29 |
| 22 | 1.37 | -0.07 | 2.88 | 1.49 | Pass | 57.14 |
| 6 | 0.76 | 3.06 | -1.65 | 1.61 | Pass | 50.00 |
| 28 | -0.68 | -3.75 | 1.09 | 1.77 | Pass | 42.86 |
| 12 | -0.84 | 2.41 | -4.68 | 1.78 | Pass | 35.71 |
| 18 | -1.36 | -2.87 | 0.05 | 1.83 | Pass | 28.57 |
| 13 | -0.07 | 3.87 | -2.82 | 1.87 | Pass | 21.43 |
| 16 | 1.62 | 1.58 | 5.02 | 1.95 | Pass | 14.29 |
| 25 | -2.4 | -0.49 | 0.38 | 2.41 | Pass | 7.14 |
| 1 | -2.73 | -0.8 | -2.53 | 2.79 | Pass | 0 |

| Sample | ΔL* | Δa* | Δb* | $\Delta E_{94}$ | Visual judgement | Cumulative frequency [%] |
|---|---|---|---|---|---|---|
| 9 | 0.49 | 3.52 | 5.07 | 1.66 | Fail | 5.56 |
| 7 | -1.58 | -1.29 | 1.62 | 1.72 | Fail | 11.11 |
| 26 | -0.26 | 4.39 | 3.90 | 1.84 | Fail | 16.67 |
| 29 | 2.00 | -0.92 | 4.63 | 2.26 | Fail | 22.22 |
| 14 | -2.29 | -1.56 | -6.57 | 2.69 | Fail | 27.78 |
| 21 | -2.05 | 4.37 | 1.30 | 2.73 | Fail | 33.33 |
| 27 | -1.48 | 5.92 | 1.88 | 2.83 | Fail | 38.89 |
| 31 | -2.01 | -4.88 | -1.83 | 2.88 | Fail | 44.44 |
| 32 | 2.68 | 0.76 | 5.95 | 2.92 | Fail | 50.00 |
| 4 | -1.74 | 5.73 | -0.37 | 3.00 | Fail | 55.56 |
| 8 | -2.32 | -4.92 | 0.78 | 3.14 | Fail | 61.11 |
| 30 | -1.49 | 5.58 | -4.69 | 3.17 | Fail | 66.67 |
| 24 | -2.28 | 3.96 | -6.50 | 3.33 | Fail | 72.22 |
| 11 | 3.31 | 0.09 | 2.55 | 3.35 | Fail | 77.78 |
| 20 | -2.61 | -4.98 | -7.66 | 3.61 | Fail | 83.33 |
| 3 | -3.91 | 0.52 | -4.12 | 4.02 | Fail | 88.89 |
| 23 | -3.38 | -4.10 | -8.75 | 4.12 | Fail | 94.44 |
| 5 | -3.58 | -10.31 | -3.70 | 5.70 | Fail | 100.00 |

### *Cumulative frequencies of colour tolerances*

For the two experimentally determined categories, the cumulative relative frequencies *P(i)* can be computed (see Table VI.2):

*Equation VI.7*
$$P_{pass}(i) = 100\left(1 - \frac{n_{i,\,pass}}{n_{pass}}\right), \quad P_{fail}(i) = 100\ \frac{n_{i,\,fail}}{n_{fail}}$$

In these equations, $n_{i,pass}$ denotes the cumulative frequency of the 'pass' rating up to the *i*-th sample pair and $n_{pass}$ its total sum. Beforehand, the experimental data set must be sorted in ascending order of residual colour difference (see Table VI.2). The quantities marked 'fail' need to be interpreted in a similar fashion. From a frequency plot as a function of stimulus intensity (here: residual colour difference), general descriptive features (descriptors) to be extracted. Figure VI.2 shows the results of the conducted psychophysical experiment. In the ideal case, all sample pairs assessed for colour difference exceeding the residual colour difference of the anchor pair would return a frequency of $P$ = 100%; similarly, all sample pairs exhibiting a colour difference less than the comparison difference would return a frequency of $P$ = 0%. This would lead to a step-shaped response function for $P = f(\Delta E_{94})$ with a discontinuity at $\Delta E_{94} = \Delta E_{94}^{(ref)}$. However, the experimental findings (see Figure VI.2) show considerable scatter among the visual assessments. For a large number of samples, the relative frequency $P$ is located between the two extreme points of the definition range (0,100). If comparison difference and test difference are roughly of the same order of magnitude, but differ in the direction of deviation as, e.g., in a comparison of lightness and hue differences, consistent visual assessment will be hard to achieve. Only samples with residual colour difference that are markedly higher or lower than the comparison difference will give rise to unique visual ratings. The scattering of the visual assessment results can be approximated by an *S*-shaped function (e.g., Equation VI.4). In the range of the comparison difference, the slope of the characteristic function increases with the ability of the observers to discriminate between colours and the higher is their level of conformity among each other.

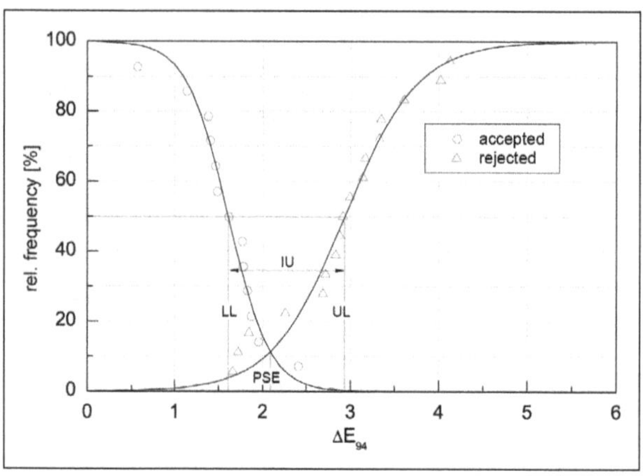


*Figure VI.2: Results of a visual assessment experiment designed to determine the acceptance threshold. If the cumulative "accepted" curve only were to be considered, the cumulative "rejected" curve would correspond to the complement at the point of intersection for 50 % frequency. In the present case, the cumulative frequency of "rejected" samples was determined separately, and deviates considerably from the complement of the cumulative "accepted" curve. The intersection point or point of subjective equality (PSE) of both curves corresponds to a residual colour difference of $\Delta E_{94} \cong 2.09$ and a relative frequency of about 11.2 %. Thus, an interval of uncertainty (IU) results from the difference between the upper (UL = 2.94) and the lower limit (LL = 1.61), i.e., IU = UL - LL = 1.33.*

The solid curves in Figure VI.2 represent logistic functions which have been matched to the discrete experimental data and which describe these to a sufficient degree of accuracy. If only the relative cumulative distribution function obtained for the rejected (fail) samples were available, the stimulus on the abscissa corresponding to the 50 % value would specify the point of subjective equality (PSE). The curve complementary to the "accepted" line would be a mirror image of

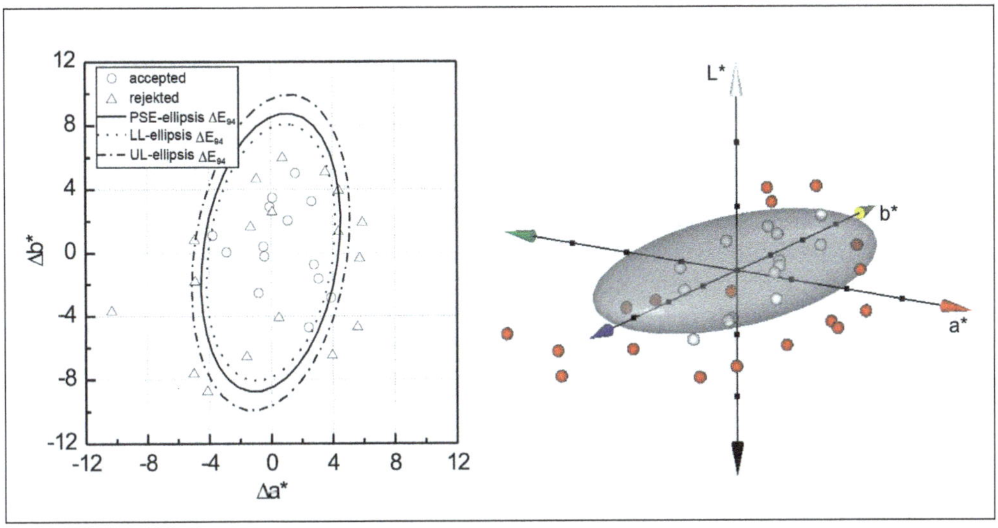


*Figure VI.3: Both diagrams include the colour measurement data shown in Figure VI.1 for 32 test samples used in the visual assessment experiment to define an acceptance threshold. In addition, the left figure shows the tolerance ellipses (80 % confidence level) derived for the PSE (= 2.09), LL (= 1.93), and UL (= 2.37) parameters using the CIE94 colour-difference model. In contrast to the case in Figure VI.2, the utilised LL and UL values correspond to a more practical acceptance level of 80 %. The ellipsoid shown in the right figure corresponds to the CIE94 metric for the PSE value. The grey and red data points represent "pass" and "fail" decisions, respectively. One division mark corresponds to a colour difference of 3 units on all axes.*

the "rejected" line, with the intersection point of both frequency distribution functions located at the 50% probability level. However, the two separately determined cumulative relative frequency distributions are not mirror functions of each other. In this case, a finite interval can be observed between the 50 % levels of both curves. The 50 % values projected onto the abscissa define the so-called interval of uncertainty (IU). If both curves are symmetric with respect to their intersection point, the middle range of IU corresponds to the stimulus of the projected PSE at the intersection point. If both curves are not symmetric, the PSE at their intersection point is certainly a better indicator for the stimulus intensity, which is neither higher nor lower than that of the reference standard. The UL and LL values represent the stimuli of the respective 50 % levels of both curves projected onto the abscissa.

### *Defining acceptance thresholds through modelling*

If the assessment results of an individual observer were reproducible and self-consistent, the "accepted" data alone would be sufficient to optimally define the acceptance threshold. However, the experimental findings in close proximity to the acceptance threshold contradict this expectation that the observer is unbiased. In such cases, it is advisable to collect "accepted" and "rejected" ratings and to factor them into the model-building so that the number of wrong instrumental decisions can be minimised. Besides the model of two assessment categories in which $\Delta E^*_{94} <$ PSE serves as the acceptance criterion, the experimental results indicate that a three-level acceptance model might be preferable: "accepted" for $\Delta E_{94} \leq$ LL, "conditionally accepted" for LL $< \Delta E^*_{94} \leq$ UL, and "rejected" for $\Delta E^*_{94} \leq$ UL.

### *Tolerance ellipse in the CIE94 colour-difference model*

The next step in data analysis consists in correlating the visual ratings and instrumental measurements with each other and constructing a **tolerance ellipsoid** about the colour position of the standard. One of many ways to determine the ellipsoid parameters is to match the logistic function to the experimental assessment data ("pass" or "fail") by minimising the weighted sum of squared residuals [7, 8]

*Equation VI.8*
$$\chi^2 = \sum_{i=1}^{N} w_i \left\{ P_i - \frac{1}{1 + e^{-(\Delta E_i - a)/b}} \right\}^2 = \min$$

where, besides the parameters $a$ and $b$, further parameters of the chosen ellipsoid equation $\Delta E_i$ have to be optimised. Formally, for example, the orientation of the ellipsoid can be defined by letting the symmetry axes point to the directions of psychologically relevant colour difference attributes of lightness ($\Delta L^*$), chroma ($\Delta C^*$), and hue ($\Delta H^*$). However, at this point, discussion of the results shall be confined to a direct comparison of the assessment results obtained with the CIE94 metric.

The tolerance ellipse corresponding to the point of subjective equality (PSE) and the corresponding tolerance ellipsoid for the CIE94 colour-difference metric are shown in Figure VI.3. With five wrong decisions (two from "pass" and three from "fail" ratings) out of a total of 32 ratings, the model is not perfect, but at least it defines a generally applicable tolerance measure in practice with 85 % correct decisions. The rejected samples also enclosed within the 2-dimensional projection are almost without exception located outside the tolerance ellipsoid and are only mapped into the acceptance region through projection onto the $a^*$-$b^*$ plane. Due to the higher number of seven and nine wrong classifications, respectively, the lower and upper limits defining the interval of uncertainty are still in need of improvement. Deviating from the limits of the cumulative frequency curves for 50 % in Figure VI.2, Figure VI.3 shows two further ellipses for the more meaningful choice of an 80 % acceptance level, with LL = 1.93 and UL = 2.37. As before, the LL ellipsoid represents a slightly more rigid criterion than the PSE ellipsoid and also excludes a visually accepted sample, while the UL ellipsoid, as a more tolerant measure, also includes visually rejected colour samples. In both cases, the new limits lead to six wrong ratings, but, for a three-level acceptance system, in which all colour samples enclosed by the LL ellipsoid are "accepted", those colour samples located between the LL and UL ellipsoids are categorised as "conditionally accepted" and need to be assessed visually, while all colour samples located outside the UL ellipsoid automatically belong to the category "rejected" and are more practice-oriented.

### *Customer and supplier specifications*

In industrial colorimetric applications, **psychophysical data** are very important when it comes to the definition of specifications between customer and supplier. Special attention must be paid to the experimental conditions and the type of stimuli employed. Both parties define the conditions under which the derived psychophysical model may be applied. Every deviation from given boundary conditions can have a negative impact on the correlation between tolerance model and visual perception. These variables include the spectral composition of the stimuli, the spatial arrangement and modulation of the stimuli (including sample and field size), the temporal variation in the stimuli, the illuminant density level (including chromatic adaptation), the noise level, differences between individuals (e.g., age, race, cultural group, etc), experience and the learning effect.

Due to the subjective nature of psychophysical measurements, experimental data always have to be interpreted with a certain caution. Practical observations have revealed that observers can vary considerably in consistency and reliability, which is why results of psychophysical experiments are subject to comparatively high scatter. Moreover, they may apply different criteria when solving the experimental task. Given this bias, the same task may produce markedly different results. In any event, it is advisable to question how well physical and **psychophysical variables** have been controlled in any experiment that provides data which have been condensed into a model. The underlying assumptions and experimental boundary conditions are frequently ignored in the practical application of such models and so the ground is unsafe.

# 6 Significance of colour measurement results

The approach to defining a tolerance that was discussed in the previous section with the aid of a practical example is subject to two significant restrictions. First, colour-difference models are frequently considered to be the perfect decision-making tool. Due to their inherent statistical nature, each colour-difference metric possesses a certain level of uncertainty which prevents perfect correlation with visual assessments. Second, tolerances are treated as "either/or" decisions. Specimens having residual colour difference values below the tolerance threshold are accepted, while those having residual colour differences above the tolerance threshold are rejected instrumentally. Every experimental collection of visual or instrumental data represents a sample, which can be comprehensively analysed and characterised with the aid of statistical tools [9, 10]. A consequence of the central limiting value theorem is that a tolerance in three-dimensional colour space is a trivariate normal distribution with a corresponding scattering ellipsoid. The closer the readings approach the surface of the scattering ellipsoid, the higher is their uncertainty. The mathematical formalism can be applied to the calculation of size and orientation of ellipsoids in colour space that correspond to the standard deviation during colour matching [11, 12] (thresholds), that correspond to instrumental and process-related testing errors [13-16] (measurement, sample preparation, etc.), or that have been defined as acceptance solids by means of visual assessments (supra-thresholds).

## 6.1 Multivariate statistics

Since a colour position can be described by a three-component vector, error analysis is more complex than in a one-dimensional case [9, 10, 17, 18]. Calculating variances of CIELAB76 coordinates $(L^*, a^*, b^*)$ is not sufficient to characterise data scatter. For instance, it is not irrelevant whether, given a large variability of $a^*$, variations of $b^*$ or $L^*$ are likewise large or small.

Generalisation from a single dimension to $n$ dimensions for the simultaneous Gaussian normal distribution $p(\mathbf{x})$ of all variables $x_i$ of a vector $\mathbf{x}=(x_1, x_2, ..., x_n)$ leads to the expression

*Equation VI.9* $$p(\mathbf{x}) = k \exp\left\{-\frac{1}{2}(\mathbf{x}-\mathbf{a})^T \mathbf{B}(\mathbf{x}-\mathbf{a})\right\}$$ with the normalisation constant

*Equation VI.10* $$k = \frac{1}{(2\pi)^{n/2} (\det \mathbf{B})^{1/2}}$$

Here, **a** is an $n$-component vector and **B** is a positive definite $n \times n$ matrix with the determinant det **B**. As will be shown below, **B** is equal to the inverse covariance matrix **V**. The functional dependence of the Gaussian distribution is governed by the quadratic equation

*Equation VI.11* $$\Delta^2 = (\mathbf{x}-\mathbf{a})^T \mathbf{B}(\mathbf{x}-\mathbf{a}) = P(\chi^2, n)$$

The quantity $\Delta$ denotes the *Mahalanobis* distance between **a** and **x**, which reduces to the Euclidean distance if **B** equals the identity matrix **I**. The *Gaussian* distribution is constant on surfaces in **x** space, for which this quadratic form is constant. For the case $n = 3$ Equation VI.11 is the equation of an ellipsoid having a centre-point $\mathbf{M} = (a_1, a_2, a_3)^T$ (see Equation VI.20).

Since $p(x)$ obviously is symmetric about $(\mathbf{x}-\mathbf{a})$, the expectation value $E[\mathbf{x}-\mathbf{a}]$ of the quantity $(\mathbf{x}-\mathbf{a})$ vanishes:

*Equation VI.12* $$E[\mathbf{x}-\mathbf{a}] = \int_{-\infty}^{+\infty} d\mathbf{x}(\mathbf{x}-\mathbf{a})\, p(\mathbf{x}-\mathbf{a}) = 0 \quad \text{or}$$

*Equation VI.13* $$E[\mathbf{x}] = \mathbf{a}$$

Hence, the vector of the expectation values is equal to **a**. Differentiation of Equation VI.12 with respect to **a** yields the expression

*Equation VI.14* $$\int_{-\infty}^{+\infty} d\mathbf{x}\, \{\mathbf{I} - (\mathbf{x}-\mathbf{a})(\mathbf{x}-\mathbf{a})^T \mathbf{B}\}\, p(\mathbf{x}-\mathbf{a}) = 0$$

where, accordingly, the expectation value of the curly braces is equal to zero, i.e.

*Equation VI.15* $$\mathbf{B}E\left[(\mathbf{x}-\mathbf{a})(\mathbf{x}-\mathbf{a})^T\right] = \mathbf{I} \quad \text{or}$$

*Equation VI.16* $$\mathbf{V} = E\left[(\mathbf{x}-\mathbf{a})(\mathbf{x}-\mathbf{a})^T\right] = \mathbf{B}^{-1}$$

Matrix **V,** the inverse of **B**, is called the covariance matrix of the variables $\mathbf{x} = (x_1, x_2, \ldots, x_n)^T$. Obviously, **V** is symmetrical, since $V_{ij}=V_{ji}$. Its diagonal elements $V_{ii}$ are the variances $\sigma^2(x_i)$ of the variables $x_i$. The covariance $\text{cov}(x_1, x_2)$ of two random variables $x_1$ and $x_2$ $(x_1 \neq x_2)$ is only equal to zero if $x_1$ and $x_2$ are independent of each other.

## 6.2 Statistics for 3-dimensional colour spaces

In the present case of a three-dimensional colorimetic problem, the vector of a colour position in CIELAB76 colour space is given by the variables $\mathbf{x} = (a^*, b^*, L^*)^T$. Inversion of the covariance matrix

*Equation VI.17* $$\mathbf{V} = \begin{pmatrix} \text{var}(a^*) & \text{cov}(a^*, b^*) & \text{cov}(a^*, L^*) \\ \text{cov}(a^*, b^*) & \text{var}(b^*) & \text{cov}(b^*, L^*) \\ \text{cov}(a^*, L^*) & \text{cov}(b^*, L^*) & \text{var}(L^*) \end{pmatrix}$$ with the elements

*Equation VI.18* $$\text{var}(x) = \frac{1}{n-1}\sum_{i=1}^{N}(x_i - \langle x_i \rangle)^2$$

*Equation VI.19* $$\text{covar}(x, y) = \frac{1}{n-1}\sum_{i=1}^{N}(x_i - \langle x_i \rangle)(y_i - \langle y_i \rangle)$$ and the mean values

$$\langle x_i \rangle = \frac{1}{n}\sum_{i=1}^{n} x_i, \quad \langle y_i \rangle = \frac{1}{n}\sum_{i=1}^{n} y_i$$

leads directly to the parameters of the Equation VI.20

*Equation VI.20*
$$b_{11}(\Delta a^*)^2 + b_{22}(\Delta b^*)^2 + b_{33}(\Delta L^*)^2 + 2b_{12}\Delta a^* \Delta b^* + 2b_{13}\Delta a^* \Delta L^* + 2b_{23}\Delta b^* \Delta L^* = P(\chi^2, 3)$$

*Table VI.3: Threshold values $P(\chi^2,3)$ of the $\chi^2$ distribution for the three-dimensional case at confidence level $(1-\alpha)$.*

| Probability $1-\alpha$ | Threshold $P(\chi^2,3)$ |
|---|---|
| 69.2 % | 3.6 |
| 93.4 % | 7.2 |
| 95.6 % | 8.1 |
| 99.8 % | 14.4 |

for the scatter ellipsoid. To illustrate the importance of the scatter ellipsoid, consider a two-dimensional example in which the ellipsoid degenerates to an ellipse (see Figure VI.4). The two positions $P_1$ und $P_2$ shown in Figure VI.4 are both located on the contour of the covariance ellipse and correspond to equally probable values of $\Omega(P_1)=\Omega(P_2)=\Omega_E$, although their distances from the centre-point of the ellipse are markedly different. Furthermore, position $P_3$ has a higher and position $P_4$ a lower probability than $\Omega_E$ $(\Omega(P_4)<\Omega_E$, $\Omega(P_3)>\Omega_E)$, although $P_4$ is located closer to the centre-point than $P_3$. An extension of this model to three dimensions is obvoius.

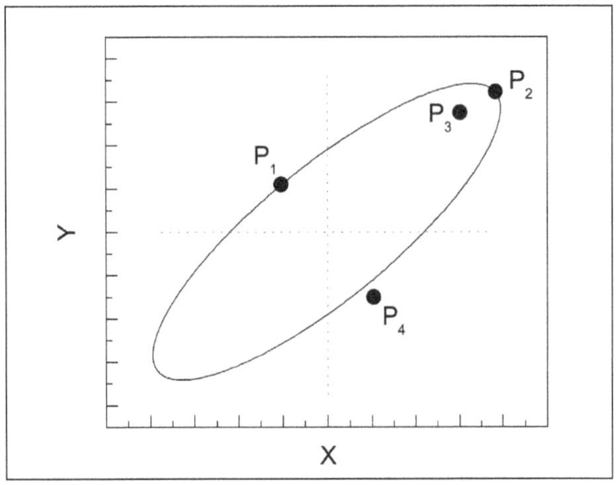


*Figure VI.4: Relative probability of various positions for a Gaussian distribution of two variables. The following relations apply:*
*$\Omega(P_1)=\Omega(P_2)=\Omega(P_E)$, $\Omega(P_3)>\Omega(P_E)$, and $\Omega(P_4)<\Omega(P_E)$).*

The constant $P(\chi^2,3)$ in Equations VI.11 and VI.20 denotes the value of the $\chi^2$ distribution for the case of three dimensions corresponding to the desired significance level and is called the probability integral. $P(\chi^2,3)$ can either be computed numerically using the formula

*Equation VI.21*
$$P(\chi^2, n) = \frac{1}{2^{n/2}\,\Gamma(n/2)} \int_0^{\chi^2} dt\, t^{\frac{n}{2}-1}\, e$$

for $n = 3$ or be taken from precomputed tables [9, 10, 18]. $\Gamma(n/2)$ denotes the Gamma function for the argument $n/2$. Some typical values relevant to practical use are shown in Table VI.3. If, e.g., $P(\chi^2,3) = 3.6$, about 69.2% of all readings are expected to be located within the scatter ellipsoid. If the readings follow a normal distribution, each reading is located within the scatter ellipsoid with a probability of 69.2% (this corresponds to a 1σ criterion for a one-dimensional normal distribution). The scatter ellipsoid is only principal axes oriented if the covariances diminish or are negligibly small.

Now that the parameters $b_{ij}$ of the scatter ellipsoid are known, all quantities are available required for calculating the critical values $t$ of the specific colour difference values and the total colour difference in a significance test:

*Equation VI.22*
$$t_{\Delta a^*} = \sqrt{\frac{P(\chi^2, 3)}{n b_{11}}}, \quad t_{\Delta b^*} = \sqrt{\frac{P(\chi^2, 3)}{n b_{22}}}$$

*Equation VI.23*
$$t_{\Delta L^*} = \sqrt{\frac{P(\chi^2, 3)}{n b_{33}}}, \quad t_{\Delta E^*_{76}} = \sqrt{\frac{P(\chi^2, 3)}{n b_E}} \quad \text{where}$$

*Equation VI.24*
$$b_{11}\alpha^2 + b_{22}\beta^2 + b_{33}\gamma^2 + 2b_{12}\alpha\beta + 2b_{23}\beta\gamma + 2b_{13}\alpha\gamma = b_E \quad \text{and}$$

*Equation VI.25*
$$\alpha = \frac{\langle \Delta a^* \rangle}{\langle \Delta E^*_{76} \rangle}, \quad \beta = \frac{\langle \Delta b^* \rangle}{\langle \Delta E^*_{76} \rangle}, \quad \gamma = \frac{\langle \Delta L^* \rangle}{\langle \Delta E^*_{76} \rangle}$$

The critical $t$ values of the colour differences can be reduced and the reliability enhanced by increasing the number $n$ of performed measurements. For the test, a certain confidence level ($1$-$\alpha$) is defined, e.g., 70%. The corresponding $\chi^2$ value for three degrees of freedom can be calculated from Equation VI.21 or taken from Table VI.3. A specific colour difference $\Delta F$ is significant, with a probability of ($1$-$\alpha$)%, if

*Equation VI.26*
$$|\langle \Delta F \rangle| > |t_{\Delta F;\,1-\alpha}|$$

is fulfilled. $\langle \Delta F \rangle$ is a place holder for each possible mean value of specific colour differences ($\langle \Delta a^* \rangle$, $\langle \Delta b^* \rangle$, $\langle \Delta L^* \rangle$, etc.).

The error theory for the three variables of a colour space can only be roughly outlined here. The formalism explained above might appear very academic at first sight, but it is indispensable for assessing the significance of colour differences. Further details and answers to questions which could not be treated in the present context can be found in the relevant literature (e.g.[13, 16]).

## 6.3 The scatter ellipsoid

Every reading is subject to more or less substantial errors which is caused by small random fluctuations (**statistical errors**) or which can be traced back to the impact of corrupting small disturbances that always push a reading in the same direction (**systematic errors**). The latter defy treatment by statistical methods. They have to be tackled by carefully choosing and calibrating a measuring instrument and by adopting other appropriate precautions which actually belong to the field of physical metrology.

### *Measurement errors*

Various error sources can influence the measurement result of effect colour shades: (i) the colour measurement instrument, (ii) the sample, and (iii) the operator. In the latter case, improper handling of the measurement system or faulty positioning of a specimen is considered to be a source of systematic error. In the second case, process fluctuations may lead to inhomogeneous colouration of a single specimen or to varying colouration of multiple specimens of the same type. The impact of these root causes on the measurement result can be analysed experimentally. The noise level of a colour measurement instrument can be determined by performing a sufficient number of measurements on a single sample under constant sample orientation and analysing the results by means of statistical methods. The mean value and standard deviation of the measurement series provide information about the quality of the measurement instrument used.

### *Homogeneity error and application error*

The measurement results will look different if the position of a measurement spot and/or sample orientation is changed for each individual measurement. In addition to the noise of the measuring device, there will be inhomogeneities in the surface coating, e.g., the azimuth effect (which can be made visible by changing the relative orientation of measuring instrument and specimen), or an irregular distribution of pigments within the pigmented layer. Metrologically, these results are strongly affected by the size of the measurement spot and the angular aperture of the spectrophotometer. Where multiple samples of the same type are measured (e.g., colour standards, components in a production line), a further cause of fluctuations in readings that must be considered is the range of production, which in turn has many causes that cannot be discussed in detail here.

The extent to which measurement, homogeneity, and application errors contribute to the total error can be analysed by comparing the corresponding spatial error distributions. The spatial distribution of the experimental data points can be illustrated by studying their projections onto the three principal planes of the coordinate system. A typical example from industrial practice is shown in Figures VI.5 and VI.6 for the effect colour shade *Light Red*. The plots show the respective error distributions projected onto the principal planes ($L^*$-$a^*$, $L^*$-$b^*$, $a^*$-$b^*$) at three viewing angles ($\varepsilon$ =25°, 45°, 70°) for the measurement error (multiple measurements in a fixed position; Figure VI.5), the sum of homogeneity and measurement error (multiple measurements at different positions; Figure VI.5), and the total error (multiple measurements of a set of similar samples at different positions; Figure VI.6). In CIELAB76 colour space, all error contributions of the investigated effect colour shade exhibit an ellipsoidal distribution and a somewhat diagonally oriented major axis. Both the orientation of the major axis in space and the proportions of the principal axes among themselves vary with the viewing angle. The major axis assumes its largest extension close to the specular angle (ε = 25°). As the aspecular angle increases, the major axis becomes shorter, so that the cigar-shaped error distribution at ε = 25° becomes rather more egg-shaped at ε = 70°. As expected, the scattering solid for the measurement error assumes its smallest (Figure VI.5, upper block) and for the total error its largest (Figure VI.6) extension in colour space; the distribution of the homogeneity error is located between these two extreme cases (Figure VI.5, lower block).

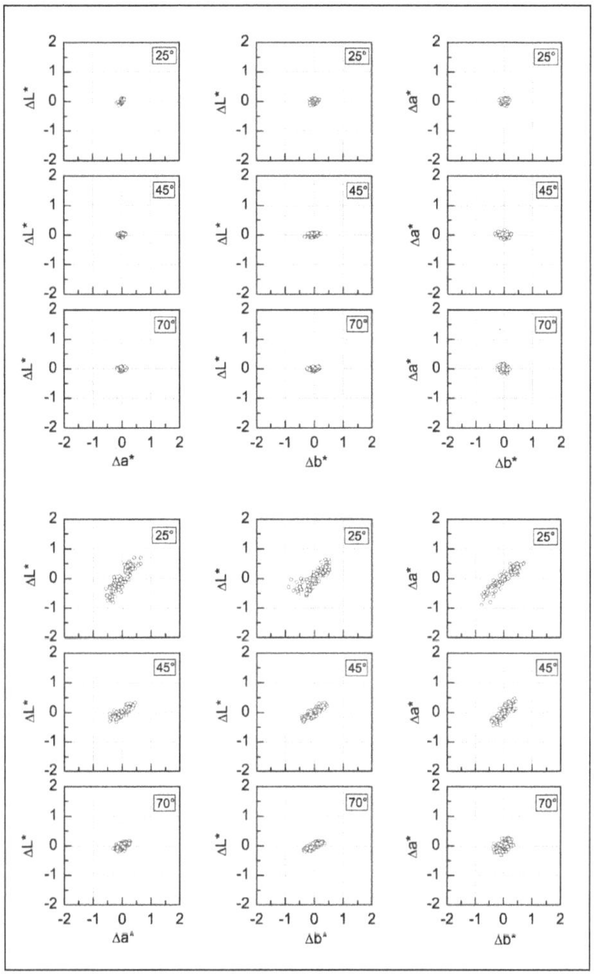


*Figure VI.5: Projection of the error distribution onto the three principal planes of CIELAB76 colour space for the gonioapparent colour shade Light Red at the three viewing angles of 25°, 45°, and 70°. The illuminating angle in each case was fixed at 45°. The results of the upper block stem from a single sample, which was measured 100 times at the same spot after the measuring head was positioned and fixed, while the results of the lower block were obtained at 100 different spots on the same sample.*

### *Correlation effect in the definition of colour tolerances*

The example discussed above clearly shows that any agreement on colour tolerances that is based on a total residual colour difference in colour space CIELAB76 does not correspond at all with the experimental facts. When $\Delta E^*_{76}$ values are set as the tolerance measure, the respective acceptance solid is spherical in shape. The experimental results clearly reveal that the scattering of the measurement data exhibits marked directionality. This correlation effect must be accounted for when colour tolerances are being defined in order that practicable and above all unique results may be obtained. Reporting the variance as a way of describing the scattering of measurement data, which is perfectly adequate in one-dimensional statistics, needs to be extended to the general case of three dimensions.

The experimental data shown in Figures VI.5 to VI.6 belong to a series of samples which were all prepared with the same base material. In accordance with

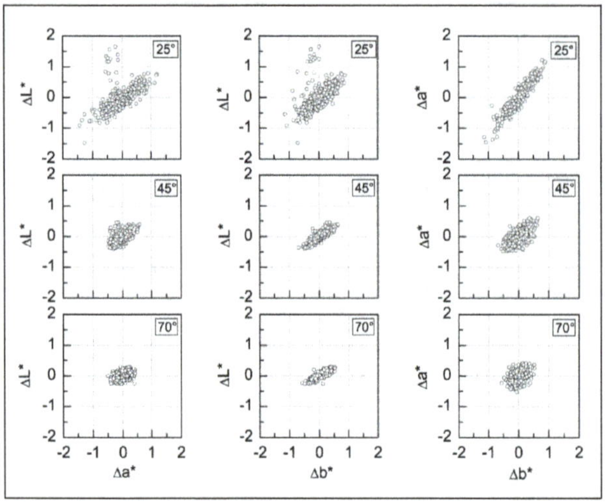

*Figure VI.6: Projection of the error distribution onto the three principal planes of CIELAB76 colour space for the gonioapparent colour shade Light Red at the three viewing angles of 25°, 45°, and 70°. The illuminating angle in each case was fixed at 45°. The results stem from a sample series of 80 similar panels which were measured at four different spots.*

*Equation VI.27*

$$\mathbf{V}_{tot.} = \mathbf{V}_{meas.} + \mathbf{V}_{hom.} + \mathbf{V}_{appl.}$$

the co-variance matrix for the total error is additively composed of the individual matrices for the measurement error ($V_{meas}$), the homogeneity error ($V_{hom}$), and the application error ($V_{appl.}$). Preparatory errors due to $n_{prep} = 1$ are excluded from the analysis. The individual matrices can be derived from the data sets shown in Figures VI.5 to VI.6. The data in Figure VI.6 (upper block) directly yield the measurement error

*Equation VI.28* $\mathbf{V}_1 = \mathbf{V}_{meas.}$

while the data set in Figure VI.5 (lower block) also contains the homogeneity error, i.e.

*Equation VI.29* $\mathbf{V}_2 = \mathbf{V}_{meas.} + \mathbf{V}_{hom.}$

The data set depicted in Figure VI.6 contains the application error $V_{appl.}$ as a further influencing parameter:

*Equation VI.30* $\mathbf{V}_3 = \mathbf{V}_{meas.} + \mathbf{V}_{hom.} + \mathbf{V}_{appl.}$

All these individual contributions can be separated by finding the differences:

*Equation VI.31* $\mathbf{V}_{meas.} = \mathbf{V}_1, \quad \mathbf{V}_{hom.} = \mathbf{V}_2 - \mathbf{V}_1, \quad \mathbf{V}_{appl.} = \mathbf{V}_3 - \mathbf{V}_2$

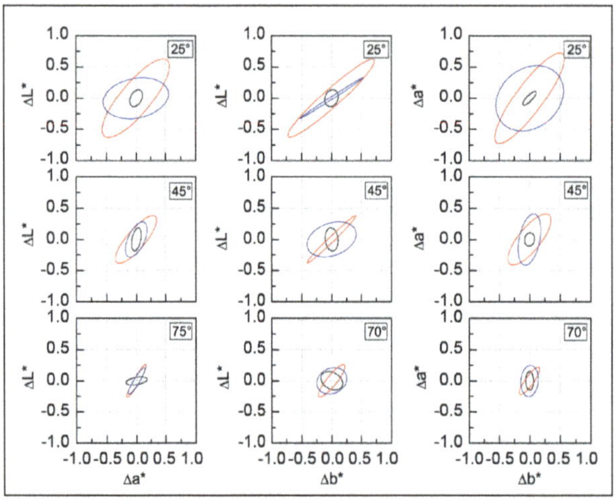

*Figure VI.7: Projection of the scatter ellipsoids for measurement error (black line), homogeneity error (red line), and application error (blue line) onto the three principal planes $L^*$-$a^*$, $L^*$-$b^*$, and $a^*$-$b^*$ of CIELAB76 colour space at the three measuring angles of ε = 25°, 45°, and 70° for the gonioapparent colour shade "Light Red". The ellipsoids for the confidence level of 70 % are based on the experimental data depicted in Figures VI.5 to VI.6.*

Inversion of the individual matrices yields the coefficients of the corresponding scattering ellipsoids. The results of all these analyses are shown in Figure VI.7 for a confidence level of 70 %. These analyses clearly reveal that data scatter is primarily caused by homogeneity and application errors; although measurement errors are not negligible, they are of minor importance, with the exception of far aspecular geometries. Particularly noteworthy is the different size, shape, and orientation of all ellipsoids. In case of the scattering ellipsoid for the measurement error, orientation and ratios of axes are only slightly angle-dependent, whereas the ellipsoids for homogeneity and application error are subject to a marked angle dependence. With increasing viewing angle, a significant volume reduction is observed in the latter two cases; at an angle of 70° all error contributions are of the same order of magnitude. This experimental finding was to be expected because, in the region far from the specular angle, the influence of the actual topology of the effect-imparting pigments on the coloristic properties is no longer critical and readings are dominated by the diffuse contribution of scattering on edges and the diffuse scattering contribution of the absorption pigments.

Application of the $\chi^2$ distribution recommended in technical standard *DIN 55600* [15] for estimating the significance of colour differences is only permissible if it can be assumed that the co-variance matrix is known. A typical application would be the production of pigments, where the variance of chromaticity coordinates can be determined from many production batches. If only a rough estimate of the co-variance matrix is available, it is not permissible to use the $\chi^2$ distribution to determine the size of the scatter ellipsoid; the $\chi^2$ distribution must be replaced by the *F* distribution [9, 10] in this case.

# 7 Tolerances for solid colours: DIN 6175 P1

In practice, the systematic deviations from uniformity in CIELAB76 are counteracted by specifying colour-region-dependent tolerance limits. Such **tolerance limits** were devised by FNF24, a technical working committee of DIN, and specified in DIN 6175 P1 [19] as a recommendation for solid colours by defining vertex points for the various regions in the chromaticity diagram (see Figure VI.8). This recommendation reflects the fact that, in the case of brilliant yellow and red colours, colour differences of $\Delta E^*_{76} \cong 1.0$ are barely perceptible whereas, close to the achromatic axis, colour differences of $\Delta E^*_{76} \cong 0.3$ are definitely perceptible. For this reason, the tolerance frame in the region of brilliant red and yellow colours was set wider than in the achromatic colour region. However, this rigid division is ultimately unsatisfactory, since inconsistencies can always occur in the border region between adjacent segments. It would be more realistic to dynamically adjust the tolerances to the position of the colours under assessment within the colour space. The proper tool for achieving this condition would be a colour difference metric in which deviations from uniformity either do not occur or are tolerable.

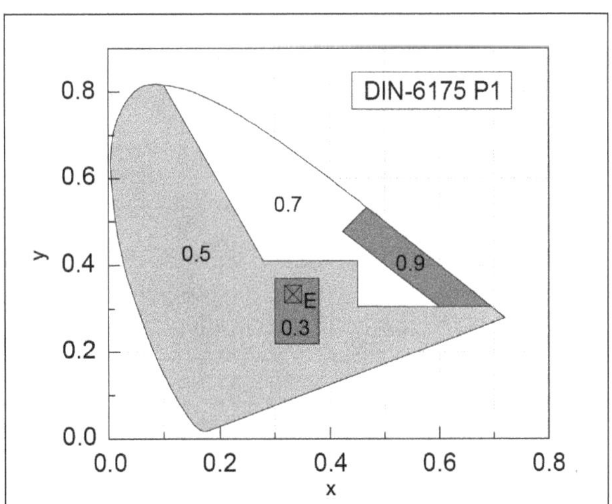


*Figure VI.8: Division of the chromaticity diagram into tolerance regions as per DIN 6175 P1. The lowest tolerance value of $\Delta E^*_{76} \cong 0.3$ is found in the achromatic region, while the highest tolerance is located in the region of brilliant yellow and red colours, where $\Delta E^*_{76} \cong 0.9$.*

### *Uniformity through DIN99 colour space?*

The revised version of the DIN standard cited above replaces the CIELAB76 colour space with colour space DIN99, which is characterised by a higher level of uniformity. Here, division of the chromaticity diagram into regions of different tolerance recommendations is replaced by a tolerance of $\Delta E^*_{76} \cong 0.3$ which is approximately independent of colour region. However, it is to be expected that there will be little experimentation in industry with new tolerance models, and that established approaches, with all their known and manageable weaknesses, will be preferred.

Although colour perception cannot express the quality of colour stimuli in numbers, it is very sensitive to colour differences of contiguous coloured areas. However, this sensitivity is considerably reduced by "breaking up" the surface coating with decorative strips, beads, and embossed edges. Area separators can raise the threshold for discriminating colour differences by a factor of 2 to 4, compared to contiguous coloured areas [20–22].

# *8 Tolerances for gonioapparent colours: DIN 6175 P2*

Acceptance limits in colour assessments of OEM coatings are increasingly approaching the perception threshold. The threshold values for solid colours are comparatively small (e.g. $\Delta E^* < 0.2$ for achromatic colours). This contrasts with the case for effect colour shades, where substantially higher values for the respective perception threshold have to be dealt with. Hence, tolerance models for solid colours cannot be applied to effect colours, which instead require modified approaches.

### *Established practice: the batch-release model*

One approach adopted in actual practice is recommended in *DIN 6175 P2* [23] as a model for releasing batches. The colour measurement must be based on the spectral method and applied at three aspecular angles: ε = as25°, ε = as45° and either ε = as75° or ε = as110°. These specifications refer to the aspecular angle. The accuracy of the angles to be set, the illumination and viewing apertures, and the colour tolerances based on colour difference values $E_{eff}$ are all defined in the technical standard.

### *Colour-difference calculation for effect colour shades*

The proposed formalism for calculating the colour difference between two effect surface coatings distinguishes the regions of achromatic or pastel colours from those of chromatic colours. The region of achromatic colours is characterised by the fact that the condition $C^*<10$ is fulfilled. The region of pastel colours is defined by the conjoint conditions $C^*<18$ and $L^*>27$. The total colour difference is then calculated with the aid of $L^*a^*b^*$ coordinates:

*Equation VI.32*

$$\Delta E'_{ab}(\varepsilon) = \sqrt{\left(\frac{\Delta L^*(\varepsilon)}{g_{\Delta L^*} \cdot S_L(\varepsilon)}\right)^2 + \left(\frac{\Delta a^*(\varepsilon)}{g_{\Delta a^*} \cdot S_a(\varepsilon)}\right)^2 + \left(\frac{\Delta b^*(\varepsilon)}{g_{\Delta b^*} \cdot S_b(\varepsilon)}\right)^2}$$

Chromatic colours are characterised by $C^* \geq 10$, with the region of pastel colours ($C^* < 18$ and $L^* > 27$) being excluded. For the purpose of calculating the total colour difference, an equation based on $L^*C^*H^*$ coordinates is adopted:

*Equation VI.33*

$$\Delta E'_{CH}(\varepsilon) = \sqrt{\left(\frac{\Delta L^*(\varepsilon)}{g_{\Delta L^*} \cdot S_L(\varepsilon)}\right)^2 + \left(\frac{\Delta C^*(\varepsilon)}{g_{\Delta C^*} \cdot S_C(\varepsilon)}\right)^2 + \left(\frac{\Delta H^*(\varepsilon)}{g_{\Delta H^*} \cdot S_H(\varepsilon)}\right)^2}$$

All specific colour difference values $\Delta L^*(\varepsilon)$, $\Delta a^*(\varepsilon)$, $\Delta b^*(\varepsilon)$, $\Delta C^*(\varepsilon)$, and $\Delta H^*(\varepsilon)$ are calculated in CIELAB76 colour space. Besides the following auxiliary and weighting functions are needed:

$$\bar{L}(\varepsilon) = \sqrt{L^*_B(\varepsilon) \cdot L^*_P(\varepsilon)}, \quad \bar{C}(\varepsilon) = \sqrt{C^*_B(\varepsilon) \cdot C^*_P(\varepsilon)},$$

$$S_L(\varepsilon) = 0.15\sqrt{\bar{L}(\varepsilon)} + \frac{31.5°}{\varepsilon}; \quad S_a(\varepsilon) = 0.7; \quad S_b(\varepsilon) = 0.7;$$

$$S_C(\varepsilon) = \max.\left\{0.7;\ 0.48\ \sqrt{\bar{C}(\varepsilon)} - 0.35\ \sqrt{\bar{L}(\varepsilon)} + \frac{42^\circ}{\varepsilon}\right\},$$

$$S_H(\varepsilon) = \max.\left\{0.7;\ 0.14\ \sqrt{\bar{C}(\varepsilon)} - 0.2\ \sqrt{\bar{L}(\varepsilon)} + \frac{21^\circ}{\varepsilon} + 0.7\right\}$$

The last two expressions show that the values of $S_C(\varepsilon)$ and $S_H(\varepsilon)$ must not drop below the limiting value of 0.7.

For different aspecular angles ε, the colours to be assessed may be located in different colour regions and so the calculation of colour differences may invoke different equations.

At the boundary between adjacent colour regions, the two colour difference equations do not yield the same colour difference values (discontinuity). Consequently, for the calculation of $\Delta E_{eff}(\varepsilon)$, the relation

*Equation VI.34* $$\Delta E_{eff}(\varepsilon) = \sigma(\varepsilon) \cdot \Delta E'_{ab} + (1 - \sigma(\varepsilon)) \cdot \Delta E'_{CH}$$

is adopted which smoothes the discontinuity by introducing a “**smearing function**”

*Equation VI.35* $$\sigma(\varepsilon) = \frac{1}{1 + e^{(C^*(\varepsilon) - C_o(\varepsilon))}}$$ where

*Equation VI.36* $$C_o(\varepsilon) = 10 + \frac{8}{1 + e^{(27 - L^*(\varepsilon))}}$$

For colours that are clearly more chromatic than the boundary between the colour regions, the smearing function $\sigma(\varepsilon)$ assumes a value of zero, so that the term $\Delta E'_{ab}(\varepsilon)$ vanishes and the total residual colour difference is given by $\Delta E'_{CH}(\varepsilon)$. In the opposite case of colours that are considerably less chromatic than the colour boundary, $\sigma(\varepsilon)$ approaches a value of 1 and the term $\Delta E'_{CH}(\varepsilon)$ vanishes. In the transition area, $0 < \sigma(\varepsilon) < 1$; here the total colour difference $\Delta E'_{eff}(\varepsilon)$ is given by a linear combination of $\Delta E'_{ab}(\varepsilon)$ and $\Delta E'_{CH}(\varepsilon)$.

When it comes to the acceptance of delivered paint material and the application of finishes at seamless contiguous areas, a tolerance frame of $\Delta E_{eff}(\varepsilon) \leq 1$ is recommended for all aspecular angles, with the appropriate *g* factors in Table 6.4 being adopted to suit the respective application. When these *g* factors are used in vehicle refinishing, a tolerance frame of $\Delta E_{eff}(\varepsilon) \leq 1.5$ needs to be applied for contiguous areas; for areas separated by beads and gaps, $\Delta E_{eff}(\varepsilon) \leq 3$ applies.

### *Discontinuity between solid and effect colour shades*

A weakness of the tolerance model propagated in DIN 6175 P2 consists in the discontinuity between solid and effect colours. In any future model, this limitation needs to be resolved in a way that solid colours can be treated as a limiting case of a more general model for effect colour shades. A further weakness of the model is the fact that only a limited number of effect colour shades could be evaluated for the purpose of developing the model parameters. As a result, the tolerance model does not perform equally well in all regions of colour space. In particular, further developments are required in the regions of pastel and dark colours. As before, the usefulness of the results needs to be validated in visual assessments.

*Table VI.4: Values of the g factors for the delivery of paint, and the execution and repair of surface coatings.*

| Specific Application | $g_L$ | $g_a$ | $g_b$ | $g_C$ | $g_H$ |
|---|---|---|---|---|---|
| Delivery of paint | 1.0 | 1.0 | 1.0 | 1.0 | 1.0 |
| Application of paint | 2.0 | 1.2 | 1.2 | 1.8 | 1.2 |
| Repair of surface coatings | 2.0 | 1.2 | 1.2 | 1.8 | 1.2 |

The complexity of the tolerance model undoubtedly also arises from the use of the non-uniform CIELAB76 metric, whose deviations from uniformity formally need to be eliminated. A tolerance model which is based on an enhanced colour difference metric, e.g. CIE94 or DIN99, would be formally easier and more successful.

# *9 Literature*


[1] Bartleson, C. J., Grum, F., publisher., *Optical Radiation Measurements. Visual Measurements*, Vol. 5, Academic Press, New York, USA, 1984

[2] Gescheider, G. A., *Psychophysics. The Fundamentals*, Lawrance Erlbaum Associates, Publishers, Mahwah, NJ, USA, 3rd edition, 1997

[3] Roehler, R., *Sehen und Erkennen*, Springer-Verlag, Berlin, Germany, 1995

[4] MacAdam, D. L., "Visual sensitivities to colour differences in daylight", *J. Opt. Soc. Amer.* 32, p. 247, 1942

[5] Berns, R. S., "Deriving instrumental tolerances from pass-fail and colorimetric data", *Colour Res. Appl.* 21, 3rd edn.459, 1996

[6] Berns, R. S., *Billmeyer and Saltzman's Principles of Colour Technology*, 3rd edn., John Wiley & Sons, Inc., New York, USA, 2000

[7] Strocka, D., Brockes, A., Pfaffhausen, W., "Untersuchung von Farbabstandsellipsoiden: Methodik und Ergebnisse an einem Beispiel, *Farbe + Design* **15/16**, p. 75, 1980

[8] Strocka, D., Brockes, A., Pfaffhausen, W., "Influence of experimental parameters on the evaluation of colour difference ellipsoids", *Colour Res. Appl.* **8**, p. 169, 1983

[9] Stange, K., *Angewandte Statistik. Eindimensionale Probleme*, Vol. 1, Springer-Verlag, Berlin, Germany, 1970

[10] Stange, K., *Angewandte Statistik. Mehrdimensionale Probleme*, Vol. 2, Springer-Verlag, Berlin, Germany, 1971

[11] Brown, W. R., MacAdam, D. L., "Visual sensitivities of combined chromaticity and luminance differences", *J. Opt. Soc. Am.* 39, p. 808, 1949

[12] Brown, W. R., "Statistics of colour-matching data", *J. Opt. Soc. Am.* 42, p. 252, 1951

[13] Völz, H. G., *Industrial Colour Testing. Fundamentals and Techniques* (2nd edition), Wiley-VCH, New York, USA, 2001

[14] Völz, H. G., "Über die Signifikanz von Farbmessungen an pigmentierten Systemen", *Farbe + Lack* **88**, p. 264, 1982; *ibid.* p. 443 (Das Prüffehlerellipsoid Teil II (Experimentelle Durchführung))

[15] "Prüfung von Pigmenten. Bestimmung der Signifikanz von Farbabständen bei Körperfarben nach der CIELab-Formel", Deutsches Institut für Normung, DIN 55600, Beuth-Verlag GmbH, Berlin, Germany, January 2004

[16] "Verfahren zur Vereinbarung von Farbtoleranzen", Deutsches Institut für Normung, DIN Fachbericht 49, Beuth-Verlag GmbH, Berlin, Germany, 1995

[17] Brandt, S., *Datenanalyse. Mit statistischen Methoden und Computerprogrammen*, Bibliographisches Institut AG, Zürich, Switzerland, 1981

[18] Kreyszig, E., *Statistische Methoden und ihre Anwendungen*, Vandenhoeck und Ruprecht, Göttingen, Germany, 1974

[19] "Farbtoleranzen für Automobillackierungen. Teil 1: Uni-Lackierungen", Deutsches Institut für Normung, DIN 6175-1, Beuth-Verlag GmbH, Berlin, Germany, 1986

[20] Danner, M., Anselm, D., "Farbtoleranzen in der Kraftfahrzeuglackierung. Teil I: Bestimmung von Krümmungsradien und Unterteilungen von Lackflächen an Personenkraftwagen", *defazet* 30, p. 433, 1976

[21] Terstiege, H., Gundlach, D., "Farbtoleranzen in der Kraftfahrzeuglackierung. Teil II: Der Einfluß zylindrischer Krümmungen auf die Farbmessung von Uni-Lackierungen", *defazet* 32, p. 70, 1978

[22] Heinrich, F., Zwernemann, K., "Farbtoleranzen in der Kraftfahrzeuglackierung. Teil III: Der Einfluß zylindrischer Krümmungen auf die Farbmessung von Uni-Lackierungen", *defazet* 32, p. 114, 1978

[23] "Farbtoleranzen für Automobillackierungen. Teil 2: Effektlackierungen", Deutsches Institut für Normung, DIN 6175-2, Beuth-Verlag GmbH, Berlin, Germany, 2001

# VII Pigment optics - physical processes

## 1 Colour-generating processes

*Uwe Hempelmann*

The colour vision in human beings, as described in the previous chapters, is a sensory perception that arises in the human brain. Nevertheless, the observed colours naturally have physical origins. This chapter will take a more detailed look at the physical processes, which, in combination with the human eye and brain, generate the perception of colour.

According to *Nassau*, it is possible to identify as many as 15 different causes of colour. While some of the details are debatable, the following table still provides a good overview of the physical and chemical mechanisms, which will then be discussed briefly. A more detailed approach is beyond the scope of this book - the reader is referred to the literature cited in Section VII.8.2.

Thermal emitter is a term used to describe flames, lamps and carbon arcs whose colour and light are primarily due to the heating or burning of a substance. Often, their colour can be described by the equations for blackbody radiation, which are well known and can be found in any physics textbook.

*Table VII.1: Causes of colour, according to Nassau.*

| Cause of Colour | | Example |
|---|---|---|
| 1 | Incandescence | Filament bulb, carbon arc |
| 2 | Gas excitations | Vapour lamps |
| 3 | Vibrations and rotations | Water, ice, iodine |
| 4 | Transition metal compounds | Inorganic pigments |
| 5 | Transition metal dotations | Ruby, emerald |
| 6 | Organic compounds | Dyes |
| 7 | Charge transfer compounds | Lapis lazuli, magnetite |
| 8 | Metals | Copper, silver, aluminium, gold |
| 9 | Semiconductors | Silicon, cinnabar, diamond |
| 10 | Doped semiconductors | Coloured diamonds |
| 11 | F-centres | Amethyst |
| 12 | Refraction and polarization | Rainbow, haloes |
| 13 | Scattering | Blue sky |
| 14 | Interference | Oil on water, soap bubbles, coating on camera and telescope lenses |
| 15 | Diffraction | Colours on diffraction gratings, haloes, some biological colours |

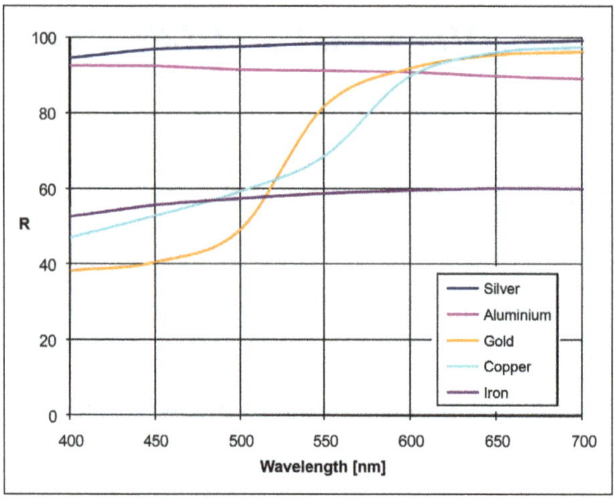


*Figure VII.1: Reflectance spectra of metals.*

Fluorescent tubes are widely used for illumination in everyday life and can simulate daylight quite closely.

### *Colours caused by absorption and emission*

With the exception of the aforementioned thermal emission sources, colours are typically the result of absorption and emission of photons by atoms, molecules and crystals. The absorption or emission of a photon is contingent on a corresponding effect being exerted on the material, which is absorbing or emitting the photon. In spectroscopy, transitions are classified as rotational, vibrational and electronic. Typically, the energy differences necessary for electronic transitions come closest to those needed for absorption in the visible range.

Exceptions to this are the faint blue colour of water and ice, and the violet colour of iodine vapour. The iodine colour stems from the dissociation of the iodine molecule in the gas phase after heating, which leaves one of the atoms in a vibrationally excited state.

The faint blue colour of very pure water and ice is due to weak absorption induced by the overtones of the various vibrational modes of the molecule. It is a matter of debate as to whether or not some of the colour is due to scattering.

Many substances, especially transition metal compounds, owe their colour to electronic transitions in partly-filled molecular orbitals (see Table VII.2).

Transition metals also act as dopants in host crystals: rubies and emeralds are well known examples. A small quantity of dopant atoms alters the crystal lattice of an uncoloured host and gives rise to a very distinct colour.

*Table VII.2: Coloured transition metal compounds used as inorganic pigments in industry.*

| | |
|---|---|
| Chromium oxide | $Cr_2O_3$ |
| Haematite | $Fe_2O_3$ |
| Goethite | FeOOH |
| Malachite | $Cu_2(CO_3)(OH)_2$ |
| Azurite | $Cu_3(CO_3)_2(OH)_2$ |

*Table VII.3: Refractive index (n(l) and k(l)) and reflectance (R(l)) of some metals.*

| Compound | n | k | R [%] |
|---|---|---|---|
| Glass | 1.50 | - | 4.0 |
| Aluminium | 1.44 | 5.23 | 82.7 |
| Sodium | 0.044 | 2.42 | 97.5 |

Organic solids used as pigments are of huge economic and industrial importance. Many of the colours we encounter every day require the use of organic dyes and pigments. Absorption by organic solids can be described in terms of the molecular orbital theory and the influence of any chromophoric groups present.

Charge-transfer processes are responsible for the familiar colours of ultramarine and blue sapphire. Magnetite and lead chromate also owe their colour to this mechanism.

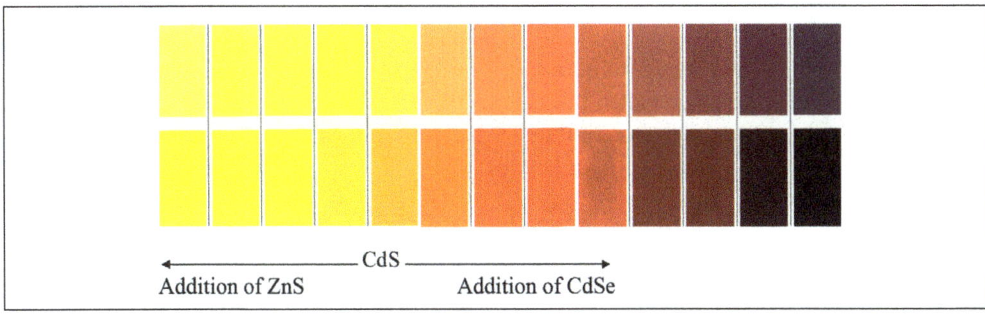


*Figure VII.2: Colour variation in cadmium sulphides.*

### *Metallic lustre*

Metals exhibit a characteristic reflectance, which is described as lustrous and is due to the relatively mobile electrons within the crystal lattice. The typical metallic colour of gold, silver and brass stems from the structure of their valence and conduction bands (see Figure VII.1)

The reflectance is described by *Fresnel* equations, which establish a link reflectance to refractive index, yielding producing a complex quantity that characterises both the absorption and the behaviour of the electromagnetic wave during phase transition.

*Equations VII.1 to VII.3*

$$R(\lambda) = 100 \frac{[n(\lambda) - 1]^2}{[n(\lambda) + 1]^2}$$

$$N(\lambda) = n(\lambda) + i \cdot k(\lambda)$$

$$R(\lambda) = 100 \frac{[n(\lambda) - 1]^2 + k(\lambda)^2}{[n(\lambda) + 1]^2 + k(\lambda)^2}$$

Table VII.3 shows some typical values of R(l), n(l) and k(l) for metals.

### *Coloured semiconductors*

Semiconductors are characterised by an energy gap between the conduction and valence bands. If the energy difference corresponds to the visible range, the semiconductor in question is coloured. Vermillion (HgS) and galena (PbS) are examples of coloured semiconductors that have served as pigments.

As the colours in this case due to transitions between energy bands, they are very pure and clear. Cadmium chalcogenides were created for this reason in the past. The exact position of the wavelength at which transition occurs can be varied by vaying the cations and anions.

For example, incorporating $Zn^{2+}$ ions into the crystal lattice shortens the distance between anion and cation, widens the band gap and shifts the transition to shorter wavelengths. Replacing sulphur by selenium has the opposite effect.

Additional variation can be obtained by doping, which is responsible for the typical colour of yellow and blue diamonds, which are semiconductors.

*Table VII.4: Coloured semiconductors and the corresponding band gap.*

| Compound | Mineral | Band gap [nm] | Colour |
|---|---|---|---|
| C | Diamond | 230 | Colourless |
| ZnO | Zinc oxide | 410 | Colourless |
| CdS | Cadmium yellow | 480 | Yellow |
| HgS | Cinnabar | 620 | Red |
| $Ce_2S_3$ | Cerium sulphide | 650 | Red |
| PbS | Galena | 3100 | Black |

### *Colour centres*

X-rays and ultraviolet radiation can induce colour in uncoloured media due to the presence of so-called “F centres” or defects in the crystal lattice. These centres have a tendency to recombine. When common salt is irradiated with x-rays for a day, it acquires an intense, albeit temporary, blue colour. Amethyst is a further, more exotic example of this effect, but is much more stable.

Besides these origins within the atomic or molecular structure, some colours are created by the interaction of light with the geometric structure of the material, and are the outcome of refraction, diffraction, interference and scattering.

## *2 Reflection, refraction, diffraction, interference*

*Uwe Hempelmann*

Various terms and definitions are used to describe what happens when light, and electromagnetic radiation in general, either is involved in a phase transition or interacts with matter.

In everyday life, light is often regarded as a beam or ray that propagates linearly. That is because the distances and dimensions involved are much larger than the wavelength of visible light (400 to 700 nm). The actual behaviour can be described by geometrical or ray optics which explains the reflection of light from a plane surface without the need to account for its electromagnetic nature.

Figure VII.3 illustrates the **reflection** of a light ray from a planar surface: the ray is reflected at the same angle with respect to the surface normal. If the surface is matte or textured, the ray will be reflected diffusely to a degree governed by the roughness of the surface. If the surface is an ideal matte, the light will be scattered uniformly in all directions (and is called a Lambertian scatterer; see Figure VII.4).

In many instances, only some of the light is reflected by the surface – the rest penetrates into the medium. The amount reflected depends on the angle of incidence and on the refractive index of the medium concerned (see below; Figure VII.7 illustrates reflection as a function of the angle of incidence $\theta$ for refractive index n = 1.5).

When light enters a medium of different density, i.e. air/water or air/pigmented coating, it undergoes **refraction**. The difference in optical density between the two media causes a change in the speed of light that varies with density and wavelength. The light ray or wave also changes direction.

Light passing from a less dense medium to a denser medium at an angle $\alpha$ from the normal is refracted towards the normal. As a result, the angle $\beta$, measured from the normal, will be smaller.

The speed of light varies with the optical density of the medium; the relationship between the speed of light in a given medium and in a vacuum is called the refractive index and is described by Snell's law:

*Equation VII.4 Snell's law:*
$$\frac{\sin \alpha}{\sin \beta} = \frac{n_\beta}{n_\alpha}$$

Naturally, this law also applies to light passing from a dense to a less dense medium. However, unlike the situation shown in Figure VII.5, a limiting case arises: when the angle of

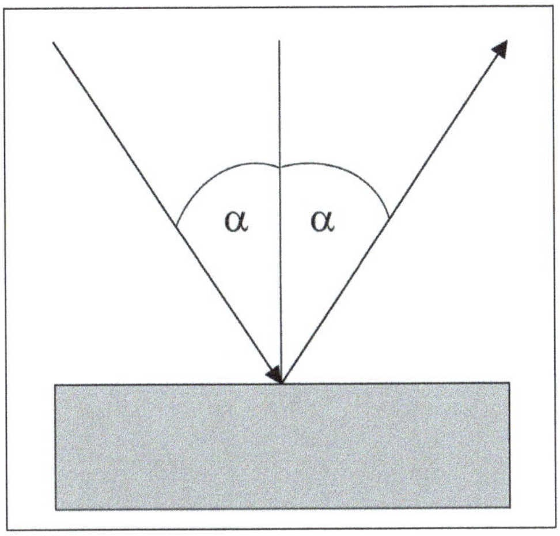


*Figure VII.3: Reflection of light from a plane surface.*

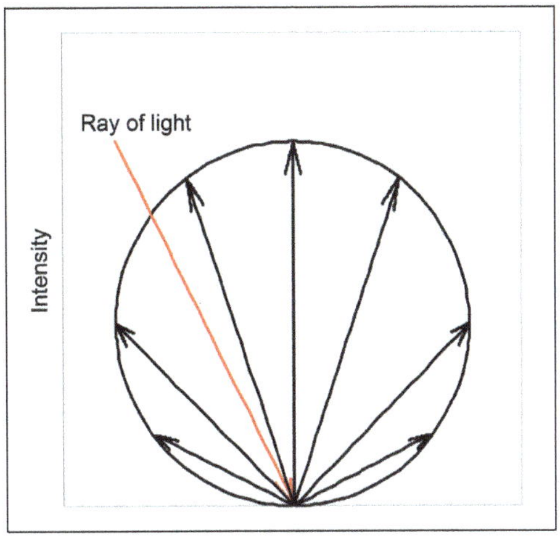


*Figure VII.4: Reflection of light from an ideal matte surface.*

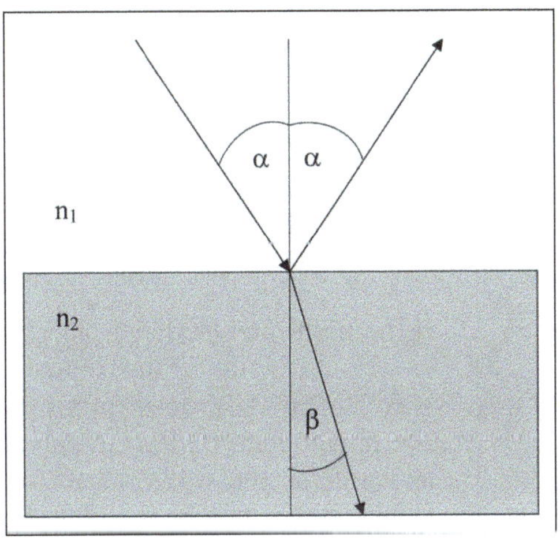


*Figure VII.5: Refraction of light by a planar surface; the refractive index of the first medium is smaller than that of the second ($n_2 > n_1$).*

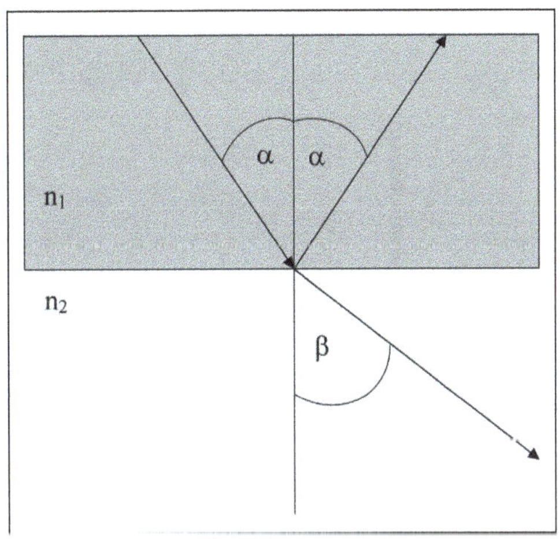


*Figure VII.6: Refraction of light at a planar surface; the second medium has a lower refractive index than the first ($n_2 < n_1$).*

the light in the less dense medium is 90° or greater relative to the normal, the light is unable to escape to the denser medium; this angle is called the critical angle.

At the critical angle, **total internal reflection occurs** and all of the light is reflected. Binocular prisms and optical fibres work on this principle. Assuming that the refractive index is approx. 1.5 for glass or an organic binder and approx. 1 for air, we can write the following Equation VII.5:

*Equation VII.5*
$$\frac{\sin\alpha}{\sin 90} = \frac{1}{1.5} \Rightarrow \alpha = \arcsin\,[0.67] = 41.8°$$

Light striking the phase boundary at this angle or higher will undergo total internal reflection (Figure VII.7).

*Fresnel* formulated laws describing the reflection of light moving between media of different refractive indices (see *K. D. Moller*, Literature in Section VII.8.1) In their full form,

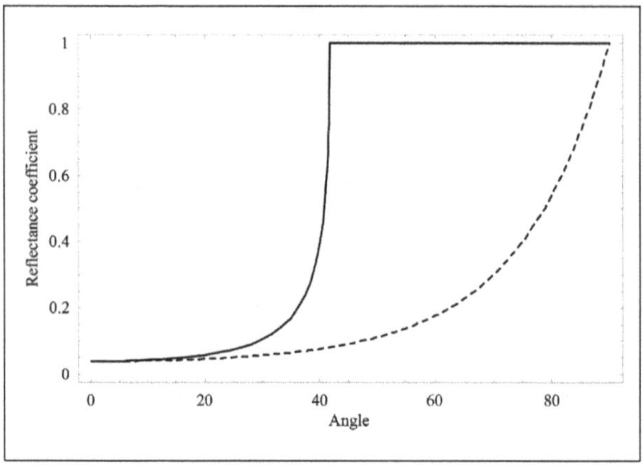


*Figure VII.7: Reflection as a function of angle of incidence: $r_{int}$: refraction from a medium of high refractive index into one of low refractive index; $r_{ext}$: refraction from a medium of low refractive index into one of high refractive index.*

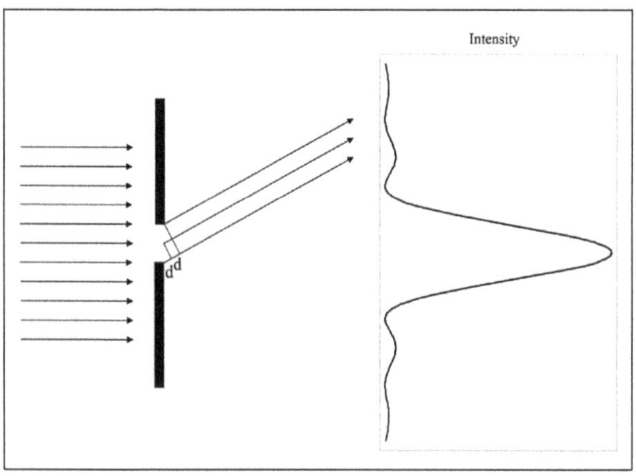


*Figure VII.8: Schematic representation of the diffraction of a light wave at a slit and the observed intensity distribution.*

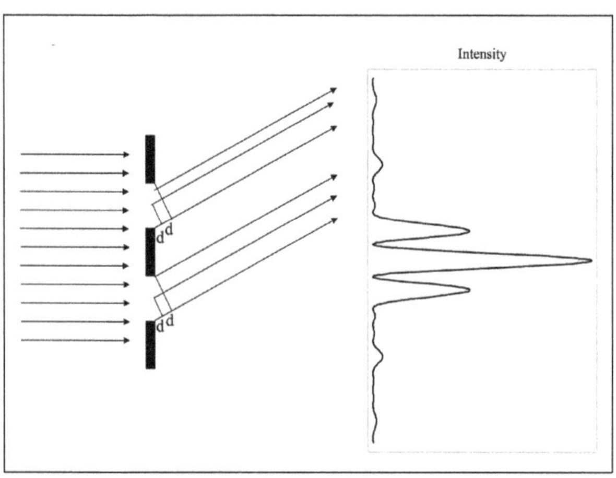


*Figure VII.9: Diffraction of light at a double slit, and intensity distribution in the far field.*

these laws account for the degree of polarisation as well as the complex refractive index. Here, we will only be concerned with the simplified forms for **transmission** and **reflectance** under normal incidence.

*Equations VII.6 and VII.7*

$$R = \left(\frac{n_\alpha - n_\beta}{n_\alpha + n_\beta}\right)^2$$

$$T = 1 - R = \frac{4\, n_\alpha\, n_\beta}{(n_\alpha + n_\beta)^2}$$

Light striking a phase boundary of refractive index of 1.5 yields the following values for reflectance and transmission:

$$R = \left(\frac{1 - 1.5}{1 + 1.5}\right)^2 = 0.04$$

$$T = 1 - R = \frac{4 \cdot 1 \cdot 1.5}{(1 + 1.5)^2} = 0.96$$

If the light interacts with objects or structures similar in size to its wavelength, any description of the phenomena that occur must account for the wave nature of light. These phenomena include diffraction and interference.

The simplest example of the diffraction of light is the effects that occur at a narrow slit. These are explained by the *Huygens-Fresnel* principle, which states that every point on the slit can be regarded as a source of a new spherical wave.

Figure VII.7 shows that, to an extent depending on the direction of propagation, the optical path difference d causes reinforcement and cancelling; the intensity observed behind the slit is therefore angle-dependent. On a screen placed behind the slit at a distance greater than the width of the slit, a pattern of stripes is observed. The spacing of the stripes is determined by the width of the slit and the wavelength of the light.

Positive overlapping occurs at angle a if the optical path difference d is a whole multiple of the wavelength l (n = 1, 2, 3, ...)

*Equation VII.8* $n\lambda = d \sin \alpha_n$

This behaviour can be extended to two or more slits. In this case, the pattern of stripes becomes more complex, as it varies with the wavelength and the widths, and the spacing of the slits (Figure VII.8 and Figure VII.9).

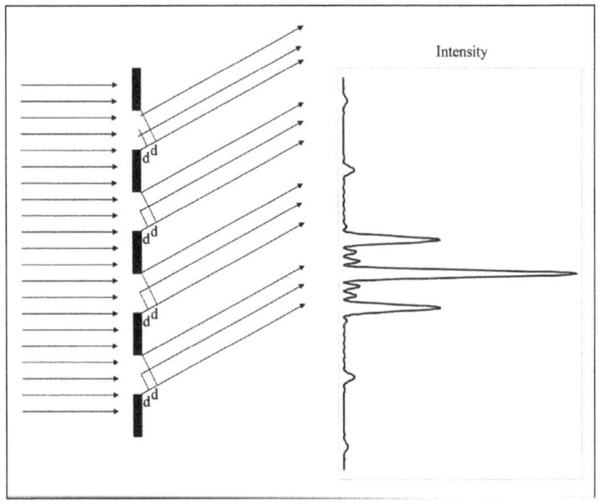


*Figure VII.10: Diffraction of light at a simple grating, and intensity distribution in the far field.*

A large number of slits creates a grating in which fine lines are positioned close to each other on a carrier. Optical elements of this kind are crucial components (monochromators) of colorimeters.

Although the physical effects are the same in principle, this behaviour is usually described as **interference** instead of diffraction. It is easy to demonstrate with the aid of a laser pointer and a polaroid plastic film found in sun-glasses. If the laser beam passes through the film, a typical pattern of points will be observed in the far field as a result of diffraction, i.e. interference.

Together, the physical effects just described form the basis of the colours often displayed by butterflies and birds. In industry, they are used to create interference pigments.

# 3 *Mie theory*

*Uwe Hempelmann*

**Scattering** of electromagnetic waves by particles is one of the most important processes in physics and the underlying theory can be used to describe a great many phenomena. *Mie* published his theory on the absorption and scattering of an electromagnetic wave by a sphere in 1908.

It is sometimes called the **Lorenz-Debye-Mie theory**, because major preliminary work had already been done by *Lorenz* and *Debye*. The equations serve as a good approximation for characterising the absorption and scattering of pigment particles within a coating or a plastic.

Mie theory describes the absorption and scattering of a planar electromagnetic wave by a single spherical particle and accounts for scattering of the incident wavelength. It does not account for **Raman** scattering (inelastic scattering of light by atoms and molecules) and similar effects. Mie theory provides solutions for the *Maxwell* equations for surface waves on a particle. It has since been extended to cylindrical particles. Meanwhile, methods have been developed that describe the **absorption** and **scattering** of arbitrarily shaped particles; unfortunately, these entail much more complex calculations.

Our brief overview of Mie theory starts with a summary of the input parameters. The results of a Mie calculation are presented as a function of these parameters, which regrettably take some getting used to, but are necessary for showing the results in a presentable form.

The first input parameter is particle size, which determines the extent of Mie scattering; for our purposes, the particle radius will be denoted by a. In later equations describing absorption and scattering, the particle radius will be denoted by α, a parameter encompassing the refractive index of the medium and the wavelength of the incident radiation.

*Equation VII.9*

$$\alpha = \frac{2\,\pi\,m_{Medium}}{\lambda}\,a$$

This equation demonstrates that scattering of light does not depend on the wavelength or on the particle size, but rather that it is the ratio of these two parameters which determines the extent of the interaction.

The second input parameter is the relative complex refractive index. For a given wavelength, the real component describes the refraction while the imaginary component describes the absorption. Unlike the case for transparent liquids, it is not possible in the case of pigments to neglect the imaginary component.

*Equation VII.10*

$$m = \frac{m_{Particle}}{m_{Medium}} = \frac{n_{Particle} - in\,\kappa_{Particlel}}{n_{Medium} - in\,\kappa_{Medium}}$$

Again, it is not the absolute value which is important, but rather the ratio of the refractive indices of the medium and the particle. The composite parameter β will be introduced later as a simplification.

As mentioned above, the complex refractive index characterises not only the behaviour during phase transition, but also the absorption (through the imaginary component). The imaginary component κ can be transformed by the factor 4 π/λ into the absorption a', where λ is the wavelength of the incident light. If the medium in question is homogeneous, the expression can be written with z as the optical path length: $I = I_0 \exp[-a' z]$, which is the familiar Lambert-Beer law (s. Equations VII.11 to VII 14) . However, it must be remembered that we are dealing here with particles that scatter light whereas the Lambert-Beer law is concerned with absorption.

*Equations VII.11 to VII.14*

$$m = n\,(1 - i\,\kappa)$$

$$a' = \frac{4\,\pi}{\lambda}\,\kappa$$

$$I = I_0 \exp[-a'z]$$

$$\ln\frac{I}{I_0} = -a'z \quad \text{vs. Lambert-Beer law: } \log\frac{I}{I_0} = -\varepsilon\,c\,l$$

When reading text books or publications on Mie theory, it is essential to look at the parameter definitions. There is, for example, an alternative definition for the complex refractive index, which simply states: m = n + i k. In this case, all equations need to be corrected by the factor -n.

The equations in **Mie theory** provides solutions to the Maxwell equations under certain conditions. These solutions require the use of *Ricatti-Bessel* functions, which are modified Bessel functions; this is also true of the *Neumann* and *Hankel* functions.

*Equations VII.15 to VII.17*

$$\psi_n(Z) = \left(\frac{\pi Z}{2}\right)^{\frac{1}{2}} J_{\left(n+\frac{1}{2}\right)}(Z)$$

$$\xi_n(Z) = -\left(\frac{\pi Z}{2}\right)^{\frac{1}{2}} N_{\left(n+\frac{1}{2}\right)}(Z)$$

$$\zeta_n(Z) = -\left(\frac{\pi Z}{2}\right)^{\frac{1}{2}} H^{(2)}{}_{\left(n+\frac{1}{2}\right)}(Z)$$

This is the core of the formalism for calculating the scattering and absorption cross-sections. The equations for $a_n$ and $b_n$

*Equations VII.18 and VII.19*

$$a_n = \frac{\psi_{n'}(\beta)\,\psi_n(a) - m\,\psi_n(\beta)\,\psi_{n'}(\alpha)}{\psi_{n'}(\beta)\,\zeta_n(a) - m\,\psi_n(\beta)\,\zeta_{n'}(\alpha)}$$

$$b_n = \frac{m\,\psi_{n'}(\beta)\,\psi_n(a) - m\,\psi_n(\beta)\,\psi_{n'}(\alpha)}{m\,\psi_{n'}(\beta)\,\zeta_n(a) - m\,\psi_n(\beta)\,\zeta_{n'}(\alpha)}$$

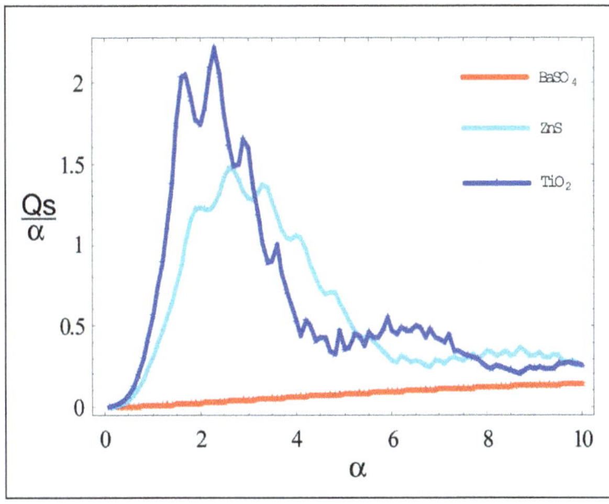


*Figure VII.11: Results for the application of Mie theory to scattering y white pigments.*

form the summands which are needed for calculating the scattering and absorption cross-sections. The poor convergence of these expressions is the reason why Mie theory only gained practical importance with the advent of computers, even though the necessary theoretical groundwork was done back in 1908.

*Equations VII.20 to VII.22*

$$Q_{ext} = Q_{sca} + Q_{abs}$$

$$Q_{sca} = \left(\frac{2}{\alpha^2}\right)\sum_{n=1}^{\infty}(2n+1)\,\{|a_n|^2 + |b_n|^2\}$$

$$Q_{ext} = \left(\frac{2}{\alpha^2}\right)\sum_{n=1}^{\infty}(2n+1)\,\{\mathrm{Re}\,(a_n + b_n)\}$$

The term **extinction** here is understood to be the sum of the effects of absorption and scattering. The respective effect is determined by the absorption and the scattering, along with the type of observation.

The results of some sample calculations for Mie theory will now be discussed. The equations given above are intended to provide the reader with the necessary insight into the parameters employed and the background to the calculations.

Figure VII.11 is a plot of the scattering cross-sections of white pigments. Absorption cross-sections and their contribution to the **total extinction** can be neglected in this case.

The quotient $Q_s/\alpha$, which normalizes the scattering cross-section to a certain amount of substance, is plotted as a function of $\alpha$, which, as already mentioned, is proportional to the particle size at a constant wavelength. At a wavelength of 470 nm and a refractive index of 1.5 in the medium, the expression $\alpha/10$ is equal to the particle radius in µm.

The plot demonstrates why titanium dioxide yields paints of superior hiding power; all other substances perform far worse under comparable conditions. The values for its scattering

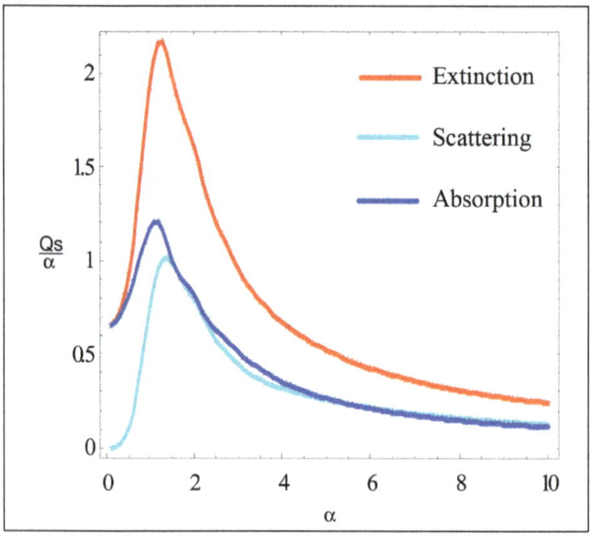


*Figure VII.12: Results of Mie calculations for coloured pigments.*

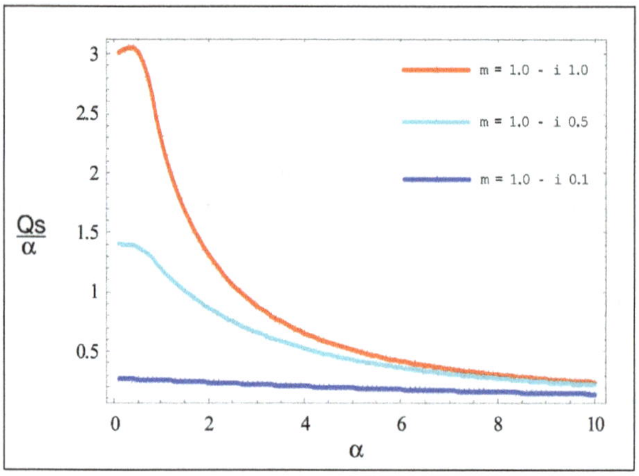


*Figure VII.13: Results of Mie calculations for particles similar to carbon black.*

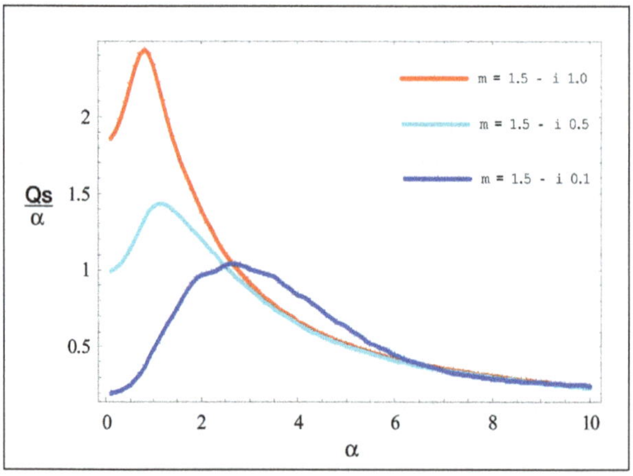


*Figure VII.14: Results of Mie calculations for organic pigments.*

cross-sections are much higher than those for other substances, such as zinc sulfide and barium sulphate.

The pronounced maximum exhibited by scattering cross-sections underscores why particle size is such an important parameter in pigment production.

The practical consequences of Mie theory can be seen for barium sulphate, zinc sulfide and titanium dioxide powders, all three of which are white. When these are incorporated into a coating in the same concentration and applied in the same thickness to white and black substrates, barium sulphate is found to have no **hiding power**, whereas titanium dioxide obliterates all signs of substrate texture.

This clear contrast between the substances' appearance as a powder and their behaviour in coatings can be explained by the fact that the relative refractive index is used in the calculations. When viewed in ambient air, barium sulphate has a sufficiently high refractive index to provide reasonable scattering intensity. In a coating, this apparently applies only to titanium dioxide.

Figures VII.12 to VII.14 show the typical behaviour of inorganic and organic pigments.

For coloured pigments (Figure VII.12), the extinction is the sum of the absorption and the scattering cross-sections. Both values are plotted: the blue line shows the absorption cross-section, the black line shows the scattering cross-section, while the red curve represents the extinction, i.e. the sum of both.

Unlike the curves shown earlier for white pigments, the cross-sections for coloured pigments are considerably smoother. The curves for titanium dioxide and zinc oxide are more jagged due to interference effects between scattered and incident light. If the absorption component is too high in the case of coloured pigments, this effect disappears because not enough light is scattered.

The curves exhibit a clear maximum at the point of optimum interaction with light. With decreasing diameter the particles do not scatter light any more and so become transparent. Absorption remains constant, however, because the light is absorbed by the whole particle, i.e. the interior and the exterior.

Once the particle size exceeds the optimum value, the absorption as well as the scattering cross sectionecline roughly as a function of $1/\alpha$ i.e. inversely proportional to the diameter, and thus exhibits a dependence on the surface area.

### *Absorption by carbon black*

Figure VII.13 show the results of calculations, which combine a small real component with absorption coefficients of different magnitudes. As expected, if the absorption coefficient is small, significant interaction with light does not occur. As the absorption coefficient increases, the extinction increases almost linearly. This linear dependence of the coefficient on the reciprocal value of the particle size is found along the entire curve. This is the type of behaviour exhibited by carbon black in practice.

### *Characteristics of organic and inorganic pigments*

When the values for the real and imaginary components of the

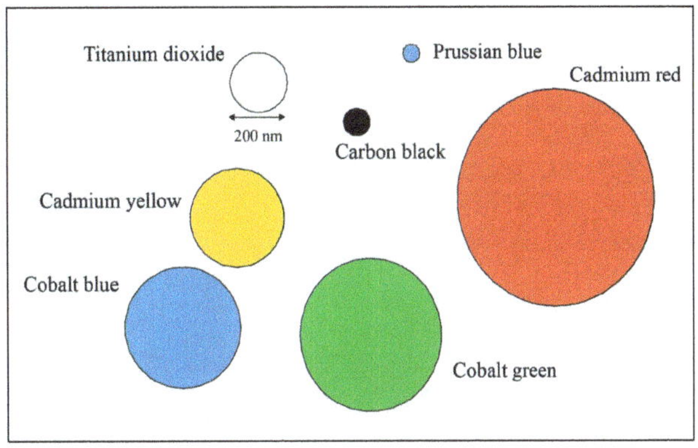


*Figure VII.15: The different particle sizes of inorganic pigments.*

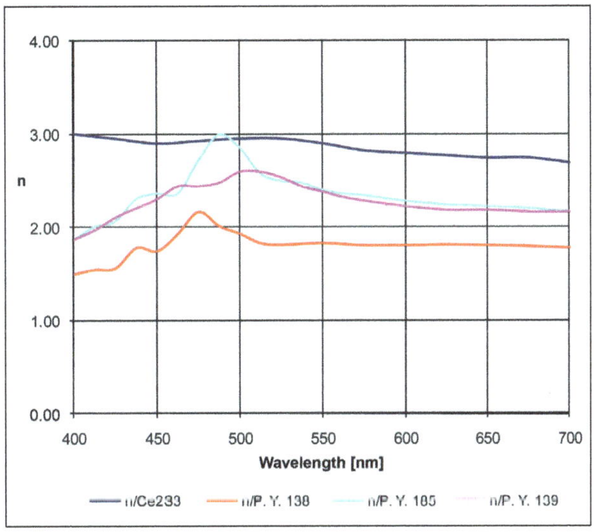


*Figure VII.16: Comparison of the real component of the refractive index of organic and inorganic pigments.*

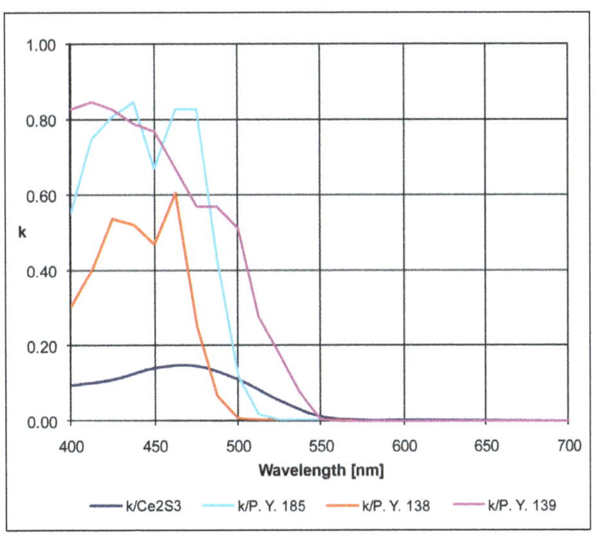


*Figure VII.17: Comparison of the imaginary component of the refractive index of organic and inorganic pigments.*

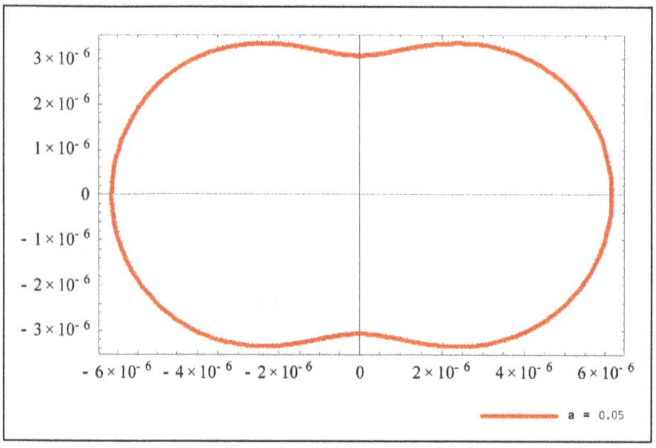


*Figure VII.18: Angular distribution of the intensity of scattered light; example λ = 580 nm, $m_{Particle}$ = 2.81, κ = 0.12 i, r = 5 nm.*

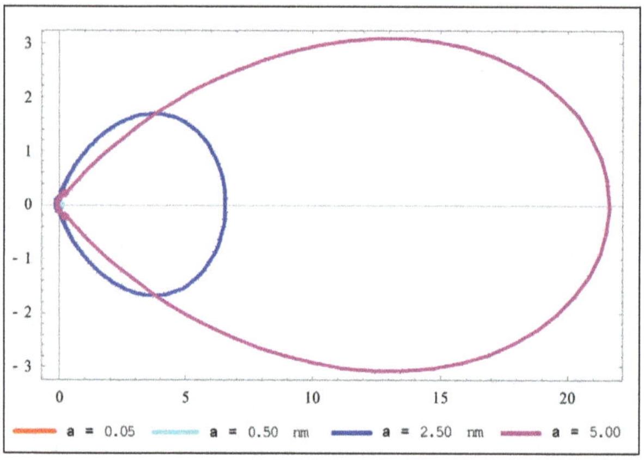


*Figure VII.19: Angular distribution of the intensity of scattered light; example λ = 580 nm, $m_{Particle}$ = 2.81, κ = 0.12 i; r = 5, 50, 250 and 500 nm.*

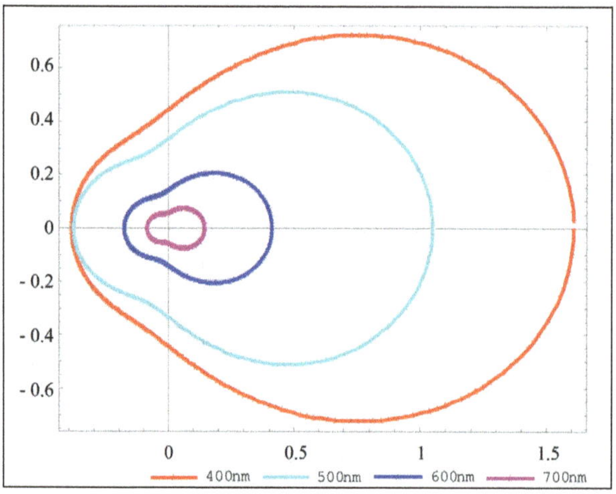


*Figure VII.20: Angular distribution of the intensity of the scattered light of haematite particles with d = 0.34 μm - wavelength between 400 und 700 nm.*

complex refractive index are increased to the level of those found in many organic pigments, a pronounced maximum for the extinction cross-section occurs. The effect of neglecting the absorption cross-section can be seen in the results for the smallest values for the imaginary component.

Figure VII.15 illustrates the relative sizes of the particles predominant in different types of pigments. It shows the different index of refraction of the substance in question leads to quite a range of optmal particle sizes.

Figures VII.16 and VII.17 show the real and imaginary components of the refractive index of some organic and inorganic pigments. The latter typically have a smaller real component than inorganic pigments. As the real component is quite often proportional to the density, this is to be expected.

In contrast, organic pigments have much higher values for the imaginary component. This behaviour is familiar to anyone who works with organic pigments every day: they have far less scattering power than their inorganic counterparts, a fact that also explains their low hiding power. However, organic pigments produce deeper colours due to their higher absorption.

So far, we have neglected the angular distribution of the scattered light in our discussion. For particles smaller than the wavelength, this angular distribution is identical with that for Rayleigh scattering. Mie theory therefore may be viewed as solution for light scattering by particles in general, while Rayleigh scattering can be considered a special case for small particles.

Figure VII.18 is a plan view of a particle located at the intersection of the x and y axes. The light wave strikes the particle from the left. The intensity distribution is familiar from the results for Rayleigh scattering.

***Angular distribution of the intensity of the scattered light***

To facilitate comparison between the values for different particle sizes, all the values are plotted in one diagram (Figure VII.19). As may be seen, the values there increase strongly, i.e. Rayleigh scattering is much less intense than Mie scattering.

As explained earlier, the parameter $\alpha$ includes the wavelength in addition to the particle size: Figure VII.20 shows, in contrast to Figure VII.19, a particle of constant size and varying wavelengths. This illustrates the point that there are natural limits to optimum particle size.

# 4 *Kubelka-Munk function for opaque layers*

*Uwe Hempelmann*

The Kubelka-Munk function, which will be derived and analysed in this section, is a highly simplified, specific instance of the equation of radiative transfer (see Section VII.7). It is also sometimes called the **phenomenological theory**. In contrast to Mie theory, its focus is not on a single particle, but rather on the collective characteristics of an entire layer.

It will become apparent in the following that some quite far-reaching assumptions are made in the derivation of the Kubelka-Munk function. The soundness of some of these assumptions can certainly be debated, but the results are generally in quite close agreement with experimental evidence. The Kubelka-Munk function forms the basis of the methods for determining the hiding power and tinting strength described in Section VII.5.

The Kubelka-Munk function starts by considering a plane-parallel layer of thickness d. For the derivation, it is assumed that the layer is of infinite size, and so edge effects can be neglected. It is further assumed that only diffuse light is involved, and not directional light.

The basic model proceeds from a continuous layer: the change in intensity of light travelling downwards in direction I is analysed, which by convention corresponds to the light falling on a layer from above. The second flux of light is in the upward direction J, and represents that light which is scattered backwards.

Often, terms such as (radiant) flux, or irradiance, are used instead of intensity. However, the results are always the same. Lest readers be confused about the nomenclature, they should bear in

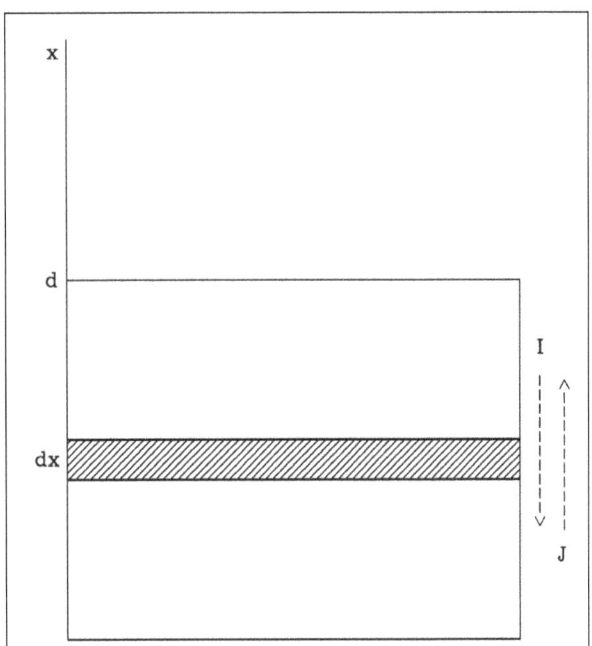


*Figure VII.21: The model for deriving the Kubelka-Munk function.*

mind that "radiometry" is often described as an acronym for "**r**evulsive, **a**rchaic, **d**iabolical, **i**nvidious, **o**dious, **m**ystifying, **e**xotic, **t**erminology **r**egenerating **y**awns".

The derivation starts by examining the changes in the upward and downward fluxes: dI/dx and dJ/dx, which yields:

*Equations VII.23 and VII.24*

$$\frac{dI}{dx} = -(K + S)\,I + SJ$$

$$-\frac{dJ}{dx} = -(K + S)\,J + SI$$

The **downward flux** I is diminished by absorption and scattering. This is expressed by the absorption coefficient K and scattering coefficient S, which are multiplied in the equation with I. As light is scattered, the upward flux J increases.

The differential dJ/dx in Equation VII.24 has a negative algebraic sign, as it moves in the opposite direction in the chosen coordinate system. The algebraic signs of the other equations follow suit.

These two differential equations can be solved for different boundary conditions. The general solution is given in Section VII.6. Of interest for practical applications is the solution for layers of optically infinite thickness, i.e. opaque or hiding.

This solution is quite simple, as demonstrated by the following derivation. Reflectance can be formulated as:

*Equation VII.25* $$R = \frac{J}{I}$$

Optically infinite thickness means that any additional thickness of the layer does not change the reflectance; this leads to the following equation:

*Equation VII.26* $$R_\infty = \frac{dR}{dx} = 0$$

The differential Equations VII.25 and VII.26 above can now be solved for these boundary conditions as follows:

*Equations VII.27 to VII.31*

$$\frac{dR}{dx} = \frac{\frac{dJ}{dx}I - J\frac{dI}{dx}}{I^2}$$

$$= \frac{(K + S)\,J\,I - S\,I^2 + (K + S)\,I\,J - S\,J^2}{I^2}$$

$$= (K + S)\,\frac{J}{I} - S + (K + S)\,\frac{J}{I} - S\left(\frac{J}{I}\right)^2$$

$$= (K + S)\,R - S + (K + S)\,R - S\,R^2 \quad = -S\left[R^2 - 2\left(\frac{K}{S} + 1\right)R + 1\right]$$

$$= R^2 - 2\left(\frac{K}{S} + 1\right)R + 1$$

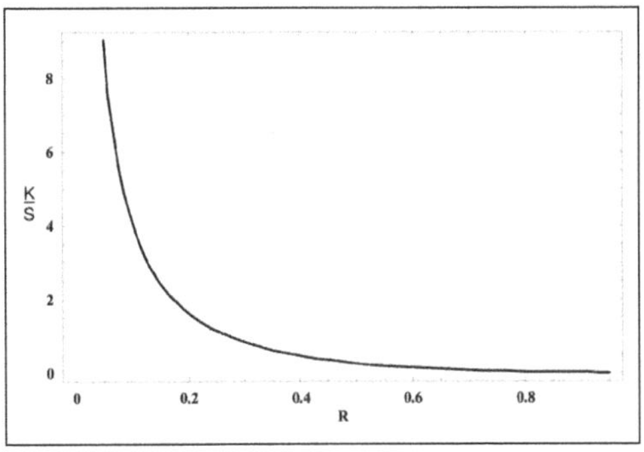


Figure VII.22: Kubelka-Munk function versus reflectance.

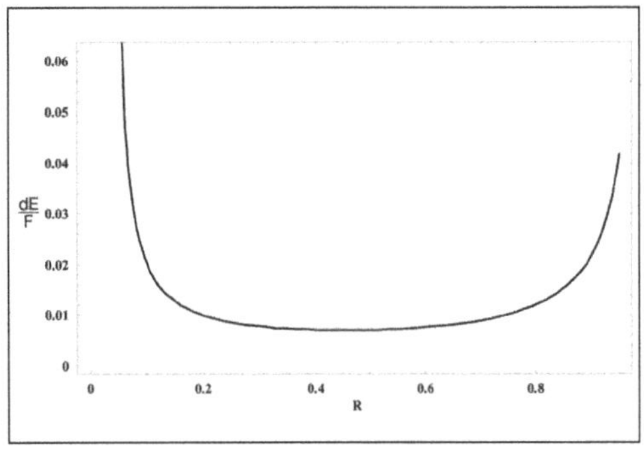


Figure VII.23: Error in the Kubelka-Munk function.

Or, to use a more common notation in which $R_\infty$ represents the reflectance of an opaque layer:

*Equation VII.32* $$\frac{K}{S} = \frac{(1 - R_\infty)^2}{2R_\infty} = F(R_\infty)$$ The inverse is:

*Equation VII.33* $$R_\infty = 1 + \frac{K}{S} - \sqrt{2\left(\frac{K}{S}\right) + \left(\frac{K}{S}\right)^2}$$

***Absorption and scattering coefficient as a function of reflectance***

This, the Kubelka-Munk function, is probably one of the most widely used equations in colorimetric applications. It is an integral component of many methods for calculating tinting strength as well as recipes. The results of **a single** reflectance determination can be used to establish the **ratio of absorption to scattering coefficient**. Figure VII.22 shows a plot of the Kubelka-Munk function versus reflectance.

The Kubelka-Munk function as described above forms the starting point for many applications. It is therefore important to establish the accuracy of its values. If we assume an error of 0.001 for the reflectance measurement, we obtain the results in Figure VII.23.

In practice, reflectance should be kept between 0.2 and 0.8 for the sake of maximum accuracy. Very light and very dark specimens should therefore not be tested, wherever possible.

# 5 Saunderson correction: how surfaces influence the outcome of reflectance measurements

*Uwe Hempelmann*

Surfaces exert a huge influence on the outcome of reflectance measurements and so merit as much attention as that given to the measuring geometry of colorimeters. The most common approach in practice is to apply the **Saunderson correction**.

In the first step, we will derive the equation for light falling perpendicularly onto a specimen with diffuse observation. This case will then be generalized and modified for other geometries.

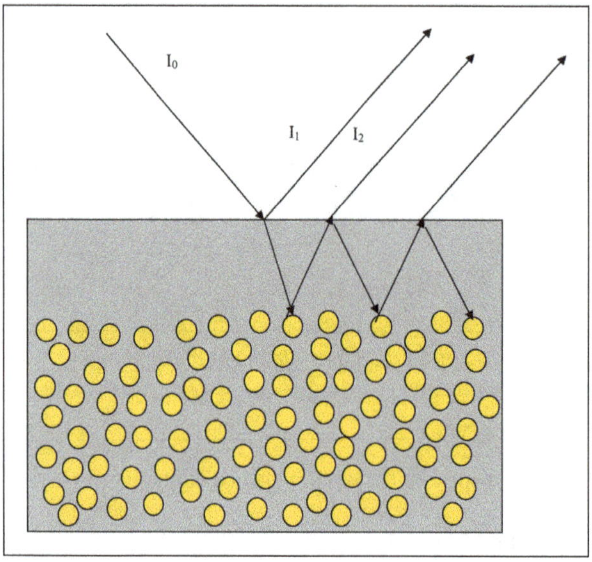


*Figure VII.24: Model of the Saunderson correction.*

The following assumptions are made: A sample contains particles (i.e. pigments) which scatter and absorb light. The refractive index of the medium is n, with the tacit assumption that it is larger than one. All derivations pertain to the air/medium interface.

In the following, the layer is assumed to be glossy and of infinite depth, i.e. opaque. Furthermore, the light scattered within the layer is taken to be ideally diffuse, leading to a **Lambert cosine** distribution at the interface where the measured intensities are independent of the angle of incidence.

The observed reflectance can be thought of as the result of several processes, which are summed up in the eye of an observer or in the detector of an instrument (see Figure VII.24; note: the light rays are drawn at an angle of 45° although the incident light is considered to be perpendicular; this is done to prevent cluttering the diagram).

If we follow the path of the light falling on the pigmented layer, we first need to correct for the light, which is directly reflected by the interface. The unreflected component is refracted into the coating, where it can interact with particles in the layer.

These particles will either absorb or scatter the light. The scattered light, assumed to be ideally diffuse, returns to the interface, where again there are two possibilities: either be reflected back into the layer, or exit the layer, where it can be seen or recorded by an instrument.

The fraction of the light which is reflected back inside the layer can again be either absorbed or scattered, triggering a third process of the same type. Thus, the detectable reflected light can be thought of as the sum of an infinite number of such processes.

### *Calculation of reflectance*

The process just outlined is easily formulated in mathematical terms, for which the following constants are needed. The reflection coefficient at the surface for light incident perpendicularly from the outside is denoted by $r_0$. For subsequent derivations, we need $r_1$ (a reflection coefficient for light incident diffusely from outside) and $r_2$ (a reflection coefficient for light incident on the surface diffusely from inside the layer). The reflection coefficients can be calculated by solving the **Fresnel equations** (see Section 8 Literature: Moller, Kortüm) and may be found in optics textbooks. For coatings (n = 1.5), the values are: $r_0 = 0.0040$, $r_1 = 0.092$, $r_2 = 0.596$.

The reflectance can be thought as consisting of $I_m$, where m = 1, 2, ... constituents, the sum of which is represented by $I_r$.

*Equation VII.34* $$I_r = \sum_m I_m$$

*Equation VII.35* $\vartheta = \frac{I_r}{I_0}$

Reflectance J in this model concept is the quotient of the sum of the reflectance arising from the various processes mentioned above and the intensity of the incident light. The first contribution comes from the light, which is directly reflected from the surface of the specimen:

*Equation VII.36* $I_1 = I_0 \, r_0$

For further assumptions, we use R to denote the reflectance of the absorbing and scattering particle, i.e. without the influence of a surface or boundary. The term $I_2$ is obtained by multiplying three factors:

$I_0 \, (1 - r_0)$: Intensity of the light entering the layer
R: Reflectance of the particles
$(1 - r_2)$: Fraction of the light not reflected back into the layer by the interface.

This leads to the following Equation VII.37:

*Equation VII.37* $I_2 = I_0 \, (1 - r_0) \, R \, (1 - r_2)$

Similarly, $I_3$ is obtained as follows:

$I_0 \, (1 - r_0)$: Intensity of the light entering the layer
R: Reflectance of the particles
$r_2 \, R$: Reflectance R of the particles multiplied by $r_2$ for the fraction of light reflected back into the interior of the layer.
$(1 - r_2)$: Fraction of the light not reflected back into the layer by the interface

*Equation VII.38* $I_3 = I_0 \, (1 - r_0) \, R \, r_2 \, R \, (1 - r_2)$

This leads to $I_4$:

$I_0 \, (1 - r_0)$: Intensity of the light entering the layer
R: Reflectance of the particles
$(r_2 \, R)^2$: Reflectance R of the particles multiplied by $r_2$ for the fraction of light reflected back into the interior of the layer; squared because it occurs twice.
$(1 - r_2)$: Fraction of the light not reflected back into the layer by the interface.

*Equation VII.39* $I_4 = I_0 \, (1 - r_0) \, R \, (r_2 \, R)^2 \, (1 - r_2)$

When Equations VII.36 to VII.39 are written beneath each other, a clear pattern emerges:

$$I_1 = I_0 \, r_0$$
$$I_2 = I_0 \, (1 - r_0) \, R \, (1 - r_2)$$
$$I_3 = I_0 \, (1 - r_0) \, R \, r_2 \, R \, (1 - r_2)$$
$$I_4 = I_0 \, (1 - r_0) \, R \, (r_2 \, R)^2 \, (1 - r_2)$$

The reflectance J is calculated by dividing the sum of the terms by $I_0$. Mathematically, the resulting sequence is an infinite geometric series. It can be presented in a closed form as:

*Equations VII.40 to VII.42*

$$\vartheta = \frac{I_r}{I_0}$$

$$\begin{aligned}&= r_0 + (1 - r_0)(1 - r_2) R \\ &\quad + (1 - r_0)(1 - r_2) R\, r_2 R \\ &\quad + (1 - r_0)(1 - r_2) R\, (r_2 R)^2 \\ &\quad + \ldots \\ &= r_0 + \frac{(1 - r_0)(1 - r_2) R}{1 - r_2 R}\end{aligned}$$

This equation enables surface effects to be separated out from particle effects in a reflectance measurement, as long as the assumptions made for the specimen are valid. For colorimetry, the inverse is needed quite often as well.

*Equations VII.43 to VII.45*

$$r_0 + R(1 - r_2 - r_0) = \vartheta - r_2 R\, \vartheta$$

$$R(1 - r_2 - r_0 + r_2 \vartheta) = \vartheta - r_0$$

$$R = \frac{\vartheta - r_0}{1 - r_0 - r_2(1 - \vartheta)}$$

We shall now derive the "inverse" measuring geometry d : 0°, i.e. diffuse illumination and observation at an angle of 0°. Although the basic premises remain the same, some assumptions first have to be made about the design and size of the detector.

In the following derivation, we assume that the detector captures all the light reflected from the sample striking at an angle j, which is deemed to be small relative to the surface normal.

This means that the detector captures the light reflected back at the angle W:

*Equation VII.46* $$\Omega = \pi \sin^2 \varphi$$

For an ideal white standard, this yields the following equation:

*Equation VII.47* $$I_W = \frac{I_0}{\pi} \Omega = I_0 \sin^2 \varphi$$

For a sample with a reflectance ρ, the equation is:

*Equation VII.48* $$I_p = \rho\, I_0 \sin^2 \varphi$$

We now follow the same procedure as before (for the 0°:d geometry) to derive the Saunderson correction for the d:0° geometry. Here, we track the path of the light in the case of an ideal glossy surface (opaque with ideally diffuse scattering particles inside).

The light striking the sample diffusely first gives rise to reflection $r_1\, I_0$ on the surface. Of this light, the detector can only capture the following fraction at the angle j:

*Equation VII.49* $$I_1 = I_0\, a\, r_0 \sin^2 \varphi$$

The magnitude of factor a depends on the geometry of the measurement instrument. Light which is not reflected is refracted inside, where it is reflected. Only the fraction reflected at the angle W' can be detected.

*Equation VII.50* $$I'_2 = I_0 (1 - r_1) R \frac{\Omega'}{\pi} = I_0 (1 - r_1) R \sin^2 \varphi'$$

The relation between the angle j' in the sample and j is:

*Equation VII.51* $$\sin \varphi' = \frac{\sin \varphi}{n}$$ Yielding:

*Equation VII.52*

$$I_2 = I_0\,(1 - r_1)\,R\,\frac{\sin^2\varphi}{n^2}\,(1 - r_0)$$

By applying the derivation as before and continuing to track the light through the sample, we can calculate the following fractions of reflected light:

*Equations VII.53 and VII.54*

$$I_3 = I_0\,(1 - r_1)\,R\,r_2\,R\,\frac{\sin^2\varphi}{n^2}\,(1 - r_0)$$

$$I_4 = I_0\,(1 - r_1)\,R\,(r_2\,R)^2\,\frac{\sin^2\varphi}{n^2}\,(1 - r_0)$$

The reflectance of the sample is calculated by dividing the sum by the contribution of an ideal white sample:

*Equations VII.55 to 57*

$$\begin{aligned}\beta &= \frac{I_p}{I_w}\\ &= a\,r_0 + (1 - r_0)\,(1 - r_1)\,\frac{1}{n^2}\,R\\ &\quad + (1 - r_0)\,(1 - r_1)\,R\,\frac{1}{n^2}\,r_2\,R\\ &\quad + (1 - r_0)\,(1 - r_1)\,R\,\frac{1}{n^2}\,(r_2\,R)^2\\ &\quad + \dots\\ &= a\,r_0 + \frac{(1\quad r_0)\,(1\quad r_1)\,\frac{1}{n^2}\,R}{1 - r_2\,R}\end{aligned}$$

Factor $a\,r_0$ is the mathematical analogue of the **gloss trap**, typically found in colorimeters that have an integrating sphere: $a = 1$ when the gloss trap is closed (all reflected light is measured) and $a = 0$ when the gloss trap is open (excluding the non-diffuse fraction of the reflected light).

The ***Helmholtz*** **reciprocity principle** in optics essentially states that results are not changed by swapping sample and detector, provided that certain conditions are observed, e.g. the sample and detector area must be the same size. Under these conditions, comparing the equations derived for 0°:d and d:0°, we get:

*Equations VII.58 and VII.59*

$$\frac{(1 - r_0)\,(1 - r_2)\,R}{1 - r_2\,R} = \frac{(1 - r_0)\,(1 - r_1)\,\frac{1}{n^2}\,R}{1 - r_2\,R}$$

$$(1 - r_2) = (1 - r_1)\,\frac{1}{n^2}$$

This means shows that the equation derived earlier can be used for the case d:0° – and is actually used more widely in practice.

For illumination at an angle less than 45° and measurement at less than 0°, we obtain the following equation:

*Equation VII.60*

$$\beta = \frac{(1 - r_0)(1 - r_0)\frac{}{n^2}R}{1 - r_2 R}$$

Factor $\bar{r}_0$ can be derived from the Fresnel equations.

# 6 *Kubelka-Munk equation for transparent layers*

*Uwe Hempelmann*

In this section, we will discuss the general solution to the differential equations of Kubelka and Munk, i.e. for transparent layers.

*Equations VII.61 and VII.62*

$$\frac{dI}{dx} = -(K + S)I + SJ$$

$$-\frac{dI}{dx} = -(K + S)J + SI$$

As deriving the solution of the differential Equations VII.62 and VII.63 is somewhat complicated, only the solution itself has been provided. For a detailed discussion, the reader is referred to the books of *Kortüm*, *Völz* and *Klein* cited in the references.

***Determination of reflectance against a white and a black background***

The solutions of the Kubelka-Munk equation are used for determining the absorption and scattering coefficients of opaque layers. Their use in practice is often hampered by the difficulty of finding a concentration at which the measured reflection exhibits significant differences against both a white and a black background.

For this reason, it is often expedient to combine a measurement over a black or white background with a measurement of an opaque layer.

For convenience, two auxiliary variables a and b will be defined. Measurement over black and white backgrounds ($R^*_{ow}$ and $R^*_{os}$ are the reflectance over black and white backgrounds; $^*$ indicates that the surface correction described in the previous section was applied) yields the following expressions for a and b:

*Equations VII.63 and 64*

$$a = \frac{(1 + R^*_w R^*_{ow})(R^*_s R^*_{os}) + (1 + R^*_s R^*_{os})(R^*_{ow} R^*_w)}{2(R^*_s R^*_{ow} - R^*_w R^*_{os})}$$

$$b = \sqrt{a^2 - 1}$$

$R^*_w$ and $R^*_s$ accordingly represent the reflectance over white and black white backgrounds, after surface correction has been applied. Substituting these results into Equation VII.66:

*Equations VII.65 and VII.66*

$$S = \frac{1}{bd}\,\mathrm{Arcoth}\,\frac{1 - a(R^*_w - R^*_{ow}) + R^*_w R^*_{ow}}{b(R^*_w - R^*_{ow})}$$

$$K = S(a - 1)$$

Instead of $R^*_w$ and $R^*_{ow}$, variables $R^*_s$ und $R^*_{os}$ can be used if they can be determined with less error.

The auxiliary variable a can also be determined by measuring an opaque layer:

*Equations VII.67 to VII.69*

$$a = \frac{1}{2}\left(\frac{1}{R^*_\infty} + R^*_\infty\right)$$

$$b = \frac{1}{2}\left(\frac{1}{R^*_\infty} - R^*_\infty\right)$$

$$S = \frac{1}{bd}\,\text{arcoth}\,\frac{1 - a\,(R^* - R^*_0) + R^*\,R^*_0}{b\,(R^* - R^*_0)}$$

$R^*$ describes the reflectance of the sample over an arbitrary background $r^*_0$. The question as to which of the possibilities described is the most favourable can only be decided on a case by case basis after errors in the respective measurements have been examined.

# 7 Multi-flux theory

*Wilhelm Kettler*

## 7.1 Criticism of the Kubelka-Munk model

Practical computer-aided recipe calculation is based on the theoretical model developed by *Kubelka* and *Munk*[1] (KM) in the 1930s. This theory replaces the scattering medium consisting of discrete scattering centres with a hypothetical continuous medium having the same scattering and absorption properties. This **continuum theory** can always be applied to discrete plane-parallel media if the number of scattering centres within the volume of interest is sufficiently high. In the mathematical treatment, only two diffuse radiation fluxes are considered: an incident flux penetrating into the medium, and a second flux escaping from the medium (Figure VII.25). In this case, scattering is defined as being the inversion of the propagation direction taken by a fraction of the radiation flux, while the process of absorption removes radiation from the considered physical system. The resulting relation between the reflectance, $R_\infty$, and the optical material parameters naturally does not exhibit any angular dependence. The metrological requirements governing the applicability of the KM-model are diffuse illumination and hemispherical observation. Such a measurement geometry cannot be realised in practice.

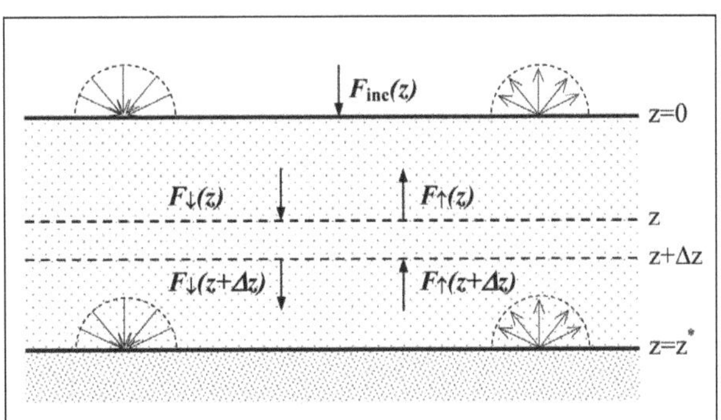


*Figure VII:25: Conservation of radiation energy, applied to a thin layer $\Delta z$ in a scattering and absorbing medium of infinite lateral extension, leads to the 2-flux equations of radiative transfer for both fluxes $F_\downarrow(z)$ and $F_\uparrow(z)$ in the positive (downwards) and negative (upwards) directions.*

In practice, the scattering particles are embedded in a matrix, whose refractive index differs from 1. Thus, measured reflectance values are additionally dependent on the refractive index of the embedding medium, as

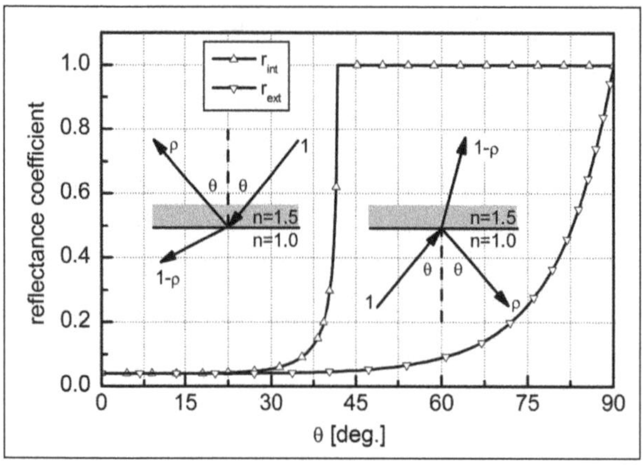


*Figure VII:26: Reflectance coefficient ρ of a matrix-air interface as a function of angle of incidence θ for internal (open triangles) and external (open rotated triangles) illumination. The refractive index of the matrix is taken to be n=1.5.*

regular reflection occurs at the phase boundary. Figure VII.26 illustrates internal and external Fresnel reflection at a matrix-air interface for a refractive index of $n = 1.5$. As can be seen from Figure VII.26, it is not unreasonable to neglect **external reflection** $r_{ext}$ (which is only about 4 %); by contrast, **internal reflection** $r_{int}$ assumes considerable values at moderate angles and, above the critical angle, gives rise to total internal reflection. It can therefore be expected that reflection at the matrix-air interface reduces the probability of light transmission through the boundary layer and that calculated absorption values without accounting for the correct boundary conditions will be systematically too low.

While it is relatively easy to determine the external reflection coefficient experimentally, the same cannot be said for the internal reflection coefficient. The reason is that, given weak scattering or thin layer thickness, radiation reflected at the inner boundary will not be uniformly diffuse. One corollary of this is that the reflection coefficient becomes a function of layer thickness.

The Kubelka-Munk theory does not deserve excessive criticism, even though *Schuster* [2], an English astronomer, essentially developed the same model 26 years earlier and his theory is nowadays labelled "crude" and "primitive" in astrophysics textbooks. A conceptual weakness of what might thus be called the Schuster-Kubelka-Munk model (SKM) is that radiation scattered into the same hemisphere into which it originally propagated is considered unscattered. This is incorrect and so, in practice, the scattering coefficient of the SKM model needs to be halved. Although this means that the SKM scattering coefficient cannot be predicted from the optical properties of a material, such as the complex refractive index and the particle shape and size, it does not lead to serious errors in the interpretation of experimental data.

Moreover, the SKM model computes the reflective properties of particulate media for a d/d measurement geometry which physically cannot be realised. It is therefore used to interpret experimentally determined reflectance spectra for **directional/hemispherical** and **bidirectional** measurement geometries.

The SKM model is generally less successful at predicting reflective properties where

- coatings are translucent,
- the coating has a very low reflectance, and
- the coating contains special effect pigments.

The first category includes yellow and red colour shades with reflective properties similar to those of optimum colours. Interestingly, all three categories fall into a transition zone in which the geometrical structure of the radiation field in the pigmented layer is not completely diffuse.

For all of the extensive theoretical and practical work on the **2-flux model,** there have been no substantial changes to the basic concept. Only *Preisendorfer*[3] has devised a way to derive a rigorous 2-flux solution from the radiative transfer equation and to unify the different variants of the approximation model.

## 7.2 Radiative transfer equation

The Schuster-Kubelka-Munk 2-flux model for radiative transfer in continuous media is readily amenable to mathematical treatment and physically transparent. It can be used to derive a physical understanding of the optical processes and all necessary insights. However, some applications require more precise results than this approach is capable of delivering. The Schuster-Kubelka-Munk 2-flux model makes no provision for calculating the full directionality of the radiation field, which dominates the optical properties of special effect coatings. This limitation arises on one hand from the restriction on the number of radiation fluxes to just two opposing directions and, on the other, to the assumption that the scattering processes are isotropic. Generalising the SKM model to $N > 2$ radiation fluxes would not solve this problem because, even in the limiting process leading to an infinite number of radiation fluxes, the outcome would only be a radiative transfer equation for isotropically scattering media.

A rigorous derivation of the radiative transfer equation is beyond the scope of this book. Interested readers will have to content themselves with a qualitative discussion of the necessary extension of the 2-flux concept. First, though, the 2-flux concept needs to be put on a more general footing.

Application of the law of conservation of radiation energy in a thin layer between $z$ and $\Delta z$ yields the difference equation [4]

*Equation VII.70* $$F_{\downarrow}(z+\Delta z) = F_{\downarrow}(z) - \kappa\,\Delta z\,F_{\downarrow}(z) - \sigma\,\Delta z\,p_{\downarrow\uparrow}\,F_{\downarrow}(z) + \sigma\,\Delta z\,p_{\uparrow\downarrow}\,F_{\uparrow}(z+\Delta z)$$

for the downwards radiation flux. The **radiation balance** at the lower boundary of the thin layer shows that the downwards radiation flux $F{\downarrow}(z+\Delta z)$ is composed of the radiation incident on the upper boundary layer and the flux within the layer after modification by absorption ($\kappa\Delta z$ is the probability of absorption) and scattering in the opposite direction. The implicit assumption, then, is that only single scattering processes need to be considered on account of the low layer thickness ($\sigma\,\Delta z \ll 1$). The quantity $p_{\downarrow\uparrow}$ indicates the probability that a photon propagating downwards will be scattered and deflected in the opposite direction. Similarly, $p_{\uparrow\downarrow}$ can be interpreted as describing the reverse process. An analogous case can be made for the upwards radiation. Dividing both sides of these difference equations by $\Delta z$ and applying the limiting process $\Delta z \rightarrow 0$ yields the coupled differential equations:

*Equations VII.71* $$\frac{dF_{\downarrow}(z)}{dz} = -(\kappa+\sigma)\,F_{\downarrow}(z) + \sigma\,(p_{\downarrow\downarrow}\,F_{\downarrow}(z) + p_{\uparrow\downarrow}\,F_{\uparrow}(z))$$

*Equations VII.72* $$\frac{dF_{\uparrow}(z)}{dz} = (\kappa+\sigma)\,F_{\uparrow}(z) - \sigma\,(p_{\uparrow\uparrow}\,F_{\uparrow}(z) + p_{\downarrow\uparrow}\,F_{\downarrow}(z))$$

whose structure deviates from that of the SKM model. This makes provision for the fact that, because photon propagation is restricted to just two opposed directions, the probabilities must satisfy the relation

*Equation VII.73* $$p_{\downarrow\uparrow} + p_{\downarrow\downarrow} = p_{\uparrow\downarrow} + p_{\uparrow\uparrow} = 1$$

Equations VII.72 and VII.73 can be consolidated into a single equation:

*Equation VII.74*
$$\mu_k \frac{dF_k(z)}{dz} = -(\kappa + \sigma)\, F_k(z) + \sigma \sum_{j=1}^{2} p_{jk}\, F_j(z), \quad k = 1, 2$$

where the quantity $\mu_k = \cos\theta_k$ is the cosine of the angle between the positive $z$ axis and one of the two directions and only assumes the values $+1$ and $-1$ in the 2-flux model.

The form of Equation VII.74 suggests a generalisation to $N$ mono-directional radiation fluxes of different channels

*Equation VII.75*
$$\mu_k \frac{dF_k(z)}{dz} = -(\kappa + \sigma)\, F_k(z) + \sigma \sum_{j=1}^{N} q_{jk}\, F_j(z), \quad k = 1, 2, \ldots, N$$

where $q_{jk} = \pm p_{jk}(\mu_k/\mu_j)$. The symbol ± indicates that the algebraic sign must be chosen such that the overall expression is always positive. The spectral irradiance $F$ (in units of $W\,m^{-2}\,nm^{-1}$) in Equation VII.75 must now be replaced by the directional spectral radiance $L$ (in units of $Wm^{-2}\,sr^{-1}\,nm^{-1}$). For the limiting case $N \rightarrow \infty$, the sum on the right-hand side of Equation VII.76 becomes an integral. The system of coupled linear differential equations thus yields an integro-differential equation for radiative transfer in plane-parallel particulate media [5, 6]:

*Equation VII.76*
$$\mu \frac{dL(z,\mu,\varphi)}{dz} = -(\kappa + \sigma)\, L(z,\mu,\varphi) - \frac{\sigma}{2} \int_0^{2\pi} d\varphi' \int_{-1}^{+1} d\mu'\, p(\mu,\varphi;\mu',\varphi')\, L(z,\mu',\varphi')$$

$L(z,\mu)$ denotes the spectral radiance of the radiation field at geometrical depth $z$ in the direction $\theta = \cos^{-1}\mu$ while $p(\mu,\varphi;\mu',\varphi')$ is the scattering function[1]. This function shows that radiation striking a particle at an angle $\theta = \cos^{-1}\mu$ is redistributed in all other directions $\theta' = \cos^{-1}\mu'$. If, as is assumed in the SKM theory, scattering occurs isotropically such that that the radiation scattered by a particle is independent of the spatial direction, then $p(\theta)=1$ applies, where $\theta$ denotes the scattering angle. The relevant theoretical parameter in the equation above is the albedo $\omega_0$ of the single scattering process, which via the equation

*Equation VII.77*
$$\omega_0 = \frac{\sigma}{\sigma + \kappa}$$

is linked to the scattering coefficient $\sigma$ and the absorption coefficient $\kappa$ of the medium[2]. Moreover, if the radiation field is broken down into diffuse ($L_d$) and collimated ($L_c$) contributions, Equation VII.77 takes the form

*Equation VII.78*

*(Net increase)* *(Extinction)* *(Self-illumination)*

$$\mu \frac{dL_d(z,\mu,\varphi)}{dz} = -(\kappa + \sigma)\, L_d(z,\mu,\varphi) - \frac{\sigma}{2} \int_0^{2\pi} d\varphi' \int_{-1}^{+1} d\mu'\, p(\mu,\varphi;\mu',\varphi')\, L_d(z,\mu',\varphi')$$
$$+ \frac{\sigma}{4\pi} L_0\, p(\mu,\varphi; -\mu_0,\varphi_0)\, e^{-(\kappa+\sigma)\, z/\mu_0}$$

*(Collimated beam)*

which generally serves as a starting point for theoretical analyses.

[1] *In the literature frequently the misleading term phase function is used which has nothing to do with the phases of electromagnetic waves. The term phase originates from astronomy where it refers to the variation of planetary brightness with phase angle* $\alpha$

[2] *Compared to the SKM model a different notation for the material parameters (scattering coefficient, absorptions coefficient) is used, since they are defined differently.*

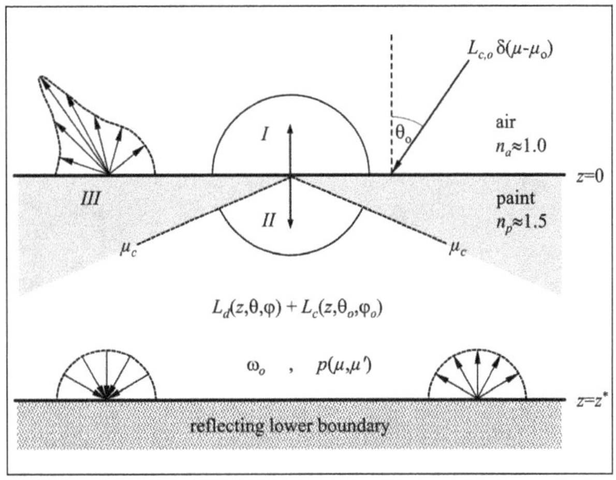


*Figure VII.27: Schematic representation of the optical phenomena in two adjacent media of different refractive indices ($n_a$, $n_p$) with planar interface, which is unidirectionally illuminated at an angle $\Theta_o$. The refractive index of the atmosphere ($n_a \approx 1.0$) is different from that of the paint layer ($n_p \approx 1.5$). The radiation reflected by the pigmented layer into the outer hemisphere (region I) stems from a cone within the paint layer (region II), whose opening angle is determined by the critical angle ($\mu_c = \cos\Theta_c$). Radiation incident at an angle exceeding the critical angle (region III) undergoes total internal reflection at the paint-air interface. The pigmented layer is bounded at the bottom ($z = z^*$) by a diffuse reflecting layer.*

So far, in formulating the radiative transfer problem, we have ignored the refractive properties of the embedding medium. For the purpose of describing the **radiation field** within the pigmented layer, the source term of Eq. (VII.79) (collimated beam) needs to be modified to account for the transmission properties of the air-paint interface. Thus, the collimated beam transmitted into the paint layer is refracted towards the axis of incidence and experiences a change in solid angle as a result of the refractive properties of the boundary layer. As can be seen in Figs. VII.26 and VII.27, the outer hemisphere is mapped onto a conical space within the pigmented layer, whose solid angle is defined by the angle of total internal reflection. In addition, the connection conditions for the spectral radiance at the optical air-paint discontinuity require further modifications to the boundary conditions. A major law of geometric radiometry, the so-called $n^2$-law of spectral radiance, holds that the ratio of radiance to quadratic refractive index of a particular medium is invariant:

*Equation VII.79*
$$\frac{L_a}{n_a^2} = t\,(a \rightarrow p)\,\frac{L_p}{n_p^2}$$

In this expression, the indices *a* and *p* denote the optical media air and paint, while *t(a→p)* is the transmission of the air-paint interface.

The equation of radiative transfer has an easy to understand basis in the principle of the conservation of energy (radiation balance), but since it is an integro-differential equation its solution is cumbersome. In mathematical physics, various precise and, for numerical solutions, approximate methods have been developed for solving it. One that lends itself particularly well to programmatic implementation is the **method of discrete ordinates** [5]. The use of a numerical quadrature rule (e.g. Gauss method) as an approximation for the integral which appears in the radiative transfer equation allows the integro-differential equation to be transformed into a system of coupled linear differential equations which can be solved by standard numerical methods. The constants of integration have to be determined from the adaptation of the general solution to the given boundary conditions. Its advantage, relative to the SKM theory, is that it accommodates more than two radiation fluxes and explicitly provides for the anisotropy of the single scattering processes in the form of the scattering function ($p(\mu,\mu')$). The quality of a radiative transfer calculation critically depends on the choice of this function. Any well-defined physical model will provide the scattering function, so there is no need to guess it. The scattering function can be calculated, for exam-

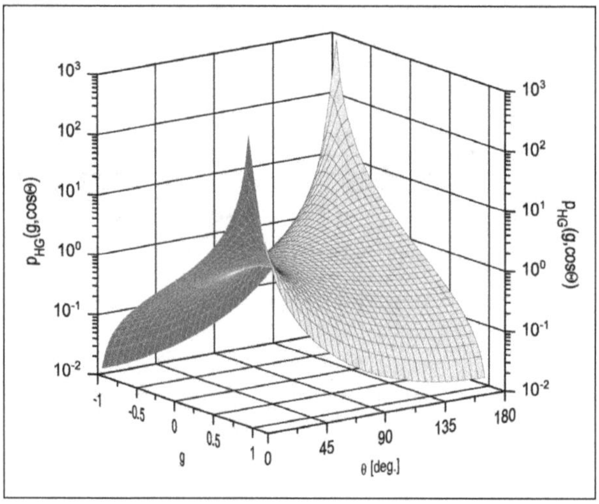


*Figure VII.28: Scattering function according to Henyey-Greenstein [8] as a function of scattering angle θ and asymmetry parameter g for |g| ≤ 0.95.*

ple, by means of Mie theory [7] (see Section VIII.3) which describes the physics of scattering by single particles when the complex refractive index and the shape and size of a pigment particle are known. These functions, which can only be expressed as infinite power series, exhibit an increasing degree of complexity with increase in size ratio $x = 2\pi d/\lambda$ ($d$ = particle diameter, $\lambda$ = wavelength). Description of the scattering function requires many expansion terms which prove unwieldy for **radiative transfer calculations** and, moreover, are computationally intensive. Only in the limiting case of Rayleigh scattering, which is satisfied when $x \ll 1$ and absorption is absent, a simple analytical expression for the scattering function can be specified:

*Equation VII.80* $$p(\theta) = \frac{3}{4}(1 + \cos^2\theta)$$

As can be seen from the polar diagram of this function, deviations from the isotropic case are relatively small and scattering contributions into the forwards and backwards domains are still the same.

For industrial pigments, the requisite optical and/or geometrical information is only available in exceptional cases [8], and so analytical scattering functions can only be “guessed”. A model function should, on the one hand, capture the essential features of the true scattering function (e.g. the degree of anisotropy), and, on the other, depend on as few parameters as possible in order that computational effort expended on the radiative transfer problem may be kept within reasonable limits. It is strongly recommended that parameterisation of the scattering function be restricted so that a numerically stable determination of all its coefficients can be obtained from a calibration echelon.

Another analytical scattering function that should be mentioned is

*Equation VII.81* $$p(\theta) = \frac{1 - g^2}{(1 + g^2 - 2g\cos\theta)^{3/2}}$$

which was proposed by *Henyey* and *Greenstein* [9] in 1941 for astrophysics studies and is frequently used in numerical simulations. This model function, which is not inspired by any physical theory, depends only on a single parameter *g*, with the definition range $-1 \leq g \leq +1$. Varying this so-called asymmetry parameter *g* allows the scattering behaviour to be continuously adjusted between pure forward scattering ($g = +1$) and pure backward scattering ($g = -1$); the case $g = 0$ simulates isotropic scattering. Figure VII.28 illustrates this situation for $|g| \leq 0.95$ (for $|g| = 1$, a δ-function is obtained).

# 8 Literature

## 8.1 General literature

This selection of literature on the topics covered above merely provides an overview for delving deeper into colorimetry. The list of publications is alphabetical and makes no claim to completeness.

Barber, P. W., Hill, S. C., "Light Scattering by Particles: Computational Methods", World Scientific, Singapure, 1990

Bohren, C. F., Huffman, D. R., "Absorption and Scattering of Light by Small Particles", John Wiley & Sons, New York, 1983

Bohren, C. F., "Multiple Scattering at the Beach", Weatherwise 1983, 36, 197-200

Bohren, C. F., "Multiple Scattering of Light and Some of its Observable Consequences", Am. J. Phys. 1987, 55(6), 524-533

Bohren, C. F., Salzman, G. C., Palmer, B. R. und Stamatakis, P., "A Multiple-Scattering Model for Opacifying Particles in Polymer Films", J. Coat. Technol. 1989, 61 (779), 41-47

Brockes, A., "Der Einfluss glänzender Oberflächen auf Remissionsmessungen", Die Farbe 1960, 1/3, 53-62

Brockes, A., "Der Zusammenhang von Farbstärke und Teilchengröße von Buntpigmenten nach der Mie-Theorie", Optik 1964, 21, 550-566

Chandrasekhar, S., "Radiative Transfer", Clarendon Press, 1950

van de Hulst, H. C., "Light Scattering by Small Particles", Dover Publications, New York, 1957

van de Hulst, H. C., "Multiple Light Scattering - Tables, Formulas, and Applications, Volume 1", Academic Press, New York, 1980

van de Hulst, H. C., "Multiple Light Scattering - Tables, Formulas, and Applications, Volume 2", Academic Press, New York, 1980

Ishimaru, Akira, "Wave Propagation and Scattering in Random Media", IEEE, 1997

Kerker, Milton, "The Scattering of Light and other Electromagnetic Radiation", Academic Press, New York, 1969

Kerker, M., Scheiner, Cooke und Kratohvil, J. Coll. Interface Sci. 1979, 71 (1), 176-187

Klein, Georg A., "Industrial Colour Physics", Springer-Verlag, Heidelberg, 2010

Kortüm, Gustav, "Reflexionsspektroskopie", Springer-Verlag, Heidelberg, 1969

Mehta, K. T. und Shah. H. S., "Simplified Equations to Calculate Mie-Theory parameters for Use in Many-Flux Calculations for Predicting the Reflectance of Paint Films", Colour Research and Application 1987, 12 (3), 147-153

K. D. Moller, "Optics", University Science Books, U.S., 1988

Mudgett, P. S. und Richards, L. W., "Multiple Scattering Calculations for Technology", Appl. Opt. 1971, 10 (7), 1485-1502

Mudgett, P. S. und Richards, L. W., "Multiple Scattering Calculations for Technology II"
J. Coll. Interface Sci. 1972, 39 (3), 551-567

Mudgett, P. S. und Richards, L. W., "Kubelka-Munk Scattering and Absorption Coefficients For Use with Glossy, Opaque Objects", J. Paint Techn. 1973, 45 (586), 44-53

Nassau, Kurt, "The Physics and Chemistry of Colour - The Fifteen Causes of Colour", John Wiley & Sons, New York, 2001

Palik, Edward D. (editor), "Handbook of Optical Constants of Solids", Academic Press, Boston 1985

Querry, Marvin R., "Optical Constants", Govern. Rep., CRDC-CR-85034, Kansas City, Missouri, MO 64110

Quinten, Michael, "Optical Properties of Nanoparticle Systems - Mie and Beyond", Wiley-VCH, Weinheim, 2011

Rybicki, George B., "Radiative Transfer", J. Astrophys. Astr. 1996, 17, 95-112

"Ullmann's Encyclopedia of Industrial Chemistry, Dyes, General Survey, A9", VCH Verlagsgesellschaft, Weinheim, 1987

"Ullmann's Encyclopedia of Industrial Chemistry, Pigments Inorganic, A20", VCH Verlagsgesellschaft, Weinheim, 1992

Völz, Hans G., "Industrielle Farbprüfung, Grundlagen und Methoden", VCH Verlagsgesellschaft, Weinheim, 2001

Völz, Hans G., "Industrial Colour Testing, Fundamentals and Techniques", VCH Verlagsgesellschaft, Weinheim, 2001

Winter, Gerhard, "Anorganische Pigmente: Disperse Festkörper mit technisch verwertbaren optischen und magnetischen Eigenschaften", Fortschr. Miner. 1979, 57(2), 172-202

## *8.2 References*

[1] Kubelka, P., Munk, F., "Ein Beitrag zur Optik der Farbanstriche", *Z. Tech. Phys.* **12**, 593, 1931

[2] Schuster, A., "Radiation through a foggy atmosphere", *Astrophys. J.* **21**, 1, 1905

[3] Preisendorfer, R. W., *Radiative Transfer on Discrete Spaces*, Pergamon Press, Oxford, UK, 1965

[4] Bohren, C. F., "Multiple scattering of light and some of its observable consequences", Am. J. Phys. **55**, 524, 1987

[5] Chandrasekhar, S., *Radiative Transfer*, Dover Publications, New York, USA, 1960

[6] Klein, G. A., *Farbenphysik für industrielle Anwendungen*, Spinger-Verlag, 2004

[7] Bohren, C. F. Huffman, D. R., *Absorption and Scattering of Light by Small Particles*, John Wiley & Sons, New York, USA, 1983

[8] Kettler. W. H., "Complex refractive index and colour of quinacridone pigments", $9^{th}$ Congress of the International Colour Association (2001), published in Proceedings of SPIE 4421, 681, AIC, (Rochester, NY, USA), 2002

[9] Henyey, L. C. Greenstein, J. L., "Diffuse radiation in the galaxy", *Astrophys. J.* **93**, 70, 1941

# VIII Practical applications

*Uwe Hempelmann*

## 1 Tinting strength

The tinting strength and hiding power discussed in Section VIII.2 quantify the amount of pigment or colourant needed to obtain a certain colour in, e.g., a paint or plastic compound. They more or less link quantification of colour impression on one hand to the physics of absorption and scattering processes on the other. The mathematical description of these processes serves as the basis for recipe calculation [1].

Tinting strength is defined in DIN 55 886 as follows: "The tinting strength of a colourant quantifies the ability of a colourant to impart colour to other substances by virtue of its **absorbing power**".

The starting point for the following considerations is a very simple experiment: the reference and sample pigments are combined with a white pigment and a binder and the resultant colours are compared. Then, the quantity of either the reference or the sample pigment is varied until sample and reference exhibit the same colour.

*Equation VIII.1*

$$P = \frac{m_b}{m}\Big|_{m_{FP} = \text{const.}}^{\Delta Q = 0}$$

where $\Delta Q$ is the match criterion and $m_{FP}$ indicates that the quantity of white pigment used is to be constant.

### Scattering power of white and yellow pigments

The same approach can naturally be used to characterise white pigments, in which event a black pigment serves as reference. The result is called the **scattering power**. In the case of light-yellow pigments, too, it is often beneficial to use a mixture containing a black pigment because it accentuates the colour difference.

### Colour difference as a characteristic

The tinting strength, as explained above, is a mass ratio for substituting one pigment for another, with the objective of minimising the difference in colour. The minimum colour difference between sample and reference is called the **residual colour difference**.

If we now seek to perform this experiment colorimetrically, we must first find a colorimetric definition of residual colour difference. It seems obvious to use the colour difference as expressed in the various equations provided by the CIE and others. However, this will not prove successful on account of their shortcomings.

A review of the literature shows that, in the case of inorganic pigments, colourists are more interested in **lightness** (Y or L* in CIELAB) whereas, in other fields (e.g. inks and textiles)

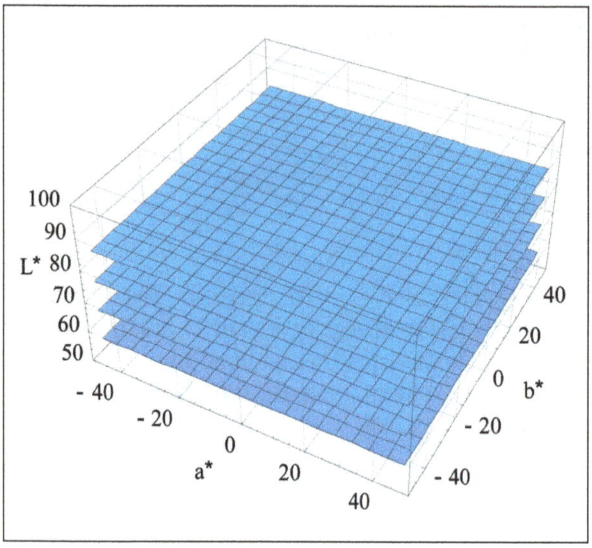


*Figure VIII.1: Planes of equal lightness in CIELAB colour space.*

and for organic pigments, **colour depth** or the **sum of the Kubelka-Munk function** over the visible range are more important.

Colour depth is a term that originated in the textile industry where it is employed in light-fastness-testing of dyes. The equivalent term in our context is colour intensity or saturation. International standards contain mathematical expressions that enable it to be applied to colorimetry.

It should be remembered, though, that the eye of the observer is the ultimate and decisive criterion in any assessment. An overview of possible tinting strength criteria is presented in Table VIII.1 [2-4].

## *Lightness as a criterion of tinting strength*

No matter which criterion is chosen it must be inherently consistent. The measured and/or calculated tinting strength must permit conclusions to be drawn about the ratio of pigment masses in the reference and sample pigments.

Naturally, all criteria yield very close results for pigments, which are similar. For more disparate samples, the differences between the criteria become more pronounced. Figures VIII.1 to VIII.3 show plots of different criteria: The lightness criterion in CIELAB colour space represents planes of equal lightness (Figure VIII.1). Further differentiation according to colour location does not take place, although nobody would think of comparing the tinting strength of a red pigment with a green one in practice.

When equal colour depth is the criterion, the **colour location** is defined too. A plot of the colour depth given by Equation VIII.2 yields the hemispheres shown in Figure VIII.2. The statement that two samples have the same colour depth therefore indicates that not only the intensity of the colour but also their coordinates in colour space are identical.

*Equation VIII.2*

$$T^* = \sqrt{a^{*2} + b^{*2} + (100 - L^*)^2} = \sqrt{C^{*2}_{ab} + (100 - L^*)^2}$$

*Table VIII.1: Tinting strength criteria.*

| | |
|---|---|
| 1 | Lightness $L^*$ |
| 2 | Colour depth $T^*$: $T^* = \sqrt{a^{*2} + b^{*2} + (100 - L^*)^2} = \sqrt{C^{*2}_{ab} + (100 - L^*)^2}$ |
| 3 | Weighted K/S sum |
| 4 | Lowest tristimulus value X, Y, Z |
| 5 | Lowest reflectometer value $R'_x$, $R'_y$, $R'_z$ |
| 6 | Lowest colour difference $\Delta E^*_{ab}$ |
| 7 | Lowest saturation difference $\Delta C^*_{ab}$ |

### *Tinting strength: lightness or colour depth?*

With regard to the question as to the correct criterion to use for tinting strength, two studies exist in which the assessments of colourists and of lay persons who had no prior training in colour assessment were compared with calculated tinting strengths: *Keifer* and *Völz* in 1977 and *Gall* in 1975.

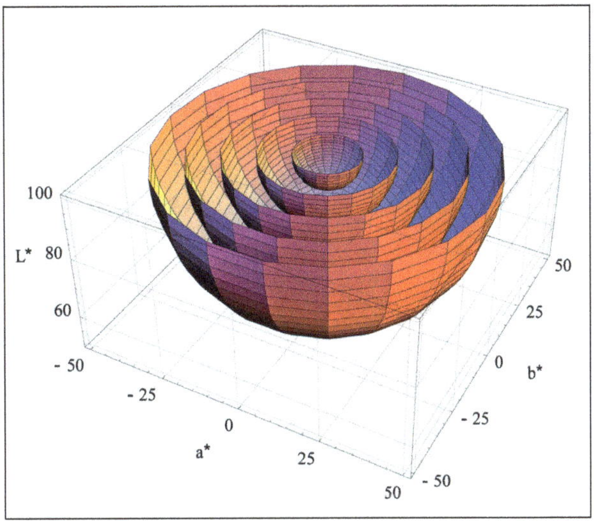


*Figure VIII.2: Planes of equal colour depth.*

However, the two studies were set up differently: Keifer and Völz were studying the substitutability of one pigment for another. They found that the **lightness criterion** yielded the best results for inorganic pigments. No significant differences were found between the colourists and the lay people.

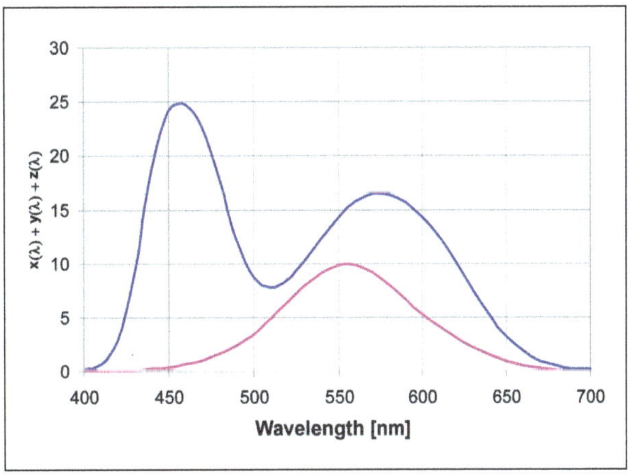


*Figure VIII.3: Spectral range of the weighted K/S sum.*

The study by Gall had a different outcome; admittedly, he was looking for samples of equal colour depth. Furthermore, his samples also consisted of organic pigments of much greater saturation. In this case, **colour depth** provided the best match for the colourists' assessment.

Thus, the tinting strength of inorganic pigments usually tends to be assessed on the basis of lightness while organic pigments are assessed according to colour depth. It should also be mentioned that these two methods tend to converge in the case of unsaturated colours.

### *Kubelka-Munk function*

Another criterion is the weighted K/S sum. Here, the reflectance spectra for the reference and sample pigment are used to calculate the Kubelka-Munk functions, which are then weighted by the sum of the tristimulus values. Figure VIII.3 is a plot of the spectral range of the function against lightness.

Once the criterion for determining the tinting strength has been chosen, the next logical step is to find a way to perform the determination without having to prepare series of samples with different masses.

The basis for all these methods is the spectral evaluation scheme first proposed by Völz [2]; it links the mass of the respective pigment to the observed reflectance spectra. The tool employed is the Kubelka-Munk theory, which formulates connection between the

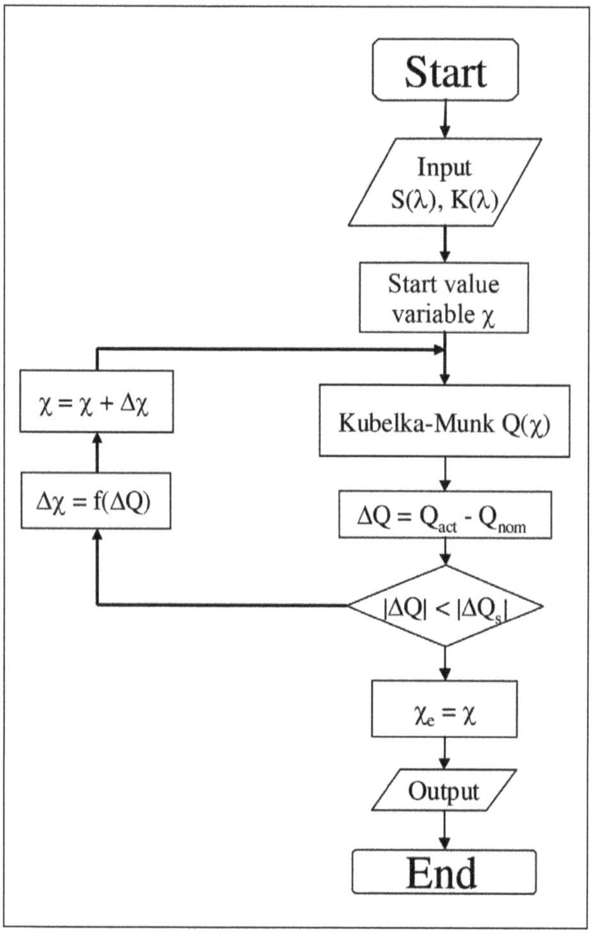


*Figure VIII.4: Flow chart for application of the spectral evaluation scheme.*

absorption and scattering processes to colorimetry. The spectral evaluation scheme has general applicability here: besides supporting calculations of tinting strength by various criteria, it can help with determining the hiding power and other properties. Its core principle is the property of the Kubelka-Munk theory to link the mass of the pigment sample to the observed spectrum.

***Residual colour difference: calculated and measured***

Figure VIII.4 shows the flow chart for one such calculation. It is a rather simple iterative calculation that can be easily performed on standard spreadsheet programs.

Let us now calculate the tinting strength for some examples to illustrate the procedure. The first example is a calculation based on the lightness criterion, which is a simplified method that does not require use of the spectral evaluation scheme. The reason for this is that the spectra of numerous inorganic pigments can be approximated by tristimulus values X, Y and Z [2]. A disadvantage is the rather large discrepancies between calculated and experimentally established residual colour differences.

The calculation involves the following steps

1 Calculation of the tristimulus values X, Y, Z of reference and sample.
2 Division by 100; application of the Saunderson correction: (X/100)`, (Y/100)`, (Z/100)`.
3 Calculation of the Kubelka-Munk function for sample and reference values F(X/100)`.
4 Calculation of the quotient: F(Y/100)` sample/F(Y/100)` reference.
5 Use of the quotient to correct the sample values.
6 Inversion of the Saunderson correction.
7 Calculation of the tristimulus values and residual colour difference.

This calculation is shown in Table VIII.2 for two iron oxide red pigments.

The first row shows the tristimulus values for sample and reference ($X_{sample}/100$ etc.). The second row contains the values obtained after division by one hundred and application of the Saunderson correction ($S_{corr}(X_{sample}/100)$ etc.). Calculation of the Kubelka-Munk function is shown in the third row ($F^{x}_{sample}$ etc.). The ratio of these values for the sample and reference yields the tinting strength (P).

To match the colorimetric properties of the sample to those of the reference, the values for the Kubelka-Munk function of the sample are then divided by the tinting strength P ($F^x_{sample}/P$ etc.). Inversion of the Saunderson correction and multiplication by 100 completes the calculation. The resulting tristimulus values show the colour that the sample would have for the corrected quantity of pigment ($100 \cdot S_{corr}(X_{sample}/P)$ etc.). These values can be used to calculate the residual colour difference. Table VIII.2 shows the difference between the calculated residual colour difference and the experimentally determined colour difference. By definition, the residual colour difference $\Delta L^* = 0$ when the lightness criterion is applied.

Let us now calculate the tinting strength by applying the spectral evaluation scheme on the basis of the colour depth criterion. As standard definitions exist for various levels of colour depth, namely: 1/3 SD, 1/9 SD, 1/25 SD (standard depth), we have an absolute, defined standard available for comparison purposes. The expression in Equation VIII.2 (see page 166) is not accurate enough for the determination, so it is necessary to apply polynomial expressions or values found in tables (see DIN 53235-1 and -2).

The determination proceeds as follows via the tristimulus values X, Y, Z for the sample.

Chromaticity coordinates

*Equations VIII.3 and VII.4*

$$s = 10\sqrt{(x - x_n)^2 + (y - y_n)^2}$$

$$\varphi = \text{Arctan}\,\frac{y - y_n}{x - x_n}$$

$a_i(\varphi)$ is taken from DIN 53235-1 and -2 or is a polynomial having the respective parameters from the same standard (where i = 1/3, 1/9, 1/25).

*Table VIII.2: Sample calculation of tinting strength based on the lightness criterion.*

| | | | | | | | | | | | |
|---|---|---|---|---|---|---|---|---|---|---|---|
| $X_{Sample}$ | 35.61 | $Y_{Sample}$ | 29.06 | $Z_{Sample}$ | 64.67 | $X_{Ref}$ | 35.04 | $Y_{Ref}$ | 28.78 | $Z_{Ref}$ | 63.83 |
| $S_{Corr}$ $(X_{Sample}/100)$ | 0.5511 | $S_{Corr}$ $(Y_{Sample}/100)$ | 0.4689 | $S_{Corr}$ $(Z_{Sample}/100)$ | 0.8111 | $S_{Corr}$ $(X_{Ref}/100)$ | 0.5444 | $S_{Corr}$ $(Y_{Ref}/100)$ | 0.4652 | $S_{Corr}$ $(Z_{Ref}/100)$ | 0.8053 |
| $F^x_{Sample}$ | 0.1829 | $F^y_{Sample}$ | 0.3007 | $F^z_{Sample}$ | 0.0220 | $F^x_{Ref}$ | 0.1907 | $F^y_{Ref}$ | 0.3073 | $F^z_{Ref}$ | 0.0236 |
| P | 0.9784 | | | | | | | | | | |
| $F^x_{Sample}/P$ | 0.1869 | $F^y_{Sample}/P$ | 0.3073 | $F^z_{Sample}/P$ | 0.0225 | | | | | | |
| $iF^x_{Sample}$ | 0.5476 | $iF^y_{Sample}$ | 0.4652 | $iF^z_{Sample}$ | 0.8092 | | | | | | |
| $100 * iS_{Corr}$ $(iF^x_{Sample}/P)$ | 35.32 | $100 * iS_{Corr}$ $(F^y_{Sample}/P)$ | 28.78 | $100 * iS_{Corr}$ $(F^z_{Sample}/P)$ | 64.40 | | | | | | |
| $L^*_{Sample}$ | 60.83 | $a^*_{Sample}$ | 24.69 | $b^*_{Sample}$ | 18.82 | $L^*_{Ref}$ | 60.59 | $a^*_{Ref}$ | 23.83 | $b^*_{Sample}$ | 18.40 |
| $L^*_{Sample/A}$ | 60.59 | $a^*_{Sample/A}$ | 24.75 | $b^*_{Sample/A}$ | 18.81 | | | | | | |
| $\Delta E^*_{ab}$ | 0.99 | | | | | | | | | | |
| $\Delta E^*_{ab}$ | 1.00 | | | | | | | | | | |

*Table VIII.3: Examples calculation of tinting strength based on the tinting strengh criterion.*

| Ratio | X | Y | Z | x | y | s | φ | W | a(φ) | $\Delta T_{s,25}$ |
|---|---|---|---|---|---|---|---|---|---|---|
| 1 : 20 | 43.67 | 33.67 | 50.87 | 0.3429 | 0.2575 | 0.7903 | 292 | 0.4761 | 3.163 | 14.53 |
| 1 : 40 | 51.28 | 42.09 | 60.89 | 0.3338 | 0.2693 | 0.6485 | 288 | 0.4393 | 3.165 | 6.38 |
| 1 : 80 | 58.74 | 51.04 | 70.00 | 0.3276 | 0.2815 | 0.5135 | 286 | 0.4156 | 3.159 | -2.01 |
| 1:100 | 61.00 | 53.93 | 72.66 | 0.3259 | 0.2854 | 0.4720 | 285 | 0.4087 | 3.156 | -4.67 |

*Table VIII.4: Comparison of the results for the determination of tinting strength based on the lightness criterion and on the weighted K/S sum.*

| Example | L* | a* | b* | Lightness criterion | | | K/S-Summe | | |
|---|---|---|---|---|---|---|---|---|---|
| | Ref. / Sample / Diff. | Ref. / Sample / Diff. | Ref. / Sample / Diff. | Rel. Tinting strength [%] | Δa* Residual colour diff. | Δb* Residual colour diff. | Rel. Tinting strength [%] | Δa* Residual colour diff. | Δb* Residual colour diff. |
| Red pigment | 60.6 | 25.7 | 16.2 | | | | | | |
| | 60.8 | 24.4 | 18.8 | 97.8 | 1.0 | 2.6 | 107.8 | 5.2 | 0.7 |
| | 0.2 | 0.7 | 2.6 | | | | | | |
| Green pigment | 75.6 | -12.7 | 9.6 | | | | | | |
| | 76.0 | -13.2 | 10.3 | 95.2 | 0.4 | -0.6 | 92.9 | -0.6 | -0.2 |
| | 0.4 | 0.5 | -0.7 | | | | | | |

*Equation VIII.5* $\Delta T_{S,I} = B_I = \sqrt{Y}\,(s\,a_I\,(\varphi) - 10) + b_I$

where $i = 1/3, 1/9, 1/25$ and $b_{1/3} = 29$, $b_{1/9} = 41$, $b_{1/25} = 56$

The results for a theoretical example are shown in Figure VIII.5. The end result is found at the intercept of the straight line with the x axis.

As the straight line has a nearly identical slope for all pigments, the colour depth can be calculated from a single sample (FIAF, Gall and Friedrichsen). This can be done easily nowadays with the aid of a spreadsheet.

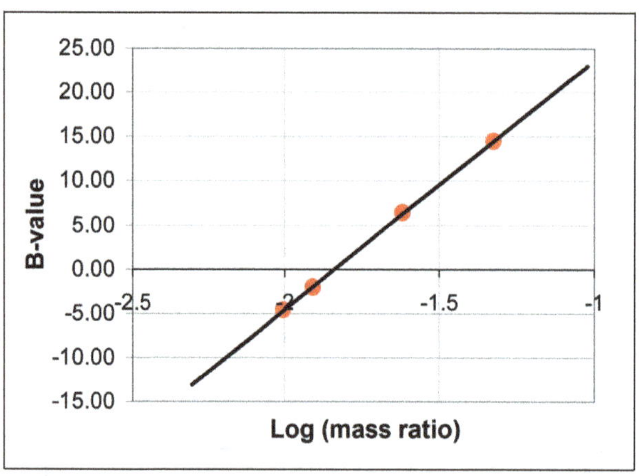


*Figure VIII.5: Determination of tinting strength based on colour depth. Plot of colour depth against the logarithm of the ratio of coloured to white pigment.*

Finally, reference is made again to the fact that the results depend on the criterion used for determining the tinting strength. Table VIII.4 shows the results obtained for a red and a green pair of pigments.

It is quite striking that the sample of red pigment has a higher tinting strength on the basis of the lightness criterion, but a lower tinting strength on the basis of the weighted K/S sum.

## 2 Hiding power

The hiding power of a pigmented substance is its ability to obliterate the colour or colour differences in the substrate (DIN 55984).

The value for the hiding power indicates in square metres the area of a substrate that can be covered in a defined contrast ratio such that after drying: $\Delta E_{ab}^* = 1$; units $m^2/l$ or $m^2/kg$

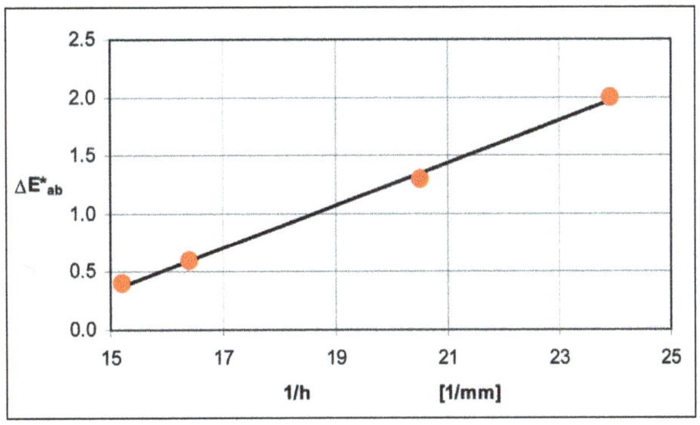


*Figure VIII.6: Determination of hiding power from a plot.*

### Hiding power: determination by means of a plot

An intuitively simple method is to determine the hiding power from a plot:

1 At least four samples are made with equal pigment volume concentration (PVC), but variable layer thickness.
2 The colour difference between the black and the white backgrounds is plotted against the inverse of the layer thickness.
3 The necessary thickness is read off at $\Delta E_{ab}^* = 1$.

The results for this method are shown in Figure VIII.6. Besides the relatively high experimental outlay, this method is rather limited in its accuracy, because the different drawdowns necessarily fluctuate in layer thickness.

A logical step in keeping with the deliberations in Section VIII.1 is to apply the spectral evaluation scheme to the hiding power. The absorption and scattering coefficients of transparent layers can be calculated with the aid of the Kubelka-Munk theory.

This presupposes a knowledge of the reflectance of the background of a non-pigmented layer of equal thickness.

As said earlier, there are two distinct methods. Either the determination is based on a calculation of two transparent layers or such a measurement is combined with the measurement of a hiding layer. Often, the second method yields more accurate results. The necessary equations are given in Section XI.3.

## 3 Literature

[1] Roderick McDonald, Colour Physics for Industry, Society of Dyers and Colourists, 1997

[2] Hans G. Völz, Industrial colour testing: fundamentals and techniques, Wiley-VCH, 1995

[3] Anni Berger-Schunn, Praktische Farbmessung, Hansen-Schmidt, 1991

[4] Juan M. Oyarzún, Pigment Processing, Vincentz Network, 2015

# IX Measuring the texture of effect finishes

*Sandra Weixel*

"First impressions count" is a statement that everyone can relate to. The visual impression which the exterior of an object conveys in the first ten seconds establishes a lasting perception of quality that transcends the design and the appearance. That is why the colour and the consistency of an effect together constitute one of the most important quality criteria of a finish and why designers are continually on the look-out for new colours.

Over the decades, metallic and special effect colours have taken on a dominant role in the colour palette of all automakers and a great many other application areas, ranging from electronic devices to household appliances to cosmetics products. Designers are looking for new colours, which not only make the product look pretty and distinct, but also underline its styling. Pigment manufacturers are developing new types of aluminium flakes to create fine or coarse-looking metallic that offer greater or lesser light-dark flop. The impression of colour, which the coating conveys depends on the illuminating and viewing angles. New, artificially produced interference pigments are broadening the range of effects that can be seen in direct or diffuse lighting conditions. These new generations of special effect pigments cannot be adequately described by traditional multi-angle colour spectrophotometers (see Section II.3) that quantify diffuse light reflection at three or five angles. So a new measurement technology has been developed for objectively describing the overall impression conveyed by effect coatings. This focuses on characterising flakes by using a camera to measure the visual impression of "sparkle" and "graininess" and to simulate how the effect changes in direct and diffuse lighting conditions.

## 1 Sparkle and graininess

Unlike the case for solid colours, the colour and the appearance of effect finishes change with change in viewing and illuminating conditions. Experience shows that two samples of effect pigments can be evaluated as a perfect match when measured with a 5-angle spectrophotometer, but still be visually unacceptable due to differences in sparkle and/ or graininess. This effect cannot be captured with conventional multi-angle colour instruments, because they measure the integral of the spectral reflection over the detected area and cannot distinguish between basecoat colour and reflection by the effect pigments.

### 1.1 Visual evaluation of sparkle and graininess

Although the light intensity is equal in all directions under diffuse illumination (cloudy sky, typical spray booth for a car), a metallic finish can create a pattern ranging from light to dark according as the flake size varies from very fine to very coarse (Figure IX.1). Common terms to describe this phenomenon are graininess, coarseness, texture or salt & pepper look. The effect is only obvious from close-up and does not change with change in viewing angle. Graininess can vary with changes in flake size, the orientation of the flakes in the coating, and with agglomeration of flakes during the application process.

*Figure IX.1: Graininess under diffuse light.*

When the light intensity comes from mainly one direction, as in the case of direct illumination (sunny sky, inspection with an intense spotlight), the same metallic or effect finish can look completely different (Figure IX.2). Small light flashes can be seen at low to high intensities. This effect is variously called sparkle, micro-brightness, glint, and diamonds, and is due to the reflectivity of the flakes – it is therefore influenced by the type of flake (cornflakes versus silver dollars, natural mica versus synthetic interference pigments or glass flakes), the concentration level of the effect pigments, the flake size, and the application method (bell/bell versus bell/pneumatic). Unlike graininess, sparkle changes with change in illuminating angle.

*Figure IX.2: Sparkle under direct light.*

The optical texture of effect coatings has been studied in two different illuminating conditions [1]. These comparative tests utilised well-defined anchor panels that consisted of eight neutral, grey metallic coating panels with a gradual change in the number and intensity of sparkle (glint). The overall colour of the panels was kept as constant as possible. A separate set of anchor panels was created for the diffuse coarseness (graininess) evaluation. For comparison, a set of more than 200 automotive colours was used with an even distribution over colour space. The sample set contained purely metallic panels, purely pearlescent panels and a mixture of both. A representative number of observers with normal colour vision was then asked to assign a numerical value to the sparkle and diffuse coarseness (graininess) of each sample panel. To improve observer accuracy and reproducibility, the observers were allowed to assign numbers from one to nine, with non-integer values of up to quarter-precision for diffuse coarseness (graininess) and half-scale values for sparkle [1]. The studies showed good results for observer repeatability and observer reproducibility. A particularly interest finding was that the amount of diffuse coarseness and impression of glint are independent parameters. A given sample could have high or low coarseness, regardless of the sparkle value. Thus, the two parameters need to be measured and quantified separately. With an eye to the future, sub-parameters, such as number and intensity of sparkle points, were discussed. These additional parameters could prove to be of help when different pigment types such as natural mica versus synthetically produced interference pigments are being evaluated.

## 1.2 Instrumental measurement of sparkle and graininess

A multi-angle spectrophotometer was developed to measure sparkle and graininess, featuring additional direct and diffuse illumination and digital-camera detection [2]. The resolution of the black and white camera correlates with the spatial resolution of the human eye. To differentiate between graininess and sparkle, the camera has to take pictures in various lighting conditions. Diffuse illumination by a white coated hemisphere is used to simulate a cloudy sky for measuring graininess. Direct illumination at three angles is used to measure the impression of sparkle in direct sunlight (Figure IX.3).

To obtain numerical values that are useful for daily process control and QC purposes, the camera pictures are analysed with algorithms compiled on the basis of visual evaluations of a variety of automotive finishes and on input from several partners in the automotive, pigment and paint industries.

### 1.2.1 Sparkle measurement under direct illumination

The impression of sparkle changes with the angle of illumination. The instrument therefore illuminates the sample at three different angles 15°/45°/75° and takes a photograph with a CCD camera located at the perpendicular (Figure IX.3) at each of these angles. The camera is equipped with a dynamic exposure time, which is dependent on the lightness of the sample. Image-analysis algorithms analyse the photographs by using the histogram of lightness levels of the individual pixels to calculate sparkle parameter values. A threshold is set and only the very bright pixels above the threshold are evaluated for sparkle. For better differentiation, the impression of sparkle is described by a two dimensional system: sparkle area and sparkle intensity at each angle (Figure IX.4).

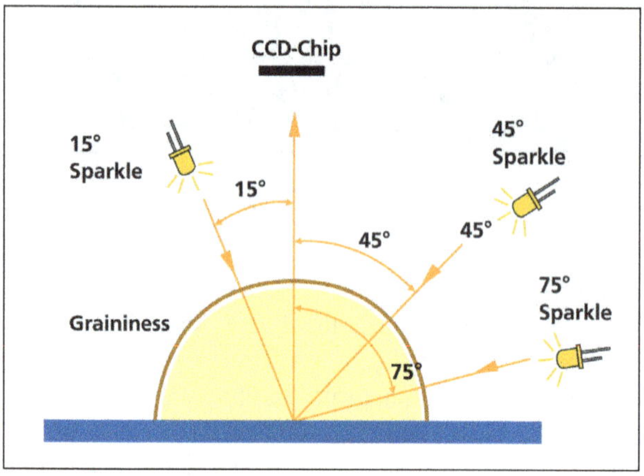


Figure IX.3: Schematic for effect measurement.

The sparkle data, sparkle area and intensity yield information on different flake types and concentration levels. Additionally, comparison of the sparkle area at 15° and 75° illumination provides information about flake orientation. The coloured hyperbola in Figure IX.4 relate to a combined value, namely sparkle

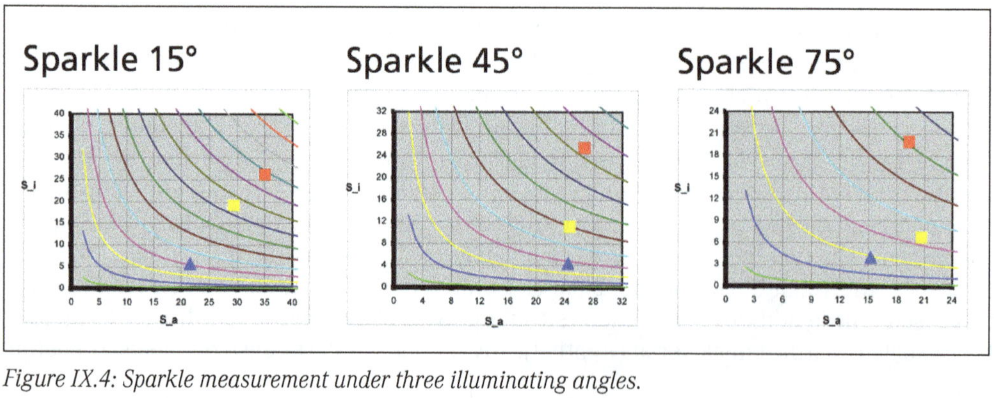


Figure IX.4: Sparkle measurement under three illuminating angles.

grade. Sparkle grade is a function of sparkle area and sparkle intensity and can be interpreted as a measure of how strongly the human observer perceives the sparkling effect of a coating. Various visual studies have shown that sparkle grade cannot be used as a major acceptance criterion for pass/fail decisions: a sample with a low sparkle area but high sparkle intensity can have the same sparkle grade as a sample with high sparkle area and low sparkle intensity, and thus appears totally different.

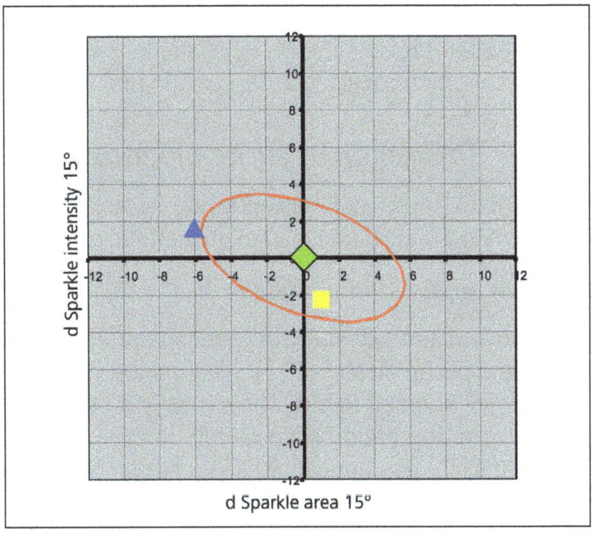


*Figure IX.5: Delta Sparkle describes the total sparkle difference between sample and standard.*

This prompted the development of a sparkle tolerance model which allows a "delta sparkle" value to be set for paint batch approval or production QC. The calculation of "delta sparkle" is a weighted value and includes an acceptance limit both for sparkle grade and for a change within the same sparkle grade. The sparkle tolerance space has the form of an ellipse, whose long axis is tangential to the sparkle grade hyperbola of the standard (Equation IX.1). It therefore corresponds to human perception, which is elliptical. Consequently, a larger deviation is allowed within the same sparkle grade than from grade to grade.

*Equation IX.1*

$$ds = \sqrt{\left(\frac{f_1(Sa_{Std}, dSa, Si_{Std}, dSi)}{Tol\_Gr}\right)^2 + \left(\frac{f_2(Sa_{Std}, dSa, Si_{Std}, dSi)}{Tol\_Gr * Tol\_GF}\right)^2}$$

where

*$Sa_{std}$ = Sparkle area of standard* — *Sa = Sparkle area of sample*
*$Si_{std}$ = Sparkle intensity of standard* — *Si = Sparkle intensity of sample*
*Tol_Gr = Tolerance grade* — *Tol_GF = Tolerance grade factor*
*dS = Delta sparkle*

### 1.2.2 Graininess measurement under diffuse illumination

Graininess is evaluated by taking a photograph with the CCD camera in diffuse lighting conditions. The photograph is analysed using the histogram of lightness levels of the individual pixels, with the uniformity of light and dark areas being combined into a single graininess value. A value of zero indicates a solid colour; the higher the value, the grainier or coarser the sample will look in diffuse light.

# 2 Sparkle and graininess applications

Sparkle and graininess are two independent parameters that can be used for classifying effect pigments and evaluating flake orientation.

## 2.1 Influence of flake size on sparkle and graininess

Sparkle and graininess data give information on flake size and concentration levels. The sample in Figure IX.6 has a silver finish with three different flake sizes (20 µm, 34 µm, 54 µm). Visually, the silver finish with the coarser aluminium pigments appears to have more sparkle in direct illumination and grainier in diffuse lighting. The greater sparkle is caused by the higher reflectivity of the larger flake. The readings correlate well with the visual assessment: sparkle area, sparkle intensity and graininess increase with flake size.

## 2.2 Influence of flake orientation on total colour impression

Besides flake types and concentration levels, comparisons of sparkle area at 15° and 75° illumination provide information about flake orientation.

### Different application methods

Automotive plants are boosting their paint efficiency by switching to 100% electrostatic basecoat application. Metallic finishes containing coarser aluminium flakes will have more non-parallel oriented flakes. The result will be a lower light-dark flop and more sparkling at a low, grazing illuminating angle. In the example shown in Figure IX.7, the basecoat of the car body was applied 100% electrostatically and the bumpers were painted with a bell/pneumatic application. The total colour difference using the mean ΔEDIN6175-2 [3] was acceptable (see Table IX.1).

Yet, visually, the car body sparkled much more than the bumper. The readings for sparkle 75° clearly confirm the visual impression. The sparkle 75° measurement evaluates the aluminium flakes, which are oriented in a non-parallel manner. The main change can, therefore, be seen in increasing sparkle area at the low grazing angle.

### Different rheology additives

Flake orientation can also be influenced by the paint formulation, e.g. the rheology additive. As fine aluminium flakes have more edges and consequently more light is scattered, orientation is more important in the case of coarser pigments. The use of an optimised rheology additive leads to a greater light-dark flop and less sparkling at lower grazing angles.

Figure IX.8 shows a waterborne system which was evaluated using three different rheology additives: a standard system, an acrylic thickener and a wax additive called "Aquatix". Visually, the three panels look the same in direct illumination at a steep angle (sparkle 15°). Comparison at a lower grazing angle (sparkle 75°), reveals less sparkling by the system containing the wax additive.

The readings correlate with the visual judgement. The sparkle area for the system containing the wax additive at 75° is smaller than for the other two systems. As sparkle 75° evaluates flakes which are oriented in a non-parallel manner, this clearly shows that the use of the wax additive "Aquatix" improves the orientation of the aluminium flakes.

*Table IX.1: ΔEDIN6175-2 is well below one for all measurement angles.*

| Sample Name | ΔEDIN6175-2 avg. |
|---|---|
| FuelDoor2 | 0.59 |
| FuelDoor1 | 0.88 |
| DPillar_R | 0.63 |
| DPillar_L | 0.56 |
| Door_R | 0.53 |
| Door_L | 0.62 |
| Bumper_R2 | 0.56 |
| Bumper_R1 | 0.40 |
| Bumper_F3 | 0.89 |
| Bumper_F1 | 0.87 |
| Bumper_F3 | 0.90 |

# 3 Conclusions

The introduction of more and more new effect pigments requires new innovative measurement technologies to capture the total colour impression. It is no longer sufficient to measure the colour impression merely at different viewing angles. As the colour perception of effect

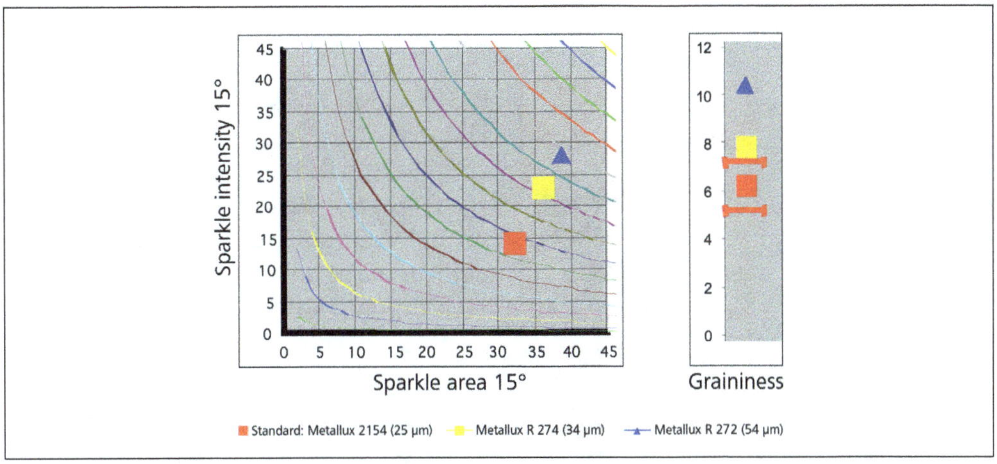


*Figure IX.6: Influence of flake size on sparkle and graininess.*

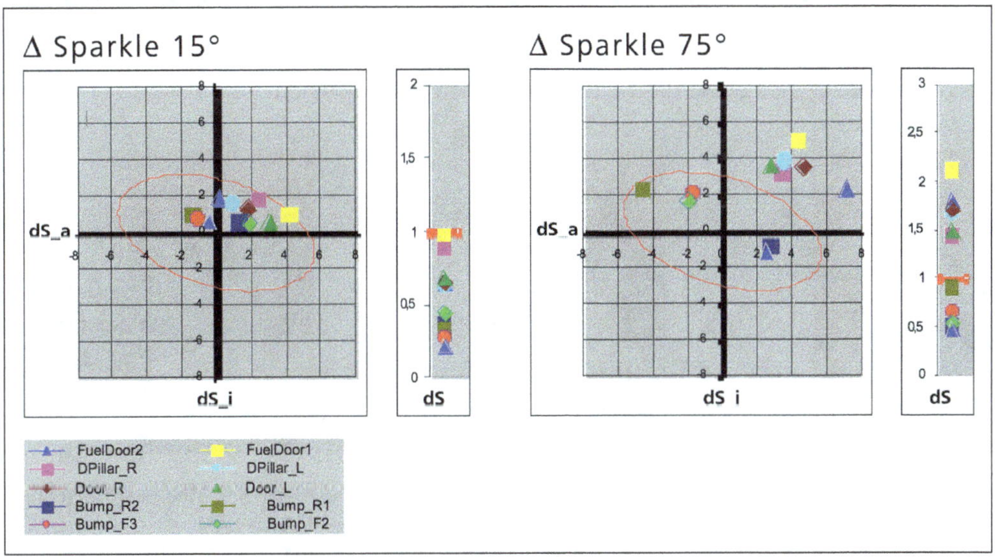


*Figure IX.7: Influence of application method on flake orientation.*

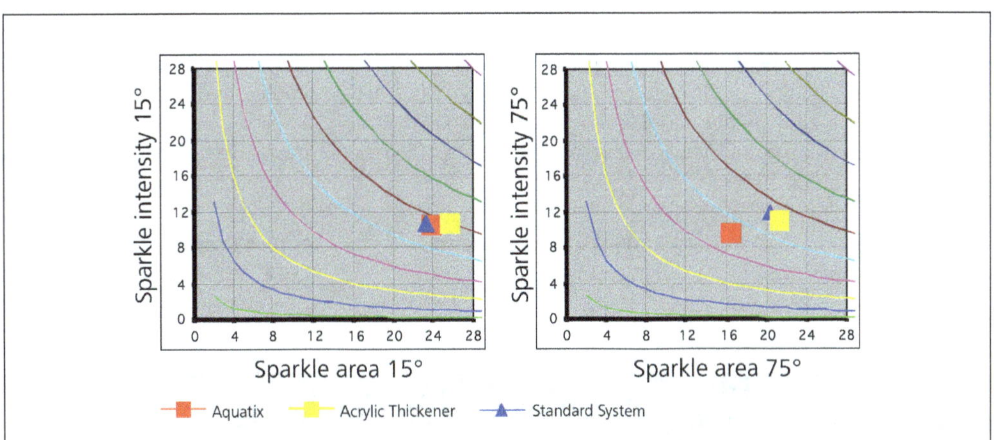


*Figure IX.8: Influence of rheology additive on flake orientation.*

pigments also changes in different lighting conditions (direct and diffuse illumination), so-called sparkle and graininess effects need to be described objectively and measured. These new parameters are valuable tools for new colour development as well as for a colour-harmony evaluation of a "multi-component" product.

# 4 Literature

[1] Kirchner E.J.J., van den Kieboom G.J., Njo S.L., Supèr R., Gottenbos R., The Appearance of Metallic and Pearlescent Materials, COLOR research and application 2006 4, 31

[2] Kigle-Böckler, G.; New measurement system for characterising the total colour impression of effect coatings; $12^{th}$ international AIC Congress, New Castle Gateshead, UK, 2013

[3] DIN6175-2. Tolerances for Automotive Paints - Part 2: Goniochromatic paints, 2001 [general] Rodrigues, A. B. J., in AIC Colour and Paints Proceedings of the Interim Meeting of the International Colour Association, Porto Allegre, 2004, P. 103-108

# X Characterisation of pigments

*Rainer Henning*

By absorbing light selectively, conventional pigments convey the impression of colour to the human eye and are therefore correctly called colourants in German standard DIN 55943. The definition therein also states that pigments are substances, which are insoluble in the application medium and it subdivides them into organic and inorganic colourants.

In the following, the most important organic and inorganic pigments for industrial applications are briefly characterised in terms of their coloristic properties. For this, pigmented coatings are adjusted with white to a standard depth (SD). For most inorganic pigments, 1/3 SD is possible, but in some cases the chroma of the full shade only allows 1/9 SD.

All the organic pigments described below can be adjusted to 1/3 SD.

### Standardised samples

The mixtures are made in accordance with a standardised procedure:

Binder: alkyd-melamine, pigment grinding on a paint-shaker machine (Skandex) for 3 h, followed by mixing of the pigment paste with the quantity of white necessary for the specific standard depth. The final paints are applied in drawdown on black and white contrast until hiding and are then oven-dried.

Colour measurement is performed with a sphere spectrophotometer in the wavelength range 400 to 700 nm; the figures quoted are net of a calculated gloss reduction of 4 %.

The chemical structure of most colourants can be found in the Colour Index. Sometimes several products from different suppliers exist under the same C.I. number; these may exhibit significantly different coloristic figures upon closer analysis.

### Colour Index

The Colour Index (C.I.) has been published by the British Society of Dyers and Colourists and the American Association of Textile Chemists and Colorists since 1925.

The index gives generic names and classifies the colorants as dyes or pigments by naming the base colour and, using a sequential number, provides information on the chemical structure as well as names of commercial products (e.g. Isoindoline has the C.I. Pigment Yellow 139 or P.Y. 139).

For more details, see Wikipedia, or visit the homepage of the American Association of Textile Chemists and Colorists.

The following sections present the coloristic differences of several chemical chromophores. In some cases, the differences within one Colour Index are shown, which may be related, e.g., to particle size or crystal shape.

# 1 *Inorganic pigments – characterisation*

*Rainer Henning*

The following descriptions focus on the most important technical properties and application areas. For more details, see reference [1].

Characterisation of the coloured pigments starts with the achromatic components white and black. They are responsible for the lightness axis in colour space and so a few remarks will be made on their specific properties.

## 1.1 *White (P.W. 6) and black (P.Bl. 7)*

Each of the achromatic pigments is of minor coloristic importance on its own, as they define the endpoint of the lightness axis. Here, chroma trends towards zero by definition.

However, there can be noticeable differences within the various titanium dioxides (P.W. 6) [2] and carbon blacks (P.Bl. 7) [3] which cannot be discussed here in detail.

In all of the plots that follow, only a neutral, white titanium dioxide in the rutile modification is used. Figure X.1 shows the colour locations of several mixtures of titanium dioxide (P.W. 6) and carbon black (P.Bl. 7).

The plot shows $b^*$ and lightness values $L^*$. Increasing quantities of $TiO_2$ lead to smaller $b^*$ values, which means the coloristic property is developing in the “blue” direction. This is caused by the specific properties of the carbon black used. Figure X.2 shows the corresponding reflectance curves.

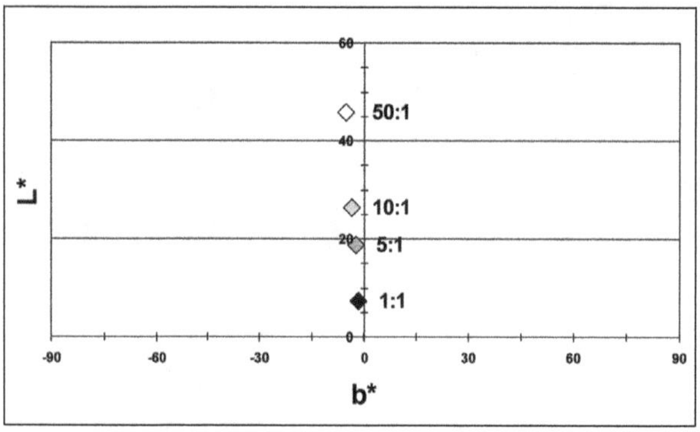


*Figure X.1: Colour locations of various mixtures of $TiO_2$ and carbon black.*

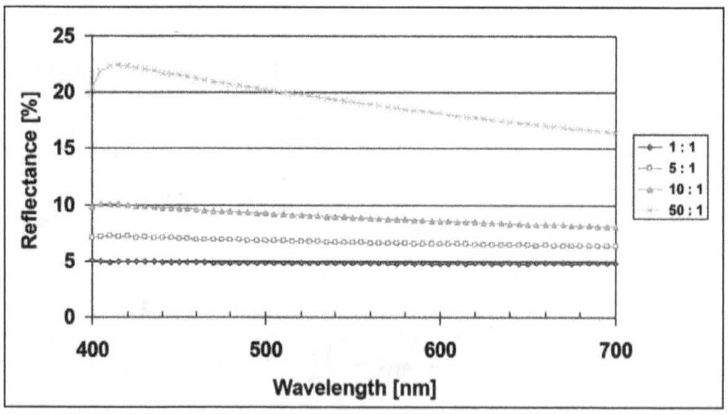


*Figure X.2: Reflectance curves of several mixtures $TiO_2$/ carbon black.*

Clearly visible is the higher reflectance at shorter wavelengths (bluish) as opposed to longer wavelengths (reddish). This becomes a maximum reflectance at higher $TiO_2$ concentrations.

The visual impression confirms the message from these measurement results: the carbon black is a bluish type.

## 1.2 *Important inorganic coloured pigments*

Figure X.3a shows the colour locations and X.3b the reflection curves of important inorganic coloured pigments.

Some chromophores exist in the red and yellow region, some of which have a rather high chroma and are also quite strong in colour for inorganic pigments. There are gaps in the blue and green area; the later discussion of organic coloured pigments will show how these gaps can be closed.

For greater clarity, the corresponding reflectance curves are divided into red, yellow and green or blue. Figure X.4 shows the reflectance curves for P.R. 101 in its opaque and transparent modification as well as for P.R. 104. Iron oxide pigments typically show slow-rising reflectance values from approx. 540 nm on; as expected, the transparent version is higher than the opaque version.

Five colour indices are important in the yellow region; their reflectance curves are shown in Figure X.5.

### 1.2.1 *Inorganic yellow and red pigments*

The reflectance curve of iron oxide yellow (P.Y. 42) is similar in shape to that of iron oxide red (P.R.101). The transparent modification again has higher reflectance values in the yellow and red wavelength ranges. Coatings with much more chroma are obtained when P.Y. 34 is used. The reflectance curve is remarkably flat up to 480 nm, and rises sharply from 500 nm on. P.Y. 184 is the most greenish-yellow member of the inorganic yellow pigment family. Its reflectance curve rises sharply from 470 nm on, remaining high at 540 nm and above. Nickel and chromium titanium yellow pigments have rather low colour strength; even the full shades do not reach 1/3 SD; so only 1/9 SD was prepared here. Consequently, the reflectance curves for P.Y. 53 and P.Br. 24 cannot be compared directly with those for the other inorganic yellow pigments. They

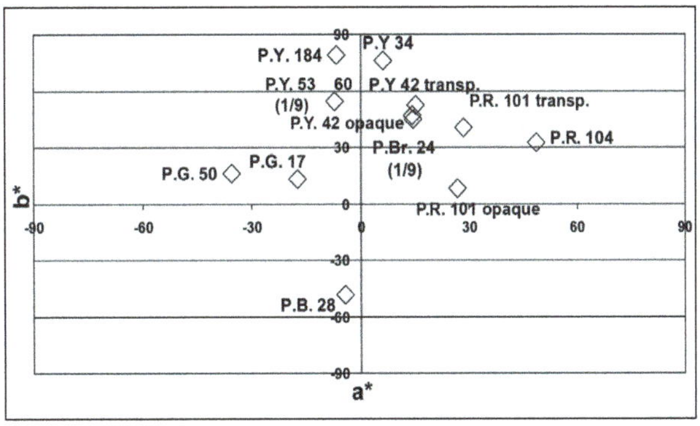


*Figure X.3a: Colour locations of inorganic coloured pigments at 1/3 (1/9) SD.*

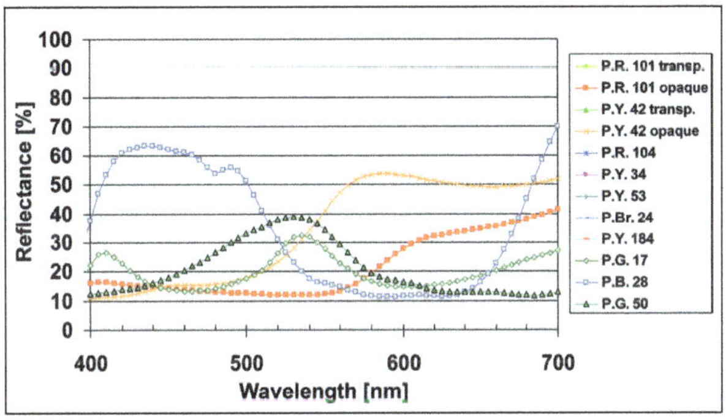


*Figure X.3b: Reflection curves of inorganic coloured pigments at 1/3 (1/9) SD.*

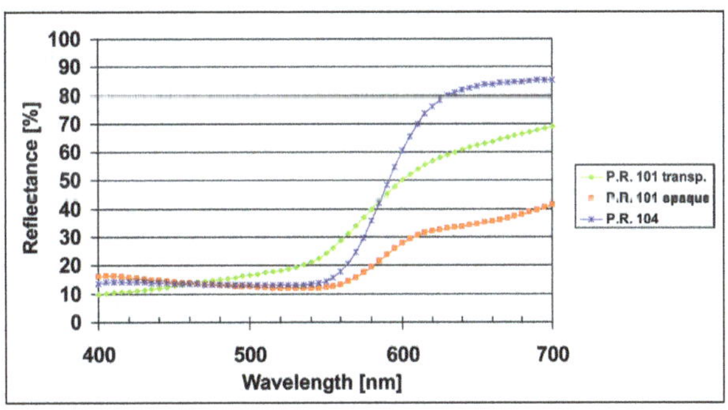


*Figure X.4: Reflectance curves of inorganic coloured red pigments at 1/3 SD.*

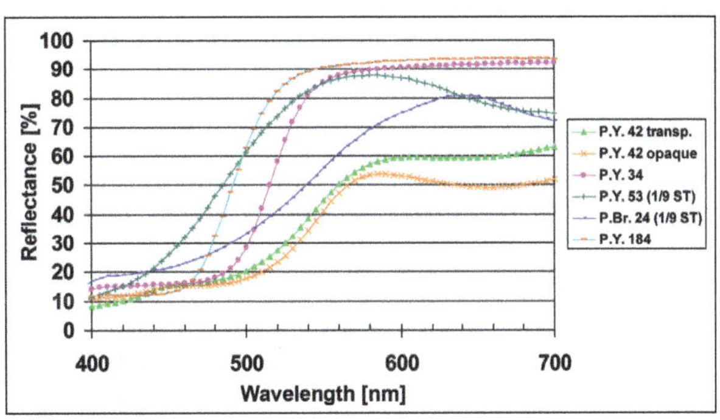


*Figure X.5: Reflectance curves of inorganic yellow pigments.*

rise slowly from 440 nm on, and pass through a broad maximum at approx. 570 nm (P.Y. 53) and 670 nm (P.Br. 24).

***Iron oxide red (P.R. 101)***
The iron in iron oxide red (P.R. 101) [4] has the oxidation state +3; together with oxygen, it forms haematite, the alpha-modification of $Fe_2O_3$. Its reflectance curve rises slowly from 580 nm on, and again at somewhat higher values in the blue wavelength range. Both the natural and the synthetic pigments are used in technical applications. Because they are inert, they are found in all paint applications. The transparent modifications also serve in wood stains and automotive effect top coats, because they scatter light much less than the opaque types.

***Iron oxide yellow (P.Y. 42)***
In aqueous solutions of iron salts, the iron oxide hydrate can be precipitated to serve as a precursor for P.R. 101. However, the precipitation can be isolated as iron oxide yellow (P.Y. 42) [4]. The reflectance curve differs from that of P.R. 101, and starts rising as early as from 520 nm on. The presence of water of crystallisation limits the heat resistance to approx. 160 °C. Above this temperature, the modification gradually changes into P.R. 101.

***Molybdate red (P.R. 104)***
Molybdate reds (P.R. 104) [5] coloristic properties are superior to those of P.R. 101 and it imparts yellowish to bluish red shades in paint formulations. The pigment is easy to disperse and has good hiding properties.

***Chrome yellow (P.Y. 34)***
The chrome yellow (P.Y. 34) [6] pigments yield paint formulations possessing good chroma, hiding and weathering fastness in the neutral-yellow to reddish-yellow range. The various commercial products, as also is the case for P.R. 104, differ in their coloristic properties. The use of lead containing pigments is being more and more restricted, because these products are being classified as problematic for creatures and the environment [6]. In the meantime, alternatives are existing with good coloristic properties; e.g. P.R. 104 can be almost completely replaced by opaque organic pigments. P.Y. 184 is the alternative for P.Y. 34; however it must be combined in some cases with organic yellow pigments.

***Nickel rutile yellow (P.Y. 53)***
Incorporating metal ions into the rutile crystal lattice leads to complex, inorganic coloured pigments (CICPs). Nickel (Ni) (P.Y. 53) [7] yields pigments that have a light, greenish-yellow hue; however, the low colour strength of the full shade prevents 1/3 SD from being obtained. The plot therefore shows the colour location and reflectance curves for 1/9 SD.

***Chrome rutile yellow (P.Br. 24)***
Instead of nickel, chromium III can be incorporated in to the rutile crystal lattice. The resulting chrome rutile yellow (P.Br. 24) [7] pigment creates a reddish-yellow hue in paint. Similar to P.Y. 53, the full shade of chrome rutile yellow is so low, that 1/3 SD is not possible, therefore the plot shows 1/9 SD only.

***Bismuth vanadate yellow (P.Y. 184)***
Bismuth vanadate yellow (P.Y. 184) [8] is a mixture of bismuth vanadate and bismuth molybdate as a 2-phase pigment which yields chromatic, neutral-yellow colourings. Its coloristic properties are superior to those of iron oxide yellow and nickel rutile yellow. It is also used as a replacement for P.Y. 34 in „lead-free“ formulations.

### *1.2.2 Inorganic green and blue pigments*

Figure X.6. shows the reflectance curves for the most important green and blue inorganic pigments.

***Chrome oxide green (P.G. 17)***

Chrome oxide green (P.G. 17) [9] is one of the few single-phase pigments. It yields green colourings of low chroma - lower even than for iron oxide pigments. Its colour strength is also extremely low. The reflectance curve exhibits maxima at 540 and 410 nm, and rises slowly from 620 nm on. The pigment is used in industrial paint formulations where brilliant shades are not required.

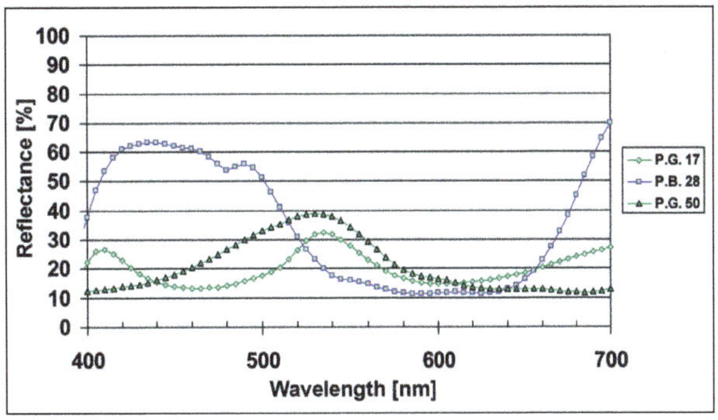


*Figure X.6: Reflectance curves of green and blue inorganic pigments.*

***Cobalt green mixed-phase pigment (P.G. 50)***

Blue to green hues are created by incorporating cobalt into the inverse titanium spinels magnesium titanate and zinc titanate. The reflectance curve exhibits a broad maximum at approx. 535 nm. Unlike the case for P.G. 17 [10], reflectance values continuously decrease towards more bluish and reddish wavelengths.

***Cobalt blue (P.B. 28)***

Formerly known as Thénard's blue, cobalt blue (P.B. 28) [10] is a cobalt aluminium spinel. It has a brilliant blue hue with a red tinge, which roughly correlates to the hue of indanthrone blue. However, its colour strength is much lower than that of P.B. 60.

# 2 *Organic pigments - characterisation*

*Rainer Henning*

Organic pigments, like the inorganic pigments, are listed in order of hue value in the CIELAB system and therefore start off with red. Only the most important technical properties and their applications can be described here; more details can be found in the literature [11].

## 2.1 *Red pigments*

Figure X.7 shows the colour locations of the organic red pigments in a white reduction corresponding to 1/3 SD.

The large number of products spans the colour space from bluish to yellowish reds; three colour indices (P.R. 52:2, P.R. 122, P.R. 202) are so bluish that their hue is at approx. 340°. They will be discussed in conjunction with the violet pigments. P.R. 188 is the most yellowish-red. Figure X.8 shows the corresponding reflectance curves at 1/3 SD.

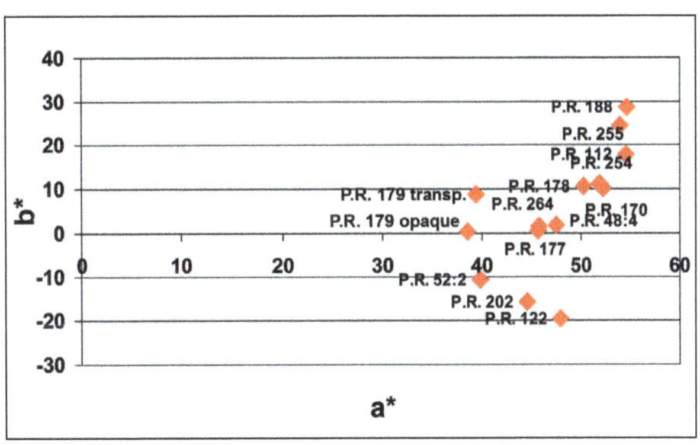


*Figure X.7: Colour location of organic red pigments.*

The steep rise from 580 nm on is typical of the reflectance curves for organic red pigments, along with more or less pronounced reflectance in the region around

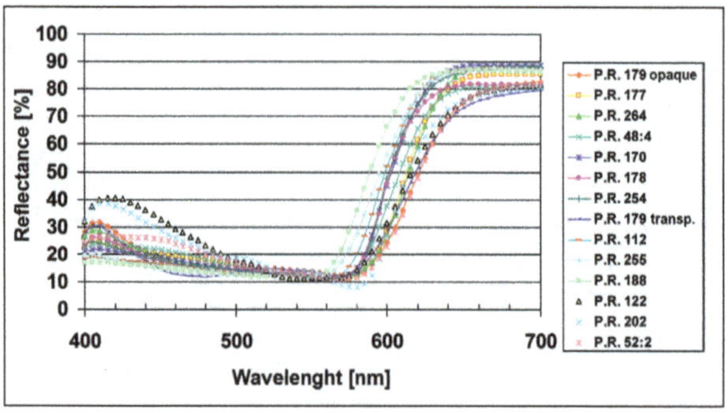


*Figure X.8: Reflectance curves of red organic pigments.*

420 nm. Bluish-red pigments are characterised by a fairly strong reflectance in that wavelength range; the higher the reflectance around 420 nm, the more bluish the hue shift (e.g. P.R. 122 and P.R. 202). Pigments with a yellowish-red hue do not exhibit a significant rise in the blue range (see P.R. 188).

***Perylene, opaque (P.R. 179)***

The opaque version of P.R.179 ranks first in the list of colour indices; at 1/3 SD, its hue is found at 0.6°, characterising the pigment as bluish-red. The transparent modification is remarkably more yellowish while having similar colour strength. Its main application lies in automotive effect colour shades on account of its very high outdoor durability. For a detailed description, see e.g. *Herbst* and *Hunger* (p. 483 ff [12]).

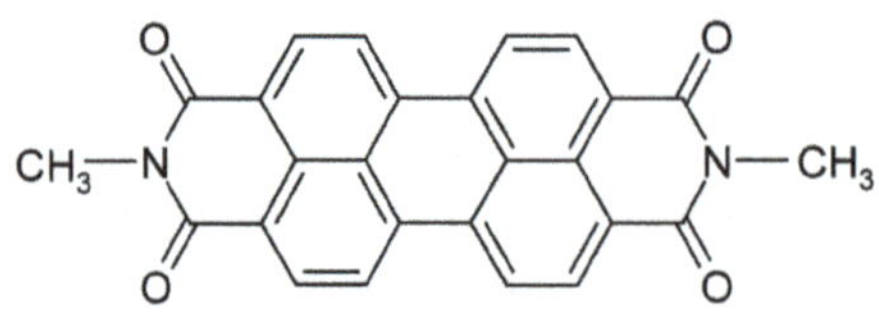


***Anthraquinone (P.R. 177)***

P.R. 177 [13] is slightly more yellowish, but much more chromatic than the opaque P.R. 179. It is highly transparent and exhibits a clean, bluish-red shade; consequently it is used in opaque, red automotive shades.

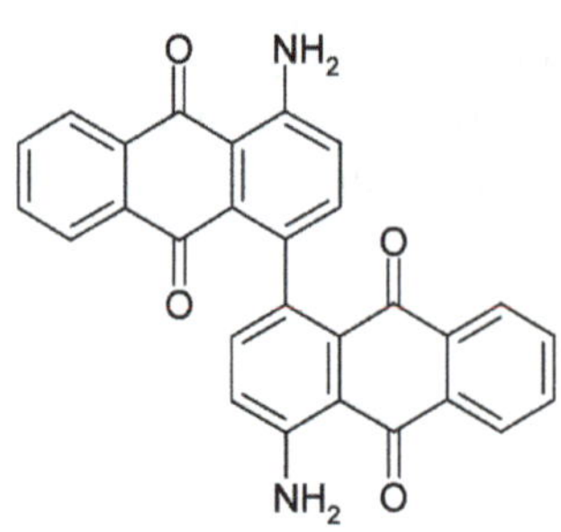


***Diketopyrrolopyrrole (P.R. 264)***

P.R. 264 [17] as a member of the diketopyrrolopyrrole (DPP) family yields a high-chroma bluish-red. It is often used in effect colour shades because of its high transparency.

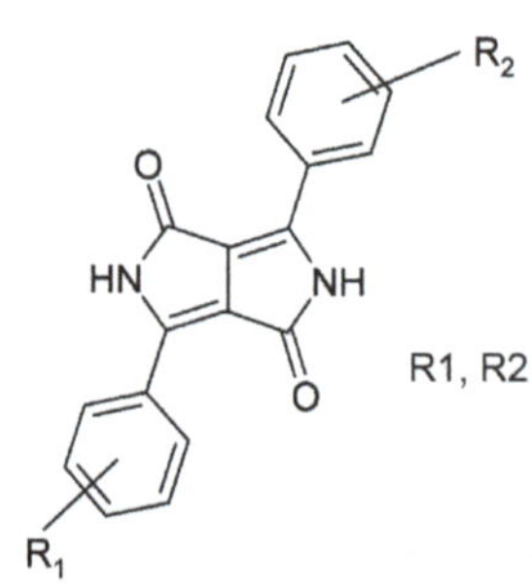


R1, R2 = phenyl

***BON pigment Mn lake (P.R. 48:4)***

This chromophore is a manganese lake [14] of monoazo-dyes with carboxylic and sulphonic acid groups. Only the manganese lakes are important for paint application, because they exhibit better fastness to light and weathering compared to salts of alkaline earth metals.

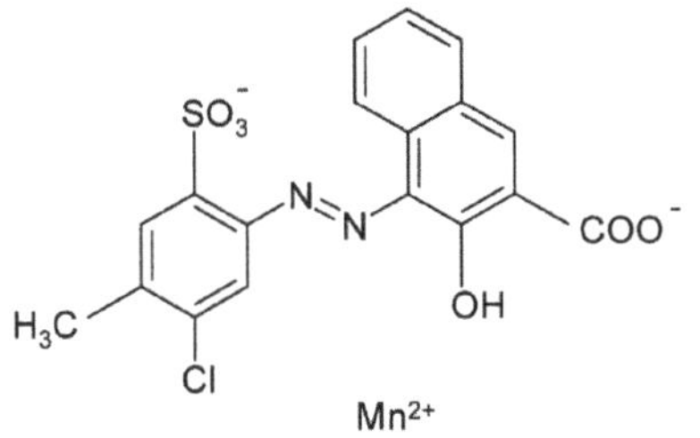


***Naphthol AS (P.R. 170)***

In its full shade, this naphthol AS (P.R. 170) [15] pigment has good fastness to light and weather; its poor solvent resistance and overcoating fastness limit its application to air-drying industrial coatings and decorative emulsion paints.

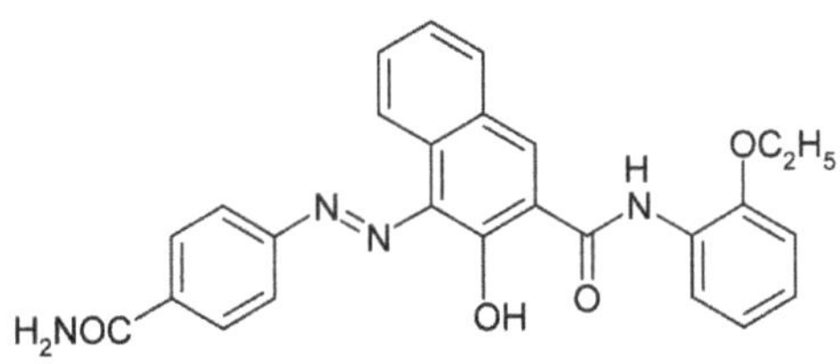


***Perylene, opaque (P.R. 178)***

Perylene opaque (P.R. 178) [16] has been the standard pigment for lead-free red automotive shades for many years. It has been used in combination with opaque iron-oxide pigments, and sometimes also with Pigment Orange 36, Quinacridones and Pigment Red 177. It has now been largely superseded by P.R. 254.

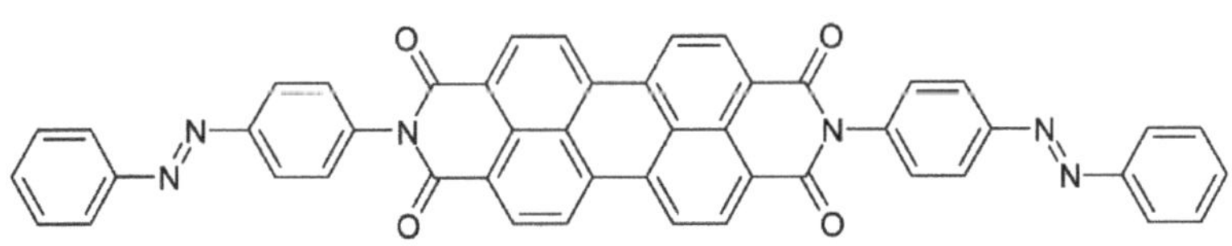

***Diketopyrrolopyrrole (P.R. 254)***

The latest high-performance organic pigments are found among the diketopyrrolopyrrole (DPP) family. P.R. 254 [17] has a very brilliant, yellowish-red hue. Due to its good opacity and high weathering fastness, it often serves in high-performance, red solid shades, especially for automotive applications.

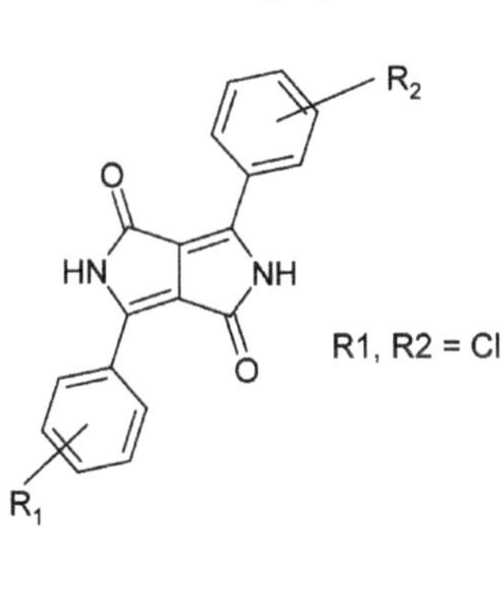

***Perylene, transparent (P.R. 179)***

Numerous finish modifications of transparent perylene (P.R. 179) [16] are available on the market; coloristically, they are much more yellowish than the opaque version. Their main application is in automotive OEM shades, because of their transparency and excellent weathering fastness.

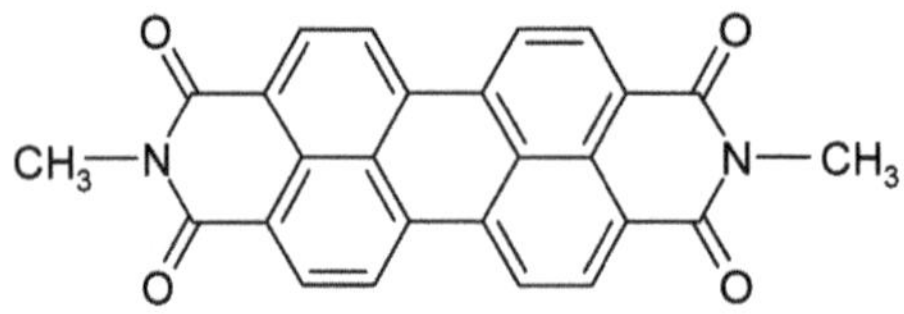

***Naphthol AS (P.R. 112)***

The naphthol AS (P.R. 112) [18] pigment has a high-brilliance, mid-red hue. It is widely used in printing inks, paints and coatings and other areas. Compared with naphthol AS (P.R. 170), it is more yellowish red, but suffers from inferior overcoating fastness.

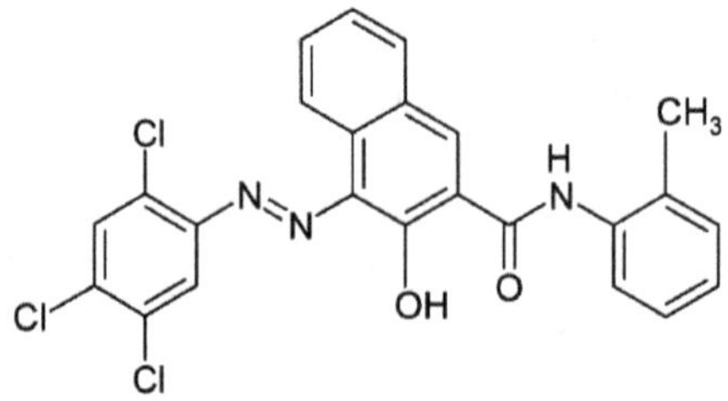

***Diketopyrrolopyrrole (P.R. 255)***

The diketopyrrolopyrrole (P.R. 255) [17] pigment is much more yellowish-red than P.R. 254 and is used in industrial and automotive coatings because of its good weathering fastness and particle-size.

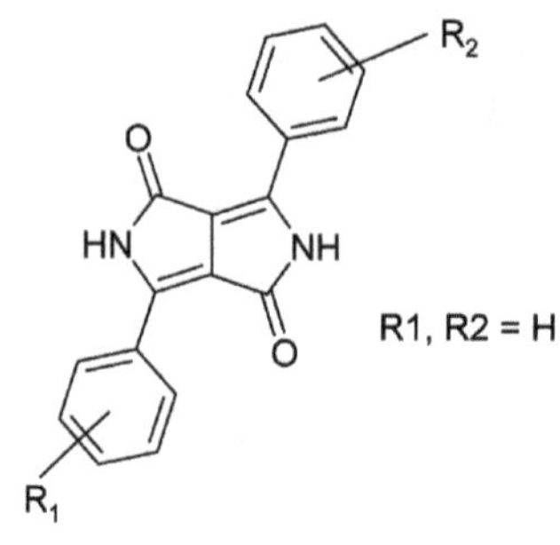

***Naphthol AS (P.R. 188)***

Naphthol AS (P.R. 188) [19] is a brilliant yellowish-red pigment with good overcoating fastness, but relatively low colour strength. Weathering fastness is good in the full shade range, and so it is used for high-performance industrial coatings.

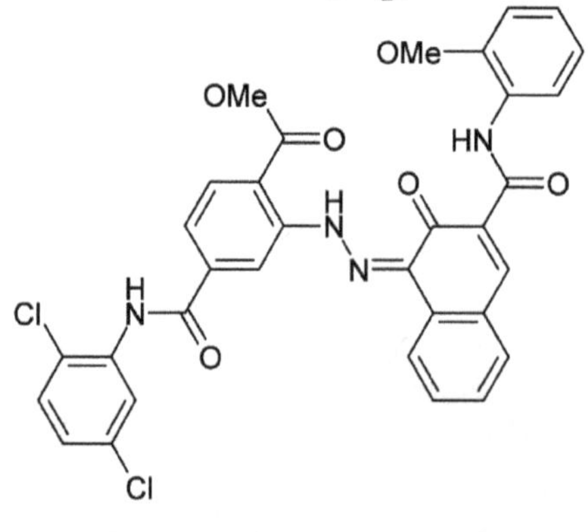

## 2.2 Orange pigments

Figure X.9. shows the colour locations for the most important orange organic pigments.

The hue of these pigments is shifted only by 10 degrees towards yellow; yet the chroma is increased by 10 CIELAB units compared to the most yellowish P.R. 188. Figure X.10. shows the corresponding reflectance curves.

All curves show a steep increase in reflectance from approx. 550 nm on; reflectance values are rather low in the shorter wavelength range, as expected.

***Diketopyrrolopyrrole (P.O. 73)***
The diketopyrrolopyrrole (P.O. 73) [17] pigment has the most reddish-orange hue and possesses the best weathering fastness of all the other orange pigments.

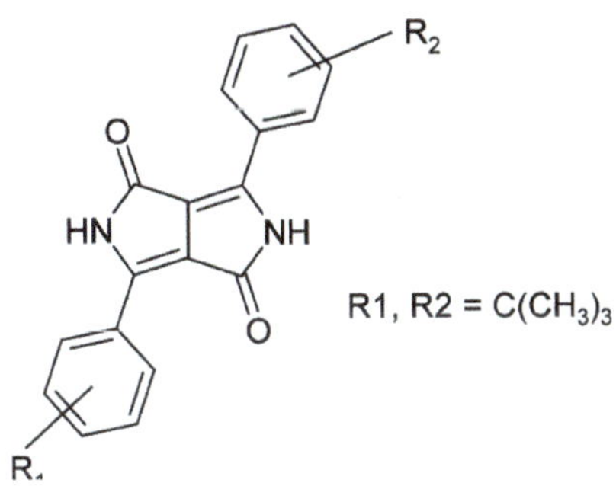


***β-Naphthol (P.O. 5)***
In its finely divided form, this monoazo pigment β-naphthol (P.O. 5) [20] has only limited fastness properties; it is therefore used in air-drying paint formulations only. A second grade on the market is coarser in particle-size, and is consequently more reddish but duller than the finely divided grade.

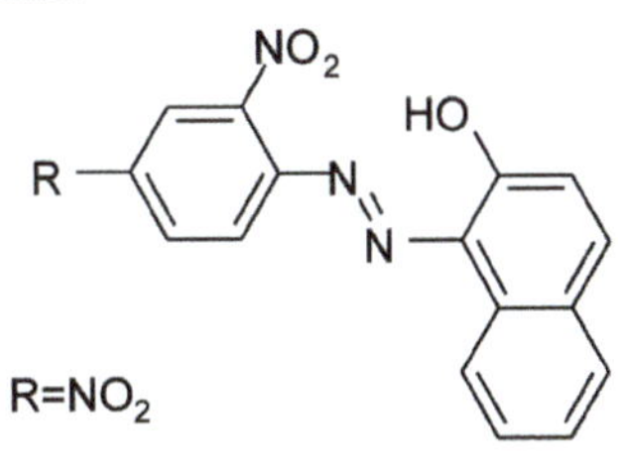


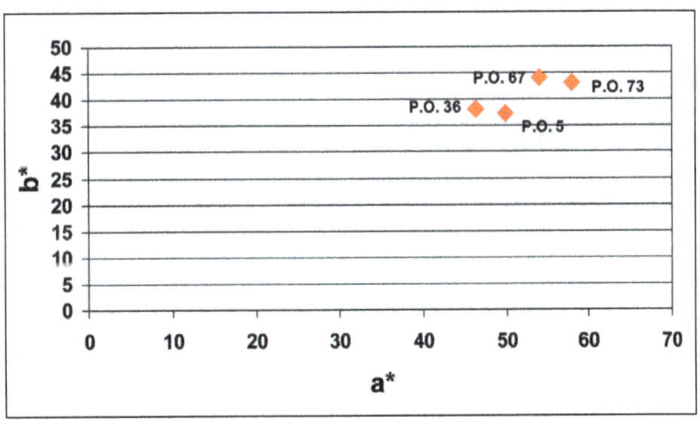


*Figure X.9: Colour locations of organic pigments at 1/3 SD.*

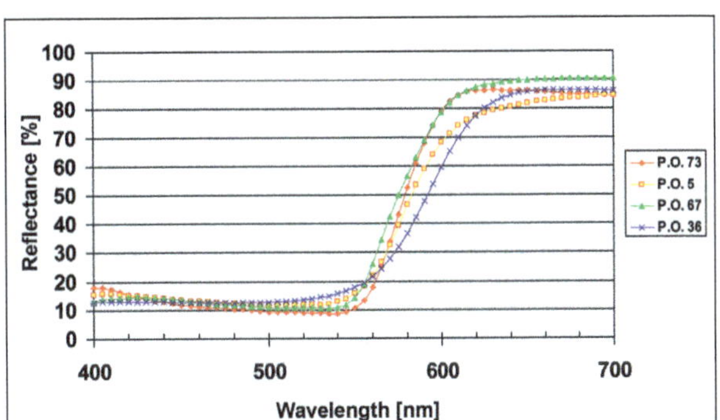


*Figure X.10: Reflectance curves for orange organic pigments at 1/3 SD.*

***Pyrazoloquinazolone (P.O. 67)***
Pyrazoloquinazolone (P.O. 67) [21] pigment has a very clean hue, but its overcoating fastness is good only below 80 °C. Its use in stoving enamels is therefore limited. Weathering fastness in the full

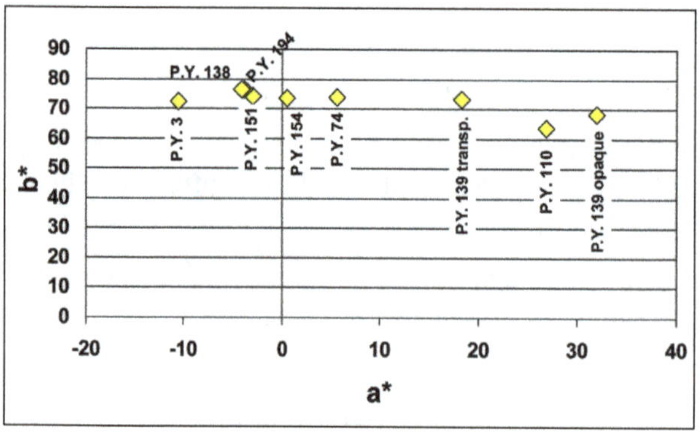


*Figure X.11: Colour locations of yellow organic pigments at 1/3 SD.*

shade range is good. Application areas lie in automotive refinishes and industrial, architectural and decorative paints.

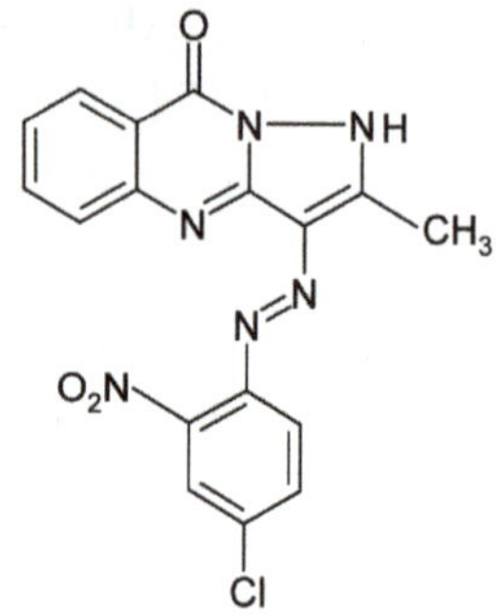

### *Benzimidazolone (P.O. 36)*

The full shade of the benzimidazolone (P.O. 36) [22] pigment offers excellent fastness to overcoating, light and weathering. It is therefore recommended for high-chroma, red RAL shades, in combination with quinacridone pigments.

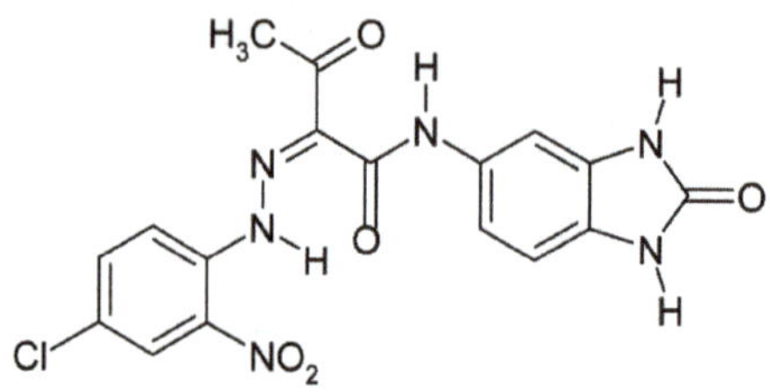

## 2.3 *Yellow pigments*

Figures X.11. shows the colour locations of the most important yellow organic pigments.

These span a range of more than 30 degrees in hue, from the opaque reddish-yellow P.Y. 139 to the greenish-yellow P.Y. 3.

The reflectance curves have low values in the blue wavelength range, rising sharply between 480 and 540 nm for various pigments. Reflectance remains at a high level in the red wavelength range.

### *Isoindoline, opaque (P.Y. 139)*

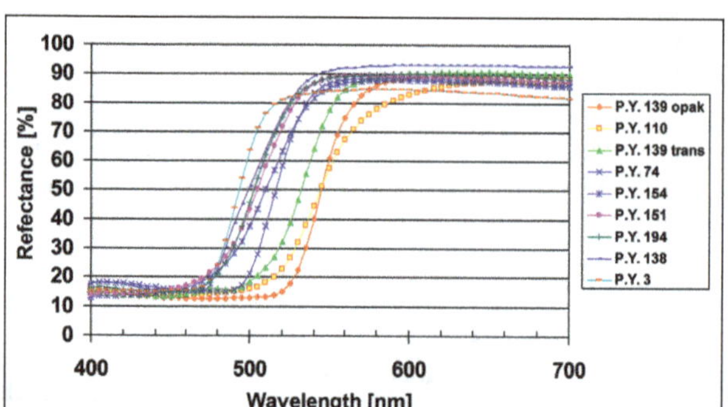


*Figure X.12: Reflectance curves for yellow organic pigments.*

This reddish-yellow isoindoline (P.Y. 139) [23] pigment has good opacity, which is unusual for organic pigments. Due to its-coarser particles, it attains only 1/3rd the colour strength of the transparent modification.

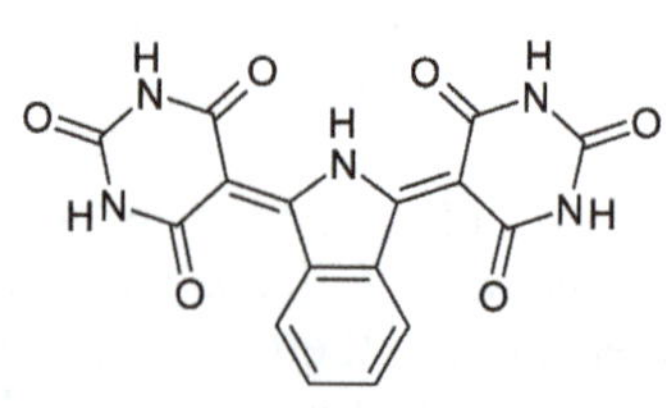

### *Isoindolinone (P.Y. 110)*

This transparent reddish-yellow isoindolinone (P.Y. 110) [24] pigment has very good weathering fastness and is therefore found in effect colour shades for automotive top-coat formulations.

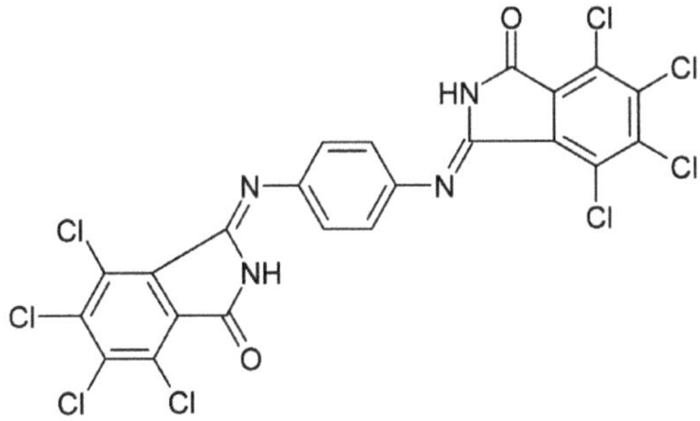

### *Isoindoline, transparent (P.Y. 139)*

The hue of this transparent modification (1/3 SD) of isoindoline (P.Y. 139) [23] is shifted by approx. 10° towards greenish yellow, compared to the opaque version. Light fastness and weathering fastness are good in the full shade range.

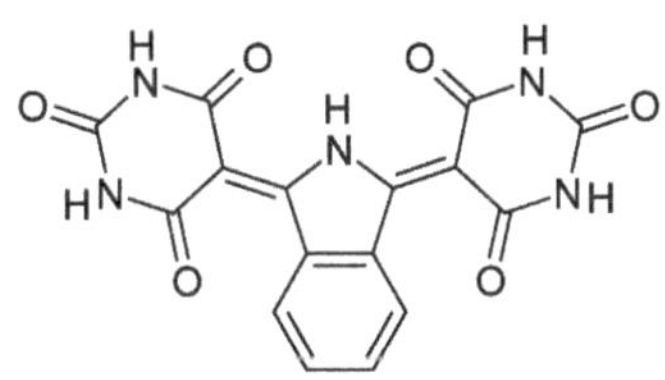

### *Monoazo (P.Y. 74)*

The monoazo (P.Y. 74) [25] pigment is found in many different applications. Its hue is comparable to that of Pigment Yellow 1, but is stronger in colour and offers superior light fastness and weathering fastness. In the full shade range, it is therefore also used for outdoor coatings.

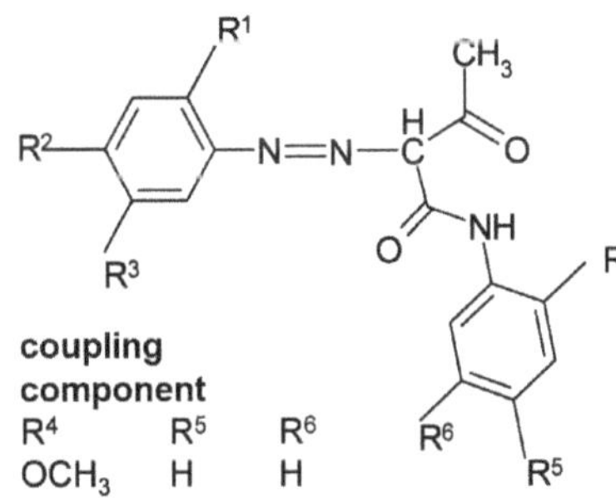

| component diazo | | | coupling component | | |
|---|---|---|---|---|---|
| $R^1$ | $R^2$ | $R^3$ | $R^4$ | $R^5$ | $R^6$ |
| $OCH_3$ | $NO_2$ | H | $OCH_3$ | H | H |

### *Benzimidazolone (P.Y. 154)*

The benzimidazolone (P.Y. 154) [26] pigment is one of the most durable organic yellow pigments; it is therefore suitable for all high-performance industrial coatings. Its weathering fastness renders it suitable for tinting.

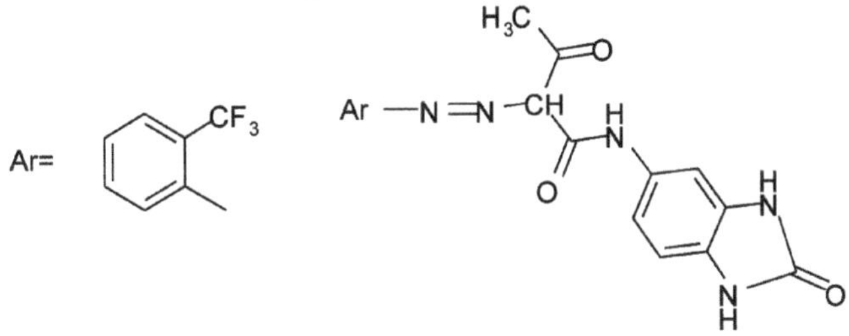

### *Benzimidazolone (P.Y. 151)*

The benzimidazolone (P.Y. 151) [27] pigment yields a clean greenish-yellow, which is slightly more greenish than P.Y. 154 but has higher tinting strength. It is mainly used in high-performance industrial coatings.

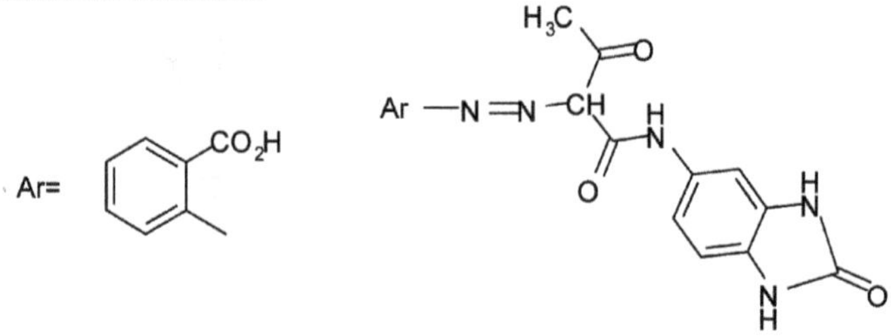


### *Benzimidazolone (P.Y. 194)*

Benzimidazolone (P.Y. 194) [22] is a greenish-yellow with medium tinting strength close to that of P.Y. 138. Its moderate overcoating fastness limits its usefulness in stoving enamels.

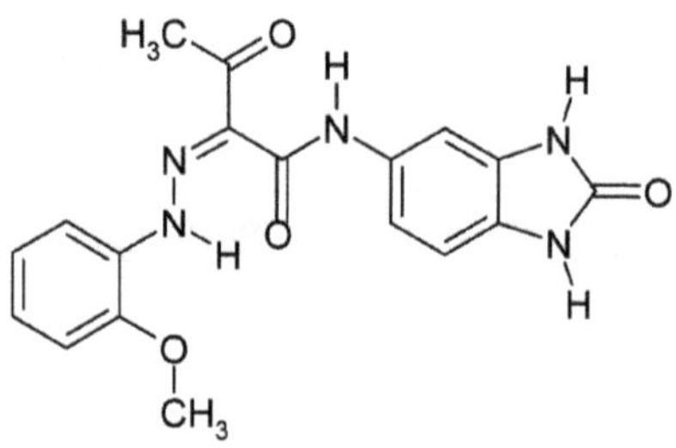

### *Quinophthalone (P.Y. 138)*

The quinophthalone (P.Y. 138) [40] pigment is a polycyclic chromophore that yields a greenish-yellow with good opacity. Due to its good weathering fastness in the full shade range, it is used in high-performance industrial coatings.

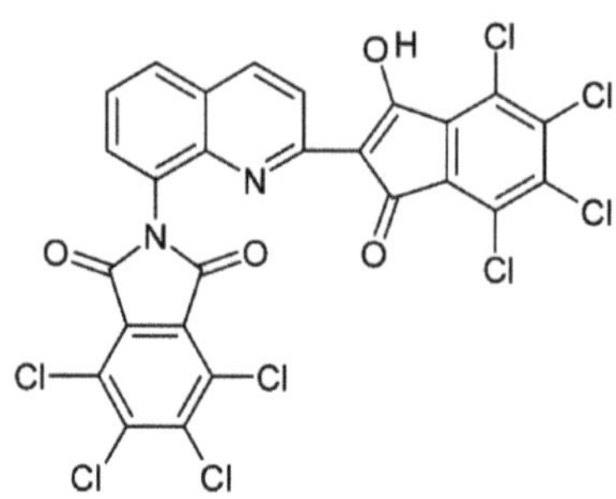

### *Monoazo (P.Y. 3)*

The most greenish-yellow pigment in this listing is monoazo (P.Y. 3) [28]. Used for creating greenish shades from mixtures with, e.g. copper phthalocyanine pigments.

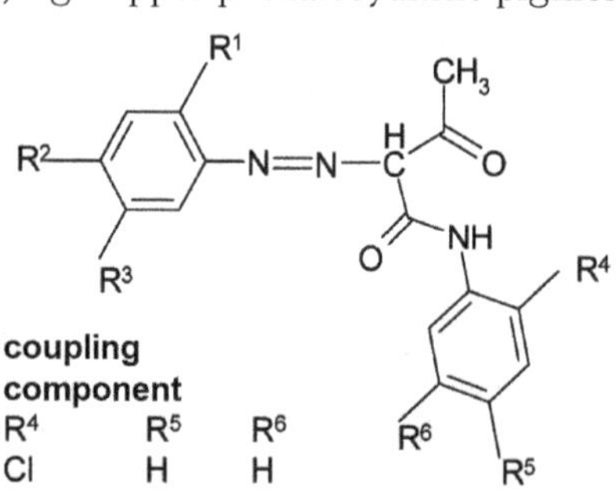

| component diazo | | | coupling component | | |
|---|---|---|---|---|---|
| $R^1$ | $R^2$ | $R^3$ | $R^4$ | $R^5$ | $R^6$ |
| $NO_2$ | Cl | H | Cl | H | H |

## 2.4 Green pigments

In the green range, only two colour indices are of technical interest: P.G. 36 and P.G. 7. Figure X.13. shows the colour location of both green pigments.

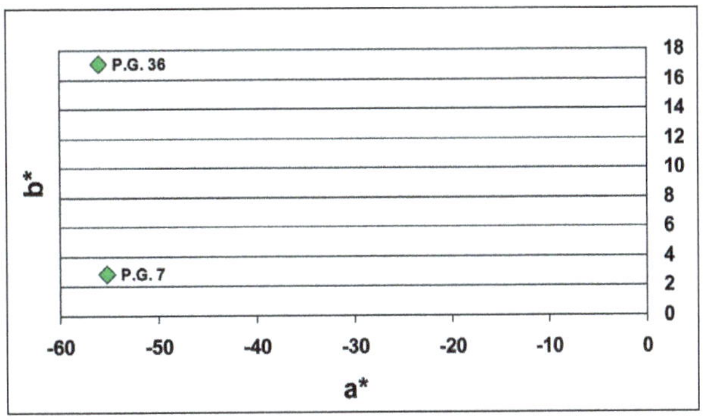

*Figure X.13: Colour location of green organic pigments.*

The colour location of P.G. 36 at 1/3 SD is more than 60 units in hue away from the greenish-yellow P.Y. 3. However, it is still significantly more yellowish than P.G. 7; this is also illustrated by the reflectance curves in Figure X.14.

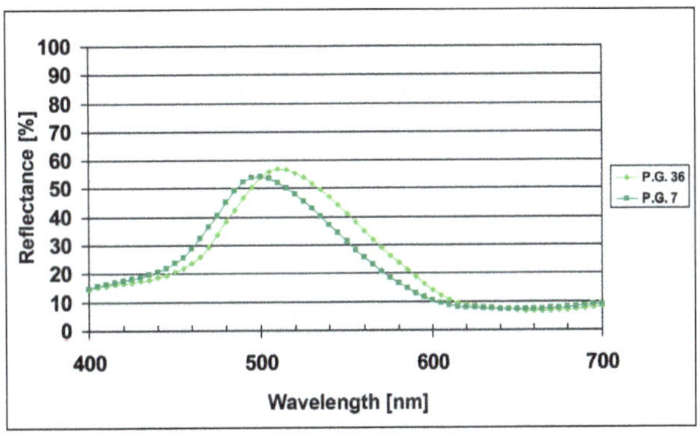

*Figure X.14. Reflectance curves for green organic pigments.*

The maximum in the reflectance curve for P.G. 36 is shifted by approx. 10 nm to higher wavelengths than is the case for P.G. 7. As the two pigments have a similar chemical structure, the shapes of their reflectance curves are similar, too.

### *Halogenated copper phthalocyanine (P.G. 36)*

The halogenated copper phthalocyanine (P.G. 36) [29] pigment is derived from P.G. 7, with the chlorine atoms replaced by bromine. As it has the same high weathering fastness as the other copper phthalocyanine pigments, it has a commensurately wide range of applications in paint formulations.

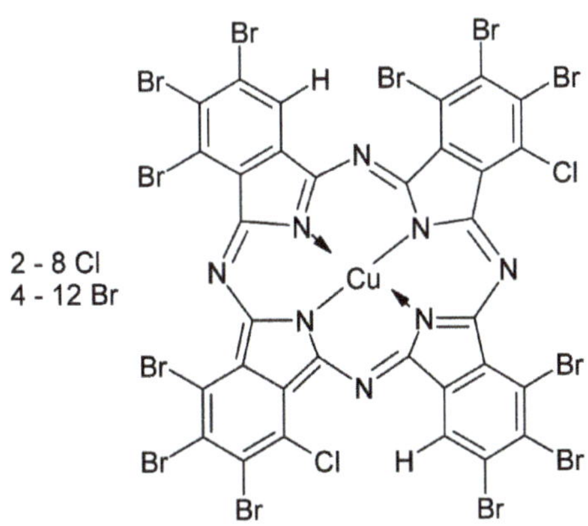


### *Halogenated copper phthalocyanine (P.G. 7)*

Replacing most of the aromatic hydrogen atoms on the phenyl substituents of the copper phthalocyanine molecule by chlorine leads to a bluish-green pigment.

This halogenated copper phthalocyanine (P.G. 7) [30] has become the standard pigment for all green paint formulations because of its high fastness properties.

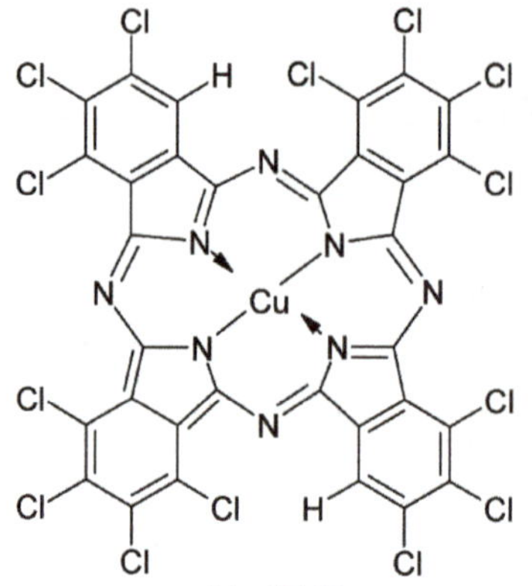


## 2.5 Blue organic pigments

The blue shade range basically contains three colour indices, ranging from greenish- to reddish-blue shades. Figure X.15 shows the colour location of the three blue chromophores.

The metal-free phthalocyanine pigment (P.B. 16) has the most greenish-blue shade, followed by the three modifications of copper phthalocyanine, with the ε modification (P.B. 15:6) being the most reddish. Pigments of the α modification are always more reddish than those of the β modification. The flocculation-stabilised products are mostly somewhat more greenish than the corresponding unstabilised types. The indanthrone pigments (P.B. 60) are much more reddish-blue.

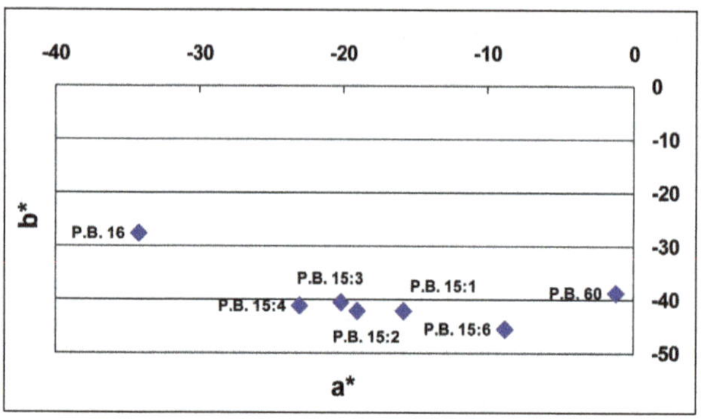


*Figure X.15: Colour locations of blue organic pigments.*

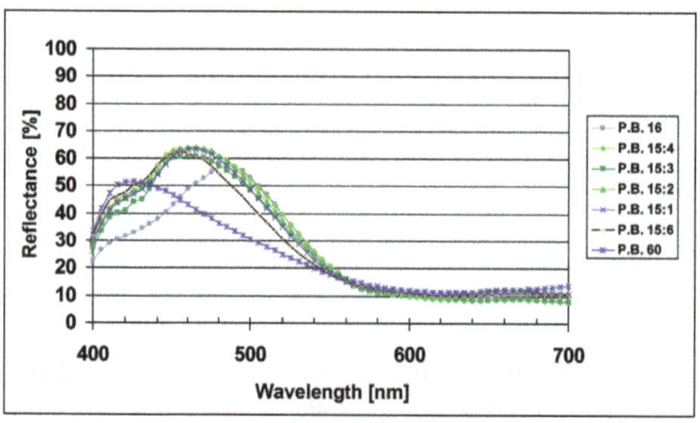


*Figure X.16: Reflectance curves of blue organic pigments.*

Figure X.16 shows the corresponding reflectance curves. The reflectance curve of the metal-free phthalocyanine pigments is still similar to that of P.G. 7; however, a shoulder can be seen between 420 and 440 nm. The copper phthalocyanine pigments have their principal maximum at approx. 460 nm and a pronounced shoulder between 420 and 440 nm.

There various modifications also differ in the wavelength range from 640 to 680 nm. Indanthrone blue typically exhibits a slow increase in reflectance from 600 nm on and a maximum at 420 nm.

### *Metal-free phthalocyanine (P.B. 16)*

The metal-free phthalocyanine (P.B. 16) [31] is a greenish-blue pigment, which tends to undergo a change in modification in some solvent mixtures. It has the best weathering fastness of all phthalocyanine pigments, even in pastel shades.

This pigment can be easily replaced for coloristic reasons by a mixture of P.B. 15:4 and P.G. 7 and so is mainly used in applications that need to be free of copper.

***Copper phthalocyanine in β modification (P.B. 15:3 and P.B. 15:4)***
These copper phthalocyanine in β modification (P.B. 15:3 and P.B. 15:4) [32] pigments yield clean, greenish-blue shades. They are used in all paint formulations because of their fastness properties. Some P.B. 15:3 grades exhibit flocculation in paint formulations, which can be overcome by using the corresponding P.B. 15:4 products. This stabilisation also helps to improve other properties in the paint formulation, e.g. flow, but in some cases solvent resistance is reduced.

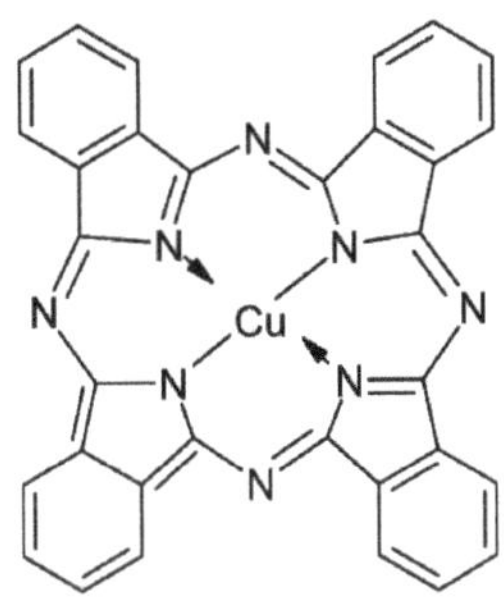


***Copper phthalocyanine in α modification (P.B. 15:1 and P.B. 15:2)***

Both modifications of copper phthalocyanine (P.B. 15:1 and P.B. 15:2) [32] are used wherever neutral blue shades are needed. Their fastness properties are the same as those of the β modification, and so they are suitable for many applications.

***Copper phthalocyanine in ε modification (P.B. 15:6)***

This copper phthalocyanine in ε modification (P.B. 15:6) [33] pigment yields the most reddish-blue shade of the copper phthalocyanine pigments. It can be used in combination with the β modification to cover the whole phthalocyanine blue range.

***Indanthrone blue (P.B. 60)***
This indanthrone blue (P.B. 60) [41] pigment belongs to the group of polycyclic chromophores and is notable for its very high fastness properties, especially weathering fastness. For this reason, it is found in pastel automotive shades, where the fastness of copper phthalocyanine is not good enough.

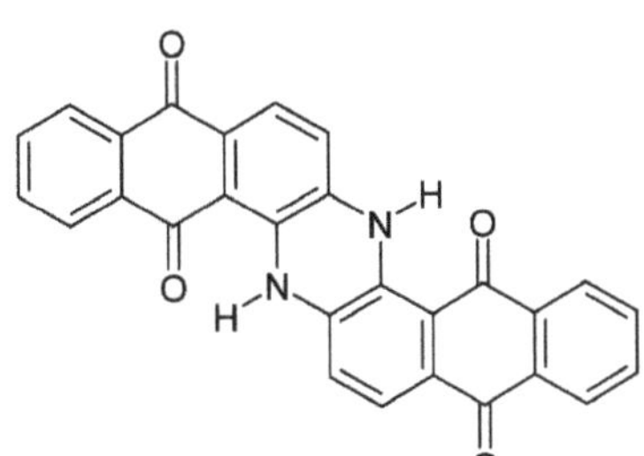


## 2.6 Violet organic pigments

In the violet range, two colour indices dominate: dioxazine violet (P.V. 23) and both modifications of quinacridone (β and γ). From the red range, three colour indices (P.R. 52:2, P.R. 122, P.R. 202) are so bluish-red that the hue angle is located at approx. 340°, and so they

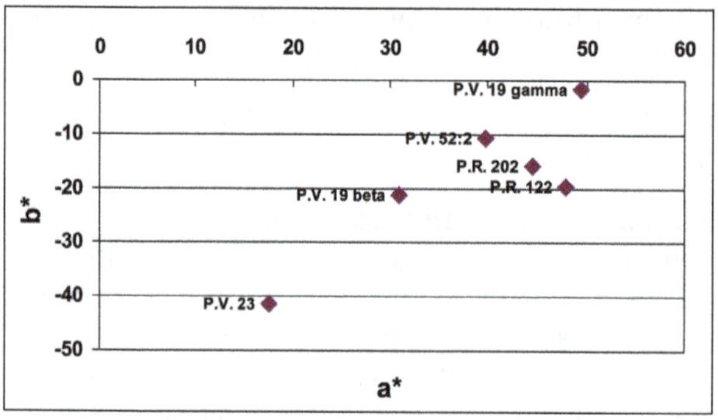


*Figure X.17: Colour locations of violet organic pigments.*

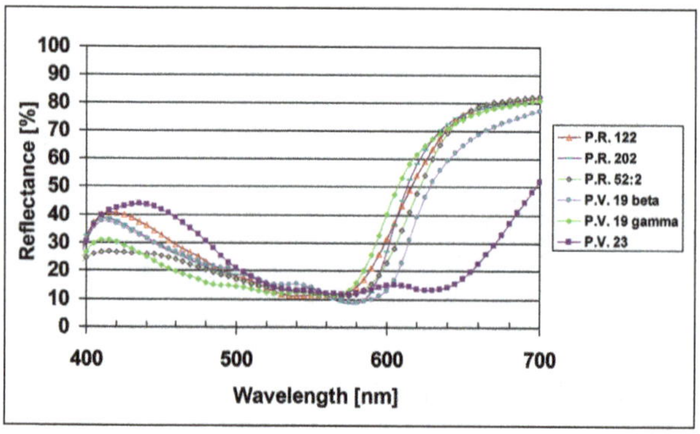


*Figure X.18: Reflectance curves for violet organic pigments.*

are characterised here. The colour locations at 1/3 SD are shown in Figure X.17.

Dioxazine violet has the most bluish-violet shade, while both quinacridones could be regarded as bluish-red pigments. This is also seen in the reflectance curves in Figure X.18.

P.V. 23 has a pronounced reflectance maximum in the short-wave range at approx. 440 nm, while the increase in the red range above 640 nm is not very strong. The reflectance curves of the other pigments are similar to those already discussed for bluish-red pigments.

### *Dioxazine violet (P.V. 23)*

The dioxazine violet (P.V. 23)[34] pigment is also known as carbazole violet and is very widely used. Its bluish-violet shade is unique. Fastness to solvents, overcoating and weathering are good. Thanks to its clean hue and high colour strength, it can be used for tinting. It is found in all paint formulations: emulsion paints, air-drying, decorative architectural paints as well as high-performance industrial and automotive refinish formulations.

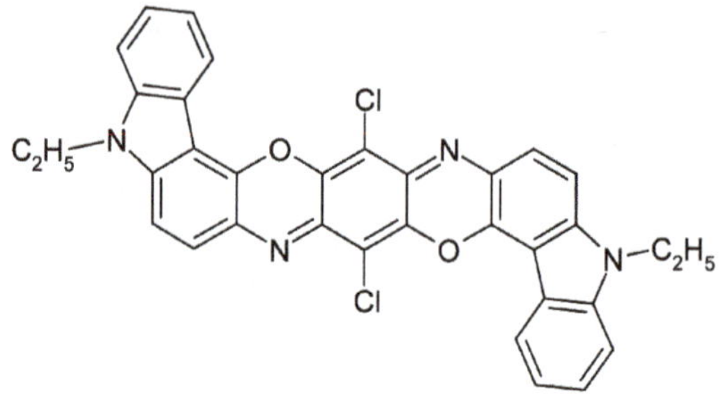

### *Quinacridone red (P.V. 19, β modification)*

The β modification of quinacridone red (P.V. 19)[35] is used in combination with iron oxide, molybdenum orange and titanium dioxide to produce deep Bordeaux and reddish-brown shades. The presence of inorganic pigments with strong scattering leads to good opacity in such formulations.

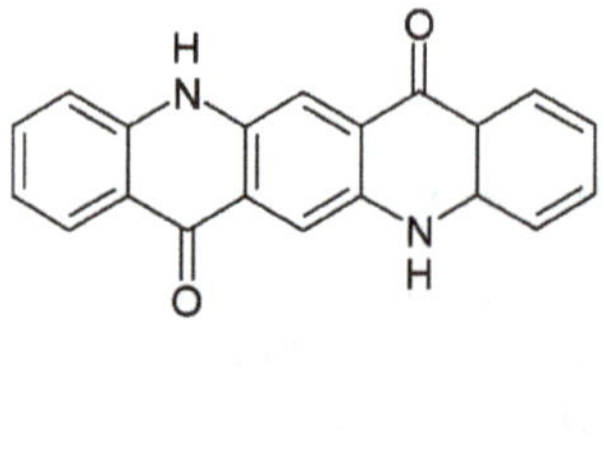

***2,9-Dimethylquinacridone (P.R. 122)***

This dimethylquinacridone (P.R. 122)[36] is the most brilliant pigment in this class and yields even better weathering fastness than the non-substituted grades. It is mainly used as a shading pigment in all kinds of paint applications, starting from metal-effect colour shades in automotive OEM formulations to all high-performance industrial coatings, powder coatings and architectural coatings.

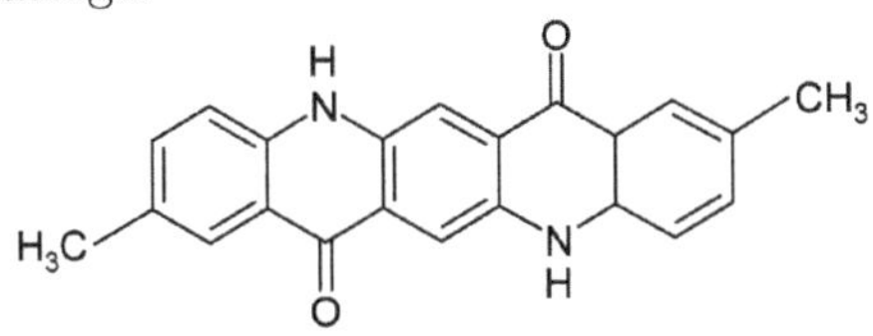


***Dichloroquinacridone (P.R. 202)***

The dichloroquinacridone (P.R. 202)[37] pigment is much more bluish and duller than 2,9-dimethylquinacridone. It is very finely divided and transparent and, when surface treated, imparts good rheology to paint. It is mainly used for automotive finishes.

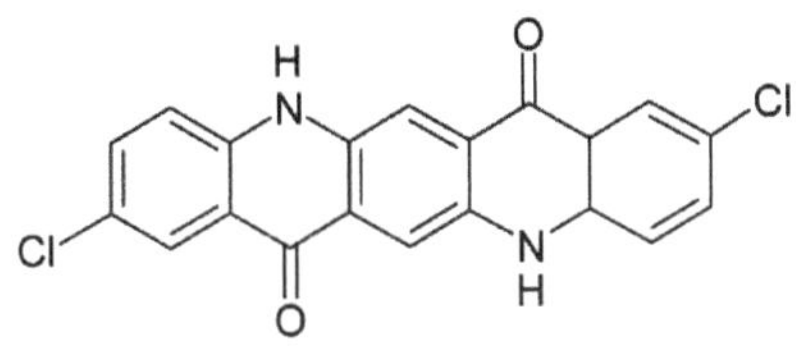


***Manganese-laked azo pigment (P.R. 52:2)***

Only the manganese salts of the azo pigment are used in paint formulations, as they have higher light fastness and weathering fastness than the corresponding alkaline-earth salts. They possess good solvent-fastness and heat stability, but their chemical resistance can be critical. The main application of these pigments (P.R. 52:2)[38] lies in combinations with other pigments of similar shade. They cannot be used together with titanium dioxide.

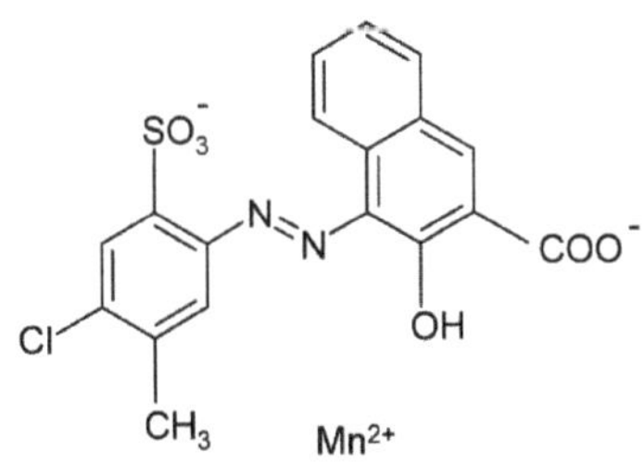


***Quinacridone violet (P.V. 19, γ modification)***

The quinacridone violet (P.V. 19, γ modification)[39] pigments yield bluish-red shades that are much more yellowish than those of the beta modification. For clean, red shades, the gamma modification is used in combination with opaque organic and inorganic pigments. Coarse particles yield better weathering fastness and cleaner full shades than fine particles. They also offer better rheological properties. This pigment closes the colour circle of the most important organic pigments, as the hue angle is located at almost 360 degrees (1/3 SD).

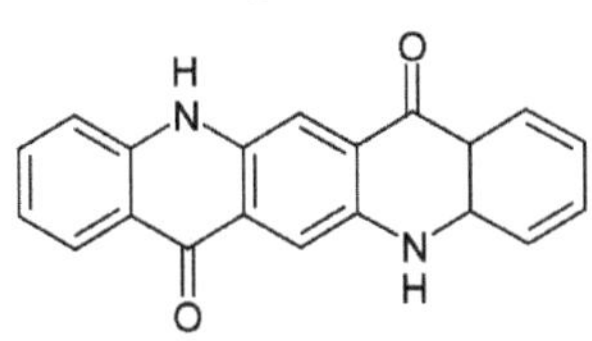

# 3 Characterisation of aluminium pigments

*Hans-Jörg Kremitzl*

Unlike solid colours, coatings containing aluminium pigments exhibit a heavily angle-dependent brightness, called the metal effect or flop, which stems from the platelet-shaped structure of the pigments. They induce an interplay between high reflection in/near the specular angle, and poor reflection far from the specular angle.

The first attempts to capture this effect metrologically were made with a goniophotometer (illuminating angle: 45°, viewing angle 38° and 0°; the ratio of the two values being a measure of the metallic effect). These devices, however, were soon replaced by spectral colorimeters capable of measuring both reflectance and colour impression.

These instruments eventually found their way into the field of automotive coatings and culminated in DIN 6175 Part 2 (03-2001) [42]: The light source is usually fixed at an angle of 45°, observation is carried out at angles of 25°, 45°, 75°, and 110° relative to the specular angle (see Section II.2.6).

Outside this standard, the $L^*$ value at 15° is still frequently used, e.g. to calculate the flop index/metal effect according to the DuPont equation [43]:

*Equation X.1*

$$\text{Metal effect/flop index} = 2.69 \cdot \frac{(L^*_{15°} - L^*_{110°})^{1.11}}{(L^*_{45°})^{0.86}}$$

## 3.1 The metallic effect and its cause

The metallic effect is caused by the interplay of

- Reflection from the pigment surface ("perfect mirror", Figure X.19a)
- Scattering at edges, irregularities, or fines (Figure X.19b).

In this context, it may be helpful to know the two basic processes for manufacturing metallic pigments:

- Milling of atomised metal powder in ball mills using lubricants such as stearic or oleic acid to prevent cold welding/agglomeration (Hametag or Hall process).
- Evaporation of aluminium under high vacuum, deposition on plastic films printed with release coat (**p**hysical **v**apour **d**eposition, PVD), stripping from the film, and shredding of the tinsel to the appropriate size.

Via the choice of manufacturing process and production parameters, such as the particle-size distribution of the metal powder, the diameter of the grinding media, and the composition of the lubricant, it is possible to produce tailor-made pigments that meet the requirements imposed on visual appearance and use. In most cases, the following rules apply.

### 3.1.1 Leafing and non-leafing properties

The wetting behaviour of the pigments can be controlled via the choice of grinding aids (lubricants): saturated long-chain fatty acids (e.g. stearic acid) usually yield pigments that are difficult to wet in most paint systems. The pigments are oriented towards the surface of the paint film (leafing pigments; Figure X.20a). Pigments milled in the presence of oleic acid or pigments produced by PVD are wetted much better by binders or solvents, and are evenly distributed in the coating system (non-leafing pigments; Figure X.20b).

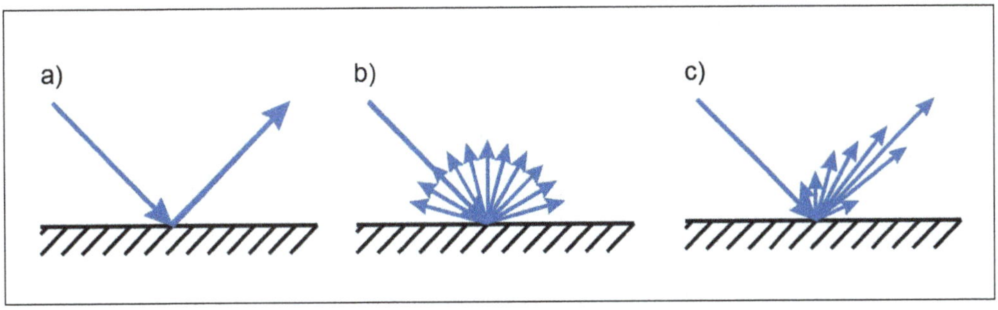


*Figure X.19a: "Perfect mirror" 100 % reflection, b: ideal matte surface 100 % scattering, c: Metallic effect (flop).*

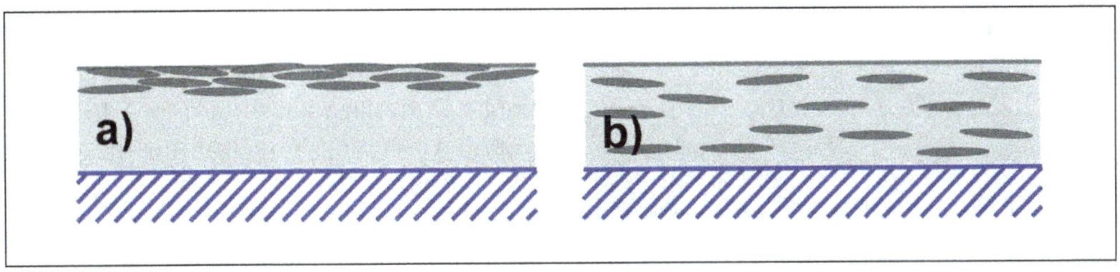


*Figure X.20: Wetting behaviour of metallic pigments; a) leafing, b) non-leafing.*

The former usually afford high brilliance and hiding power, but often come with poor mechanical and chemical stability. The latter, because the binder lies above the pigments, yield an inferior metallic effect, but are much better protected against environmental influences and can be combined with coloured pigments for tinting purposes.

Apart from controlling leafing via the milling process, it is also possible to make downstream adjustments to the orientation behaviour, e.g. by employing additives that contain phosphorus.

### 3.1.2 Particle-size and diameter

The appearance of conventionally produced (milled) pigments is significantly influenced by their particle-size. Within comparable pigment series, finer pigments are usually darker near the specular angle, but have better hiding power due to the higher degree of grinding. Coarser grades appear more metallic and more brilliant. When the pigments exceed a certain size (approx. 20 µm), the human eye is capable of resolving them individually, perceiving them as sparkles.

### 3.1.3 Particle shape, thickness, and topography

First-generation metallic pigments were (and still are) produced from relatively coarse, atomised aluminium powders by being repeatedly comminuted to particles with uneven edges, a high content of fines, and a broad particle-size distribution. Due to their characteristic shape, these pigment grades are called "cornflakes" (Figure X.21a). From the 1980s on, atomising technology became more sophisticated and yielded increasingly finer aluminium powders that only needed to be rolled out during the milling process. This prevents the particles from breaking and from tearing at the edges. The eventual outcome is "silver dollar" pigments (Figure X.21b), which offer significantly improved flop indices, but - due to higher pigment thickness and lack of fines - have less hiding power.

*Figure X.21 a-d: SEM images of various pigment generations.*

Almost perfect mirror-like surfaces became available in the 1990s with the advent of PVD technology.

Pigments produced in this way are characterised by an extremely smooth surface.

The thickness of the pigments themselves can readily be controlled and ranges from 30 to 50 nm; this reduces scattering effects at the edges to the greatest possible extent (Figure X.21c).

Just a few years ago, a wet-grinding method was developed that yields pigments whose properties imitate those of PVD pigments. Optimised aluminium powders and patented production techniques now support the production of flakes whose thickness extends well into the range of PVD pigments (Figure X.21d) and which, given adequate coating systems, can rival them in appearance.

### *3.1.4 Orientation of pigments in the paint film*

Not only the pigments, but all the other components of the coatings system exert an influence on the optical properties of effect pigment paints. Optimised effects using high-end pigments can only be obtained if the binder, additives and even the substrate permit uniform, lamellar orientation of the flakes. Thus, shrinkage of the paint film causes flakes to be better aligned in solventborne paints of high VOC content than in high-solids systems. Additionally, extenders, binder admixtures and other additives can exert a strong influence on the orientation, and hence on the metallic effect [44].

The aforementioned principles primarily apply to solventborne and waterborne systems. In UV-curing systems, there is hardly any decrease in layer thickness during curing and so the optical properties have to be adjusted via the wetting behaviour, particle-size, or by using PVD pigments.

Similar rules apply to powder-coating systems: As these pass through the liquid and highly viscous states very briefly during stoving, only small pigments have any degree of mobility, with the result that bright visual effects are more likely to be obtained with fine leafing grades. PVD pigments are of no relevance here.

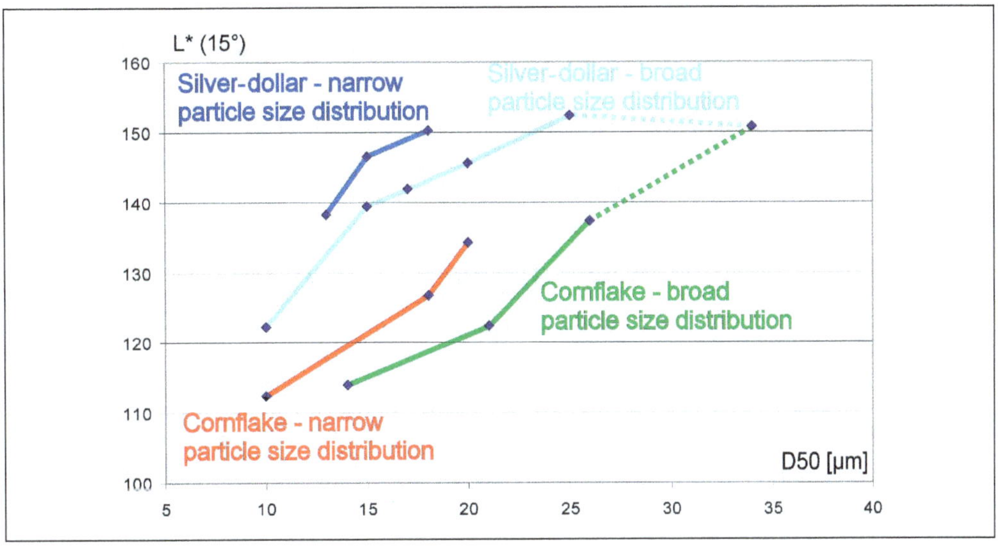


*Figure X.22: Brightness values L* (15°) of various pigment series (milled, non-leafing) as a function of mean particle-size diameter (in a solventborne medium-solids system, 2-coat application).*

## 3.2 Comparison of various pigment grades

Figure X.22 illustrates several crucial points concerning the selection of conventional non-leafing pigments:

Generally, the brightness close to the specular angle increases with increase in mean particle-size ($D_{50}$). At the bottom end of the brightness range are located "cornflake" pigments with a relatively broad particle-size distribution, followed by grades with a narrower particle-size distribution (PSD). A significant increase in brilliance becomes evident in the transition to "silver dollar" pigments. Further narrowing of the PSD ultimately leads to pigment series featuring extremely high differences in brightness between 15° and 110°.

## 3.3 Chemical and mechanical properties

Grinding is not the only step during which the additives are heavily stressed in the production process. Even in the finished product, they undergo gradual change, e.g. through polymerisation of unsaturated fatty acids, saponification reactions at the metal surface, and oxidation. This process is accelerated by unfavourable storage conditions, such as elevated temperatures and access of oxygen or humidity.

For these reasons, standards and reference samples for comparison with new batches must be monitored, and replaced as necessary, at regular intervals.

Furthermore, the negative electrochemical potential of aluminium renders it prone to oxidation. If processed in water-borne media (paints), this may lead to gassing (evolution of hydrogen).

Different strategies have been devised to counter this undesired reaction. The most common are:

- Inhibition using organophosphorus additives
- Inorganic coatings, usually based on $SiO_2$
- Organic coatings, usually based on acrylates.

This does not relieve users of these pigments of the obligation to assess their stability in the respective system. They must also bear in mind that such changes can significantly influence the wetting behaviour, and thereby the appearance.

Finally, preparation and application of effect-pigment paints is best performed under the least-possible mechanical stress in order that damage to the platelet-like structure and generation of additional fines may be prevented.

# 4 Characterisation of pearlescent pigments and special effect pigments

*Peter Gabel and Gerhard Pfaff*

The development of new effect pigments [45] has accelerated in the last 10 years.

For more than 3,000 years, people have marvelled at the brilliant and iridescent colours found in nature [46]. These colours are not limited to pearls and clam shells, and can be seen in many fascinating examples in the world of birds, fish, gemstones, minerals and insects. Production of such pigments is based on various optical principles.

DIN 55943 [47] provides a good overview of the different pigment classes. Pearlescent pigments are defined as lustrous pigments consisting of transparent or semi-transparent platelets of metal oxides or other materials of high refractive index. These are applied in a parallel orientation, which creates the characteristic pearlescence through multiple reflections. Pearlescent pigments that exhibit additional colours generated by interference of light are called interference pigments. Interference pigments as well as pigments of non-transparent flakes are by definition effect pigments whose colour is generated completely or predominantly by the phenomenon of interference of light (see Section VII.2 [47]).

This has rendered classification a more complex affair, and has led to the creation of the new term "special effect pigments." Under this categorisation, special effect pigments are pearlescent pigments and transparent and non-transparent interference pigments which, irrespective of their composition and structure, lead to pearlescent or non-pearlescent effects in combination with interference phenomena in the application medium [48].

Crucial to this development was the capability to manufacture transparent synthetic platelets. These new substrate platelets, in addition to mica, can be coated with metal oxide layers. Pearlescent and interference pigments may be natural or synthetic, and transparent or partially light-absorbing. Interference pigments based on platelets of non-transparent materials have been developed that further broaden the class of effect pigments. Various new effect pigments consisting of platelets of transparent materials have been developed in recent years and further broaden the class of effect pigments.

These new substrate materials include platelets of silicon dioxide (silica flakes), aluminium oxide (alumina flakes), and borosilicate (borosilicate or glass flakes). Additionally, synthetic mica (fluorophlogopite) is now used instead of natural mica (muscovite) for the production of new effect pigments. New developments in the range of decorative effect pigments include multi-layer pigments containing Fabry-Perot structures, pigments based on liquid crystal polymers as well as pigments with textured surfaces. The state of the art is summarised in Table X.1.

The particle-sizes of effect pigments predominantly lie in the range 5 to 100 μm, and are higher in some cases. This means that their average platelet diameter is significantly larger than the wavelength of visible light. The thickness of the platelets, with some minor excep-

*Table X.1: State of the art and historical development of special effect pigments.*

| Pigment | Substrate | Preparation | Optical Principle | Year |
|---|---|---|---|---|
| Fish silver | None | Extraction | Interference | 1655 |
| Basic lead carbonate | None | Hydrolysis | Interference | 1960 |
| Pearlescent pigment | Mica | Thermal hydrolysis<br>Sulphate process | Interference | 1963 |
| Bismuth oxychloride | None | Hydrolysis | Interference | 1970 |
| **Pearl lustre pigments** | **Mica** | **Titration Chloride process** | **Interference** | **1970** |
| Titanium dioxide | Mica | Titration Chloride process | Interference | 1970 |
| Iron(III) oxide combination pigments | Mica | Titration Chloride process | Interference | 1970 |
| Multi-layer | Mica | Titration Chloride process | Interference | 1970 |
| Metal oxide | Mica | Titration Chloride process | Interference | 1970 |
| **Effect pigment** | **Al flake** | **CVD** | **Interference** | **1987** |
| Metal oxid metal flake | Metal flake (Al,Cu,Zn..) | CVD | Fabry-Perot structure | 1999 |
| **Effect pigment** | **$Fe_2O_3$ flake** | **Hydrothermal process** | **Adsorption** | **1994** |
| Multi-layer $Fe_2O_3$ flake | $Fe_2O_3$ flake | CVD | Interference | 1998 |
| Effect pigment | $TiO_2$ flake | Thermal hydrolysis | Interference | 1997 |
| Effect pigment-LCP | Liquid crystal polymers | Polymerisation | Interference | 1995 |
| Effect pigment, multi-layer | Al film | PVD | Fabry-Perot structure | 1995 |
| Effect pigment | Synthetic Mica | Titration Chloride process | Interference | 1996 |
| Effect pigment | $Al_2O_3$ flake | Titration Chloride process | Interference | 1999 |
| Effect pigment | $SiO_2$ flake | Titration Chloride process | Interference | 2000 |
| Effect pigment | Ca borosilicat flake | Titration Chloride process | Interference | 2002 |
| Effect pigment, diffractive | Al film | PVD | Diffraction grating | 2002 |
| Effect pigment, holographic | Thermoplastic film | Thermoplastic film | Holograms | 2004 |

tions, is typically below 1 μm, which is small enough to interfere with visible light. Effect pigments therefore have high aspect ratios (ratio of diameter to thickness), with values of up to 200.

## 4.1 *Manufacture, properties and types of special effect pigments*

This section deals in more detail with the metal oxide mica pigments in Table X.1. We will also discuss innovative effect pigments, such as platelets of silicon dioxide (silica flakes), aluminium oxide (alumina flakes), borosilicate (borosilicate or glass flakes), multi-layer pigments containing Fabry-Perot structures, and pigments based on liquid crystal polymers. Information about other types of effect pigment, such as bismuth oxychloride, can be found elsewhere [45].

### 4.1.1 *Metal oxide mica pigments*

The most important industrial class of pigments consists of flakes of natural mica coated with a thin film of transparent metal oxide. While titanium dioxide is undoubtedly a key metal oxide, the use of oxides such as iron oxides, chromium(III)-oxide and mixed oxides has added to the popularity of artificial pearlescent effects.

#### *Structural principles of metal oxide mica pigments*

The simplest metal oxide mica pigment consists of three layers (see Figure X.23) of different refractive indices $n_1$, $n_2$, $n_3$; where $n_1 = n_3$ (Table X.2). There are four interfaces in this case ($P_1$ to $P_4$):

- $P_1$ and $P_4$: surrounding medium/metal oxide,
- $P_2$ and $P_3$: metal oxide/mica.

Consider titanium dioxide mica pigment. Here, it can be easily seen that interference phenomena occur through reflection from all six combinations of interfaces. Some combinations such as $P_1P_2$ and $P_3P_4$ or $P_1P_3$ and $P_2P_4$ create identical interference phenomena. The average thickness of the mica platelets fluctuates about the statistical mean. The interference effects resulting from the interface between metal oxide coating and mica substrate ($P_2$, $P_3$) combine to create a white background reflection. The interference colour resulting from a large number of pigment particles therefore depends only on the thickness of the upper and lower metal oxide layer [49–51].

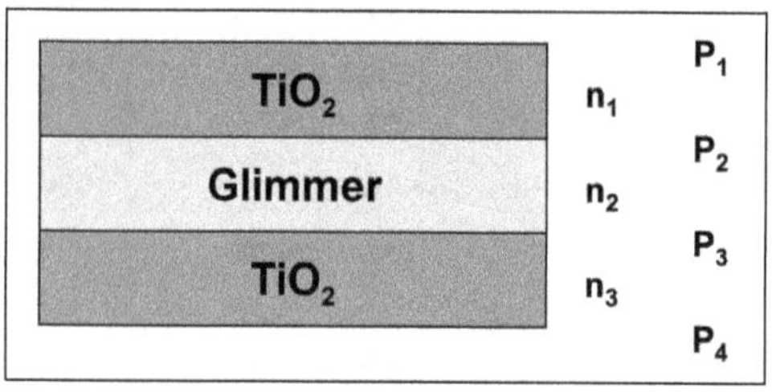


*Figure X.23: Schematic structure of titanium dioxide mica pigment with the four existing interfaces ($P_1$-$P_{4v}$).*

*Table X.2: Refractive indices of various materials.*

| Medium | Refractive index |
|---|---|
| Vacuum, air | 1.0 |
| Water | 1.33 |
| $MgF_2$ | 1.37 |
| $SiO_2$ | 1.42 to 1.48 |
| Plastics, paints, printing inks | 1.4 to 1.7 |
| Mica | 1.5 to 1.6 |
| $Al_2O_3$ | 1.7 |
| Guanine, hypoxanthine | 1.85 |
| Basic lead carbonate | 2.0 |
| Bismuth oxychloride | 2.15 |
| $Fe_3O_4$ (magnetite) | 2.4 |
| $TiO_2$ (anatase) | 2.5 |
| $TiO_2$ (rutile) | 2.7 |
| $\alpha$-$Fe_2O_3$ (hematite) | 2.,9 |

Mica is an aluminosilicate with a layer structure (Figure X.24) of inter-connected tetrahedral and octahedral layers. The tetrahedral layers consist of $SiO_4$ and $AlO_4$ tetrahedra in a ratio of 3 : 1 which are linked with layers consisting of octahedral $AlO_6$ units (with some Al in the octahedra replaced by Mg, Fe, or Li). The octahedral layers are then followed by tetrahedral layers. There are comparatively large distances between the tetrahedral and octahedral layers within the mica structure. Na, K, or Ca ions are located in the corresponding interstices. The interstices are also the areas of lowest mechanical strength in the crystal structure. They therefore act as cleavage planes when the mica is placed under mechanical stress, e.g. during grinding. Ultimately, the interstices are the reason why it is possible to cleave large mica blocks into small, thin platelets. These platelets can be coated with thin metal oxide layers to yield effect pigments.

Transparent muscovite is commonly used for effect pigments, but in some cases synthetic phlogopite serves as the carrier. Muscovite is found throughout the whole world, but only a few deposits yield suitable starting materials for metal oxide mica pigments. Muscovite has long been used as a cheap filler [52, 53].

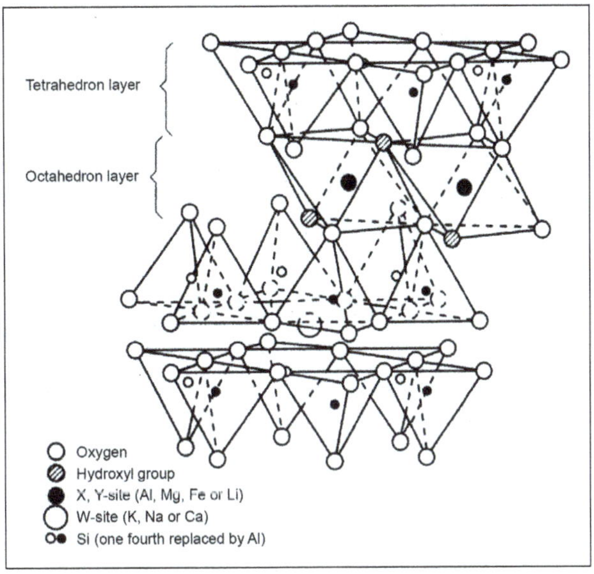


*Figure X.24: Three-dimensional structure of muscovite mica.*

### *Manufacturing process*

Manufacture of metal oxide mica pigments commences with preparation of the natural mica. The rough mica blocks are ground and then classified into a range of particle-size distributions. The subsequent coating step involves a wet chemical process to precipitate metal hydroxides or oxide hydrates onto the surface of the fine mica particles. This coating step is followed by filtration, drying and precision calcination. All these steps are crucial to the quality of the pigments. The entire manufacturing process is shown schematically in Figure X.25.

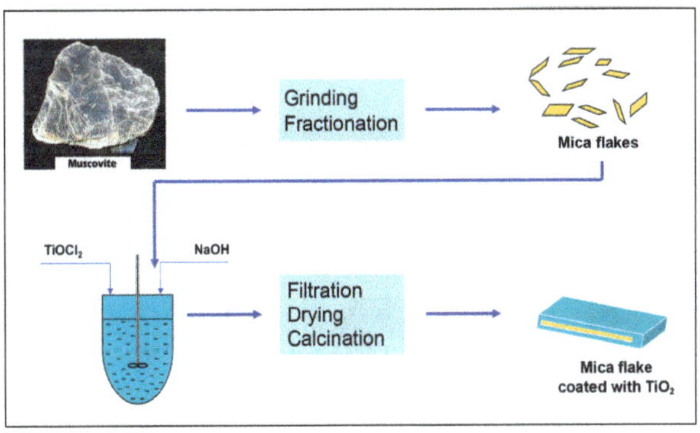


*Figure X.25: Process scheme for metal oxide mica pigments.*

### *Pigment particle dimensions*

Selection and preparation of the mica are the most important steps governing the quality of the final effect pigments. The ratio of diameter to thickness (aspect ratio) of the pigment particles is directly proportional to the particle-size distribution of the starting mica platelets. The mica particles are usually between 300 and 600 nm thick. The optimum diameter varies with the intended application.

*Figure X.26: Mica platelets with different particle-size distribution.*

A frequently employed classification distinguishes five classes of particle-size distribution (Figure X.26):

*Figure X.27: Scanning electron micrograph of mica platelets.*

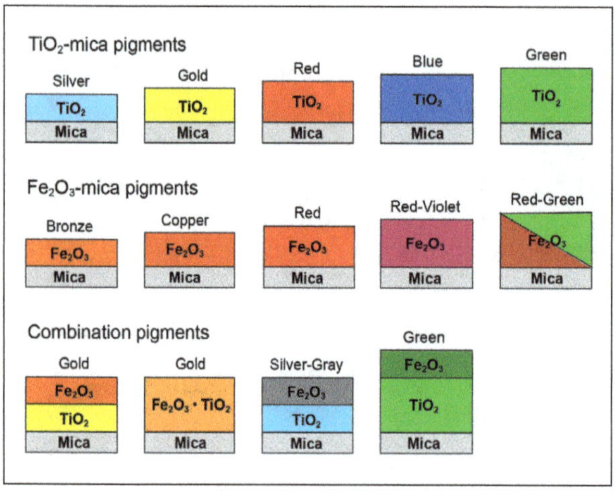


*Figure X.27a: Schematic illustration of different metal oxide mica pigments.*

M:1 to 15 µm
F: 5 to 20 µm
N: 10 to 50 µm
S: 10 to 130 µm
L: 40 to 200 µm

Lustre and hiding power of the effect pigments in the application system are the two key properties influenced by the particle-size distribution of the chosen mica. As a rule of thumb, large mica particle diameters lead to increased lustre while small mica particle diameters generate increased hiding power. In summary:

- large particles = large smooth surfaces, uniform reflection of incident light, leading to excellent lustre and brilliance,
- small particles = smaller, flat surfaces, more edges and corners, increased scattering and refraction, and therefore better hiding power.

Figure X.27 shows a scanning electron micrograph of mica platelets produced by cleaving muscovite. It can be seen that the resulting platelets are not of uniform thickness. They contain grinding-induced steps, but have a very smooth surface overall. This is corroborated by atomic force microscopy (AFM) of thin mica platelets produced by cleaving rough mica blocks.

The development of mica pigments commenced with those containing $TiO_2$. The earliest pearlescent pigments were silvery white. These do not exhibit interference effects, as the $TiO_2$ layers are too thin. Figure X.27a gives an overview of the major metal oxide mica pigments that have been developed since then. As the $TiO_2$ layer thickness on the mica increases, the interference colour develops more and more, varying from gold to green.

### *4.1.1.1 Titanium dioxide mica pigments*

The era of metal oxide mica pigments began with products based on $TiO_2$ coatings on a muscovite substrate [46, 48, 62, 65, 87]. The $TiO_2$ is usually coated onto the mica by hydrolysing titanium salt solutions in an aqueous mica suspension. The first synthesis of $TiO_2$ mica pigments was published back in 1942 [54]. Commercial use did not occur until DuPont initiated it in the early 1960s [55, 56]. Further developments were mainly driven by Merck [57] and Mearl [58]. In these products, mica acts as a template for the metal oxide layers deposited during the synthesis and as a mechanical support for the thin optical layers formed in the final pigment. Two manufacturing methods have become established:

***Homogeneous hydrolysis***

A measured quantity of titanyl sulphate is added to an aqueous mica suspension. This is heated slowly to hydrolyse the titanium salt to insoluble titanium dioxide hydrate, which is steadily deposited as a homogeneous coating on the mica particles. When all of the titanium salt has been hydrolysed, the coated particles are filtered off, dried and finally calcined at 700 to 900 °C in a kiln. At these high temperatures, water is eliminated and a thin compact coating of titanium dioxide is formed around the mica. The equations for the main reactions are as follows:

a) Coating

$TiOSO_4 + mica + n\,H_2O \rightarrow TiO_2 \ (n\text{-}1)\,H_2O/mica + H_2SO_4$

b) Calcination

$TiO_2 \ (n\text{-}1)\,H_2O/mica \xrightarrow{-(n-1)\,H_2O} TiO_2/mica$

***Titration method***

The advantage of this method over hydrolysis is that the processes are easier to control. An aqueous, strongly acidic $TiOCl_2$ solution is continuously added to a mica suspension at a pH of about 2 and at temperatures ranging from 60 to 90 °C. The titanium dioxide hydrate precipitates out of solution at a precise acidic pH. The pH is held at a constant value by continuously adding sodium hydroxide solution. A highly simplified equation for this reaction is:

$TiOCl_2 + 2\,NaOH + mica + n\,H_2O \rightarrow TiO_2 \ (n\text{+}1)\,H_2O/mica + 2\,NaCl$

Further processing of the pigment particles proceeds in the same way as for those made by homogeneous hydrolysis.

***Rutile-anatase modification***

At pH values above 1.5, the titanium dioxide precipitated onto mica platelets yields only the anatase modification. Even annealing at temperatures around 1000 °C fails to produce a rutile phase in the $TiO_2$ layers.

The anatase layers on the mica platelets have a preferred crystal orientation if they are very thin, e.g. less than 40 nm [59, 60]. This preference disappears with increase in layer thickness. Moreover, the uncalcined $TiO_2$ layers show a deviation in lattice parameters, compared to anatase, in the compact and dense solid form (single crystal/pressed pellet). These differences are reduced by calcining [60]. These phenomena are evidence that muscovite mica used as a substrate has a strong template effect in the coating and calcination processes during pigment manufacture.

The rutile modification, by comparison, offers advantages such as a higher refractive index (rutile: n = 3.10; anatase: n = 2.52) and better light stability. This fact inspired studies aimed at ways of generating rutile on the mica. It was discovered that pre-coating the mica with small quantities of tin dioxide drives the reaction in the desired direction:

$$SnCl_4 + 4\,NaOH + mica \rightarrow SnO_2/mica + 2\,H_2O + 4\,NaCl$$

$$SnO_2/mica + TiOCl_2 + 2\,NaOH \rightarrow TiO_2/SnO_2/mica + 2\,NaCl + H_2O$$

$$TiO_2/SnO_2/mica \xrightarrow{>700°C} TiO_2\ (rutile)/SnO_2/mica$$

Tin dioxide crystallises with a rutile structure and so forces the precipitated titanium dioxide to form this phase as well.

Figure X.28 shows an SEM image of a $TiO_2$ mica pigment. Both the size and shape of the various pigment particles can be clearly seen. Figure X.28 (right) and Fig. X.28 (left) show cross-sections through individual pigment particles. The crystalline titanium dioxide layer

*Figure X.28: Scanning electron micrograph of titanium dioxide mica pigment (left: overview micrograph, right: cross-section through a particle).*

formed on both sides of the mica platelet during coating and calcination can be clearly seen. This $TiO_2$ layer covers the substrate platelets completely. Its thickness can be adjusted with nanometer precision during the manufacturing process. Precise control over this layer thickness is essential for manufacturing special effect pigments that possess reproducible colour properties. This applies especially to the silvery white and interference pigments described in the following sections.

Titanium dioxide mica pigments are often divided into silvery white pearlescent pigments and interference pigments. An aqueous mica suspension does not have great appeal before addition of the titanium salt solution and subsequent hydrolysis. There is no lustre whatsoever, which is not surprising because the refractive indices of mica (n = 1.6) and water (n = 1.33) are quite similar. However, when a titanium salt is hydrolysed by one of the methods described above, the suspension acquires a silky-matte appearance initially, with a more intense lustre developing as the reaction continues. If the reaction is stopped at this point and the particles are drawn off, silvery white, lustrous pigments are obtained. At this stage, the pigments have a titanium dioxide coating 40 to 60 nm thick.

The pigments possess a neutral, silvery white lustre and represent the most important group of pearlescent pigments. The particle-size distribution determines whether they have a silky-matte, brilliant or glittering lustre. These pigments can be combined with other transparent colourants to impart a silvery, metallic lustre to any shade. They are often used without additional colourants, however. Adding a little carbon black produces a silver-grey effect, similar to that of aluminium effect pigments.

***Interference colours***

An interesting optical phenomenon occurs during titration if coating of the mica with titanium dioxide is not stopped at the silvery-white stage. Upon further addition of titanium dioxide to the reaction mixture, the pigment particles start to turn golden-yellow without losing their lustre. As more titanium dioxide is added, this colour disappears and is replaced by copper-red, and then lilac. This is followed by a lustrous blue, which then turns turquoise and finally green. If even more titanium dioxide is added, this fascinating colour display starts all over again.

What is the explanation for this phenomenon? The titanium dioxide layers have reached a thickness at which, in accordance with the optical principles for thin platelets, light interference occurs. Figure X.29 demonstrates how the layer thickness of $TiO_2$ determines the interference colour.

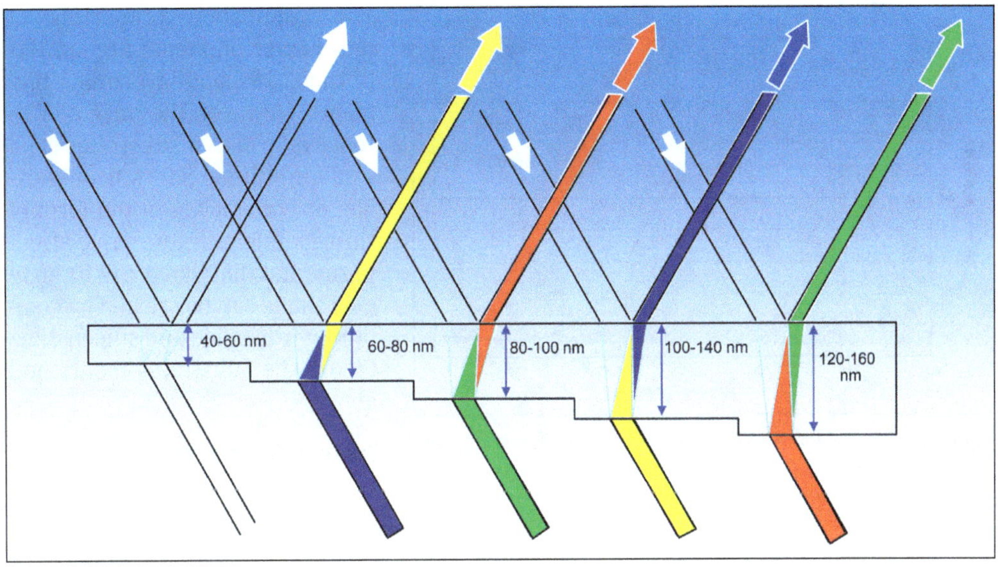


*Figure X.29: The thickness of the titanium dioxide layer determines the interference colour.*

Clearly, precise adjustment of the $TiO_2$ layer thickness in the nanometer range is crucial to achieving a defined interference colour. Even deviations of just a few nanometers lead to other colours and effects that are visible to the human eye. The manufacture of metal oxide mica pigments truly can be described as nano-technology, with the goal being to produce nano-layers, and not nano-particles [61].

In 2007, a new class of "high chroma" mica interference pigments led to improvements in physico-chemical parameters, such as particle-size and shape. Just eight pigments are needed to obtain a chroma maximum in colour space. This new colour-space concept covers more than 93 % of mica-based interference colours [64].

### *4.1.1.2 Titanium dioxide mica pigments with multi-layers*

Mica pigments bearing one type of metal oxide have recently been joined by new types of effect pigments that have more than one metal oxide layer on the substrate platelet [62, 63]. This development is based on the recognition that the intensity of interference colours is heavily dependent on the difference between the refractive indices of neighbouring optical layers and on the number of such layers. A large number of layers of alternating refractive indices (preferably alternating layers of high refractive index and low refractive index) increases the fraction of light reflected from the pigment surface and therefore boosts the intensity of the interference colour.

A typical layer structure on mica is a sequence comprising a high-refractive-index $TiO_2$ layer, a low-refractive-index $SiO_2$ layer and an additional high-refractive-index $TiO_2$ layer. Such a structure creates a three-layer optical system on both sides of the mica platelets.

Pigments with a more intense interference colour or a more pronounced colour travel can be made by carefully choosing the thicknesses of the $TiO_2$ and of the $SiO_2$ layers. Compared with interference pigments having just one $TiO_2$ layer, multi-layer mica pigments yield an interference colour of much better intensity and brilliance and excellent transparency. This is the outcome of the improved reflectance profile of the individual pigment particle-s shown in Figure X.30.

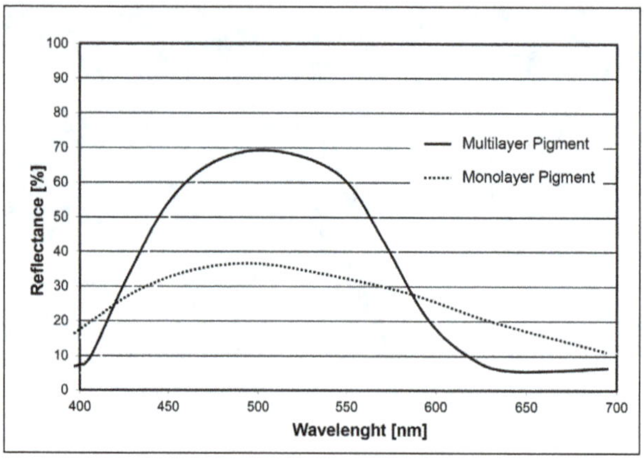


*Figure X.30: Calculated reflectance spectra of one-layer and three-layer systems with reflectance maxima in the green range of the spectrum.*

The application properties of multi-layer pigments are similar to those of mono-layer mica pigments. However, the mica can be replaced by other substrates such as aluminium oxide, silicon dioxide, or borosilicate in the form of thin platelets. Interesting effect pigments that have an intense gold shade can be created by combining iron oxide and silicon dioxide on the substrates mentioned above [65].

***Iridescent effects***

Interference pigments exhibit significant colour travel (iridescence, flop). By this is meant that the colour varies with the angle at which the pigment is observed (see Section II.3.3). The interference colour can only be seen from a certain angle, the gloss angle. The colour seen from the other positions depends on several factors. Interference pigments follow the additive mixing theory.

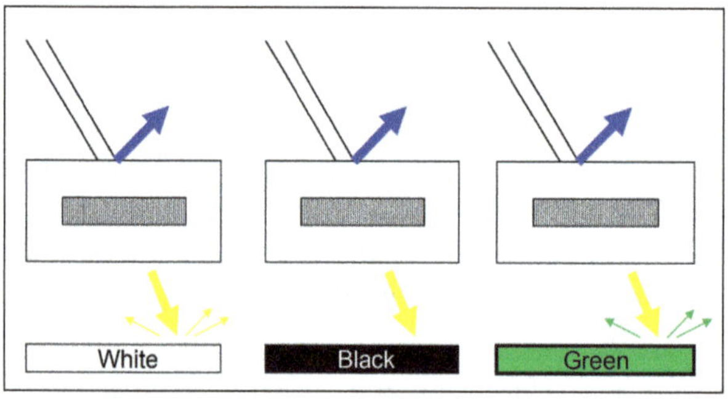


*Figure X.31: Interference pigment blue against white, black and green background.*

Consider a blue interference pigment printed on a white, a black, and a coloured, e.g. green, background (Figure X.31). Interference, as mentioned earlier, can be considered to be the selective reflection of light rays. The interaction of light with the blue pigment may be described as follows:

***White background***

White light incident at the gloss angle is reflected as blue light, whereas the remaining light passes through the transparent pigment. This fraction of the light is complementary to blue and so is yellow. The transmitted yellow light is diffusely scattered by the white background in all directions. At the gloss angle, the intense interference colour hides the yellow light scattered diffusely by the background. From all other directions, the object appears pale yellow.

***Black background***

Black absorbs nearly all incident light, including the transmitted yellow light that passes through the blue interference pigment. At the gloss angle, the eye again sees a lustrous blue object. Since light is scattered at the corners and edges of the pigment particles, the surface of the object, when seen from other angles, appears to be more or less grey-blue rather than totally black.

***Coloured background***

Due to the transparency of pearlescent pigments, the background colour shines through. At the gloss angle, though, the interference colour outshines the background colour. Interference colour and background colour can be identical or different, a fact which makes all sorts of flop effects possible. Particularly attractive effects can be achieved on curved surfaces, where both colours can be seen at the same time.

### *4.1.1.3 Iron(III) oxide mica pigments*

Haematite, the a-$Fe_2O_3$ modification of iron oxide, with its high refractive index of 2.9, good hiding power, and adequate chemical and thermal stability, makes an ideal effect pigment.

***Manufacture and properties***

Mica is coated with transparent haematite layers in a wet chemical process similar to that employed for titanium dioxide mica pigments. To this end, an iron(II) or iron(III) salt solution is added drop-wise to a mica suspension at a constant pH (as in a volumetric titration). The resultant hydrated iron oxide is then dehydrated at temperatures between 700 and 900 °C. The following reactions occur:

$$Fe^{3+} + \text{mica} + 5\ H_2O \rightarrow FeO(OH)/\text{mica} + 3\ H_3O^+$$

$$2\ FeO(OH)/\text{mica} \xrightarrow{-H_2O} Fe_2O_3/\text{mica}$$

The granular layer structure sinters during annealing at 700 to 900°C to yield a less porous, seemingly vitreous, yet crystalline, coherent phase [60].

Another manufacturing method for iron oxide-based products is chemical vapour deposition (CVD). This may consist in oxidising iron pentacarbonyl in the gas phase and simultaneously depositing the resulting iron oxide onto mica or other platelet-shaped substances [66, 67].

The mica platelets are fluidised in nitrogen gas at temperatures of about 450 °C. Then the reactants $Fe(CO)_5$ and oxygen are injected into the fluidised bed. They need to be extensively diluted in the inert (nitrogen) gas in order that proper coating may occur. The thickness of the iron oxide layer formed on the mica varies with the length of the reaction. Carbon monoxide is formed as a by-product, and can be catalytically oxidised to carbon dioxide [66]:

$$2\ Fe\ (CO)_5 + 1.5\ O_2 + \text{mica} \rightarrow Fe_2O_3/\text{mica} + 10\ CO$$

The colour and the lustre of the pigments made by this method depend upon two optical properties: absorption and interference. The layer thickness of iron oxide can be controlled in the manufacturing process to superimpose an interference colour on the red or red-brown absorption colour of haematite. Bronze, copper, and various shades of red are the result. Relatively thick layers will yield reddish-violet or even reddish-green iron oxide mica pigments. For example, a layer thickness of 75 nm $Fe_2O_3$ yields a bronze colour while a thickness of 85 nm yields a red. In the case of bronze, copper, and red shades, the reddish-brown absorption colour of the pigments is enhanced by virtually identical interference colours (yellow, copper, red). This yields very intense colours at the gloss angle. The reddish-violet and reddish-green pigments have a violet or green interference colour at the gloss angle, in addition to the reddish-brown absorption colour of the haematite, due to their thicker iron oxide layer.

Like $TiO_2$ mica pigments, pigments made with $Fe_2O_3$ layers must have defined thicknesses in the nanometer range. Even tiny deviations in the thickness of the iron oxide layer will produce noticeable differences in colour and lustre.

***Special features***

Iron oxide mica pigments are not photoactive and therefore possess excellent light stability. They also offer high brilliance due to their higher refractive index. Furthermore, the red absorption colour bestows comparatively high hiding power, a fact which can prove a decisive advantage in certain applications.

#### 4.1.1.4 Combination pigments based on metal oxide and mica

Combining transparent absorption pigments with transparent effect pigments creates new colour effects that cannot otherwise be produced by simply blending the two types. The effects result from the interplay of absorption and interference when pure titanium dioxide mica pigments are coated with an additional layer of an absorbing colourant (pigment, dye). Of course, any desired colour combination can be obtained by simply blending interference and absorption pigments.

***Structure and functional principles***

In such mixtures, the interference colour of the pearlescent pigment dominates at the gloss angle, whereas the colour of the absorption pigment is seen at all other angles. If, however, an extremely thin layer of nano-size particles is additionally applied to the interference pigment, the colour flop is more abrupt and more brilliant. A further advantage of this approach is that it eliminates dispersion problems arising from the use of transparent absorption pigments, which have large specific surface areas on account of their small particle-size [68].

The thickness of the titanium dioxide layer determines the brilliance and the interference effect at the gloss angle, whereas the transparent colourant dominates at all other angles. A colourant and interference pigment of identical colour produce a strongly coloured effect pigment with a lustre flop that can be seen at all angles. An example of this is the coating of a blue interference pigment with Prussian blue.

If the colours of the colourant and interference pigment are chosen carefully, a clear, two-colour effect can be achieved. One such example is a combination of interference blue with carmine red. At the angle of reflection, the interference blue can be seen, whereas the red absorption colour of carmine dominates at all other angles [68, 69].

***Possible combinations***

The metal oxide most commonly combined with $TiO_2$ on mica is iron(III)oxide (haematite). This combination yields a wide range of brilliant gold shades. The spectrum ranges from pale yellowish-gold to brilliant reddish-gold to greenish-gold, which looks amazingly like real gold.

Precipitation of iron(III)oxide onto a titanium dioxide mica pigment produces pale-gold to reddish-gold pigments, in accordance with the $TiO_2$ layer thickness. As $Fe_2O_3$ also has a high refractive index, the resulting interference colours are due to the combined effects of $TiO_2$ and $Fe_2O_3$ layers. The iron(III)oxide absorption pigment determines the mass tone colour.

The simultaneous precipitation of iron and titanium oxide hydrates onto mica leads to a highly refractive iron titanium oxide after calcination. The greenish-gold colour arises from highly refractive yellow $Fe_2TiO_5$ phases, which can be identified [70].

It is not only golden shades that can be obtained by combining titanium dioxide and iron(III) oxide. Variations in the layer thickness of both oxides yield a wide spectrum of colours and interference effects, from yellow to red, red-brown, violet, blue and finally green. Theoretically, the range of possible colours is virtually unlimited.

Other metal oxides can be used as a second colour component for combination pigments, provided that their refractive index is high enough. Bluish green to moss-green pigments result from a combination of titanium dioxide and chromium oxide.

Combinations of titanium dioxide and other inorganic pigments, such as Prussian blue, cobalt blue, $Fe_3O_4$ and carbon black also create impressive colour effects on the substrate. The last two pigments are particularly interesting because the interference colour is perceived as the absorption colour. The optical properties are similar to those of pigment

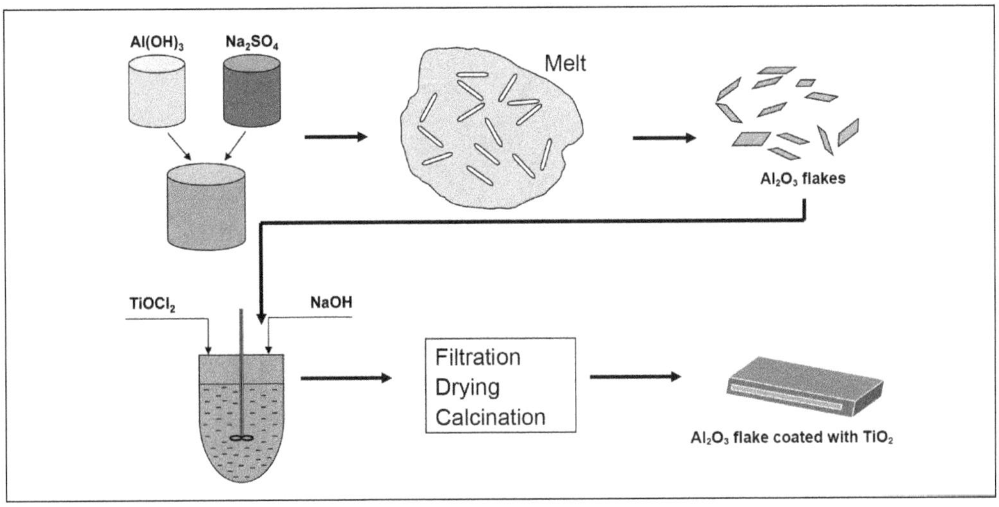


*Figure X.32: Production scheme for $Al_2O_3$ flake pigments.*

mixtures which contain carbon black and in which the transmitted fraction of the light is absorbed [71].

There are two ways to coat $TiO_2$ mica pigments with organic colourants:

- Precipitation onto the $TiO_2$ surface in an aqueous suspension in the presence of complex-forming surfactants.
- Fixing the colourant as a mechanically stable layer by using suitable additives in conjunction with physical mixing.

Combination pigments are mostly used in printing inks and cosmetic formulations. In cosmetics, reds and two-tone colours are particularly popular. However, there are regulatory restrictions on the number of organic colourants, such as carmine red or thioindigo, that can be used. The manufacturing processes are comparatively complex and must be carefully controlled to meet high purity and regulatory requirements.

### 4.1.2 Effect pigments based on alumina flakes

With a view to expanding the range of attainable colour effects, further special effect pigments have been developed that use substrate materials such as aluminium oxide, silicon dioxide, and borosilicate instead of mica [72, 73]. Effect pigments based on aluminium oxide flakes (alumina flakes) are produced in a special, technologically sophisticated manufacturing process [72, 73].

The first, crystal-growth stage yields $Al_2O_3$ flakes with a very smooth and perfect surface. The flakes have the structure of corundum (a-$Al_2O_3$). The desired flake shape is achieved by controlling crystal growth in the axial and equatorial directions. The extraordinary colour purity and transparency of effect pigments obtained by coating $Al_2O_3$ flakes with metal oxides stems from the fact that the synthesis yields thin, monocrystalline flakes [74-77].

Coating the alumina flakes with metal oxides of high refractive index, especially titanium dioxide and iron (III) oxide, yields effect pigments that are extraordinarily reflective. These pigments possess a very intense glittering effect known as sparkle. The coating process is analogous to that used for metal oxide mica pigments, except that it starts with an aqueous suspension of $Al_2O_3$ flakes. The entire manufacturing process for pigments based on aluminium oxide flakes, comprising flake production and coating with metal oxides, is shown in Figure X.32 [75].

Figure X.33: Scanning electron micrographs of pure (left) and $TiO_2$-coated (right) $Al_2O_3$ flakes.

The effect pigments just described are sold as high-chroma crystal-effect pigments [78]. For outdoor applications, they are modified with an additional weatherable surface treatment. Figure X.33 shows electron micrographs of pure and of $TiO_2$-coated $Al_2O_3$ flakes.

Silvery-white effect pigments and interference pigments based on $Al_2O_3$ flakes can be produced – in the same way as pigments based on mica platelets – in different colours, e.g. gold, red, blue or green, by varying the thickness of the titanium dioxide layer. Coating with iron oxide yields bronze, copper or red effect pigments, the precise colour depending on the layer thickness. The commercial pigments have a narrow particle-size distribution of 5 to 30 µm and a high aspect ratio. The sparkle effect mentioned above originates in the optimised thickness of all the layers in the pigment structure, including the $Al_2O_3$ substrate. Analyses show that the thickness range of the alumina flakes and the resultant pigments can be controlled very precisely [79, 80].

The lustre effects of the $Al_2O_3$-based pigments are additionally intensified by direct illumination, e.g. sunlight. Such sparkle has never before been observed in fine effect pigments. This has important implications for automotive coatings, because it was not possible in the past to obtain such intense sparkle with coarser mica pigments without substantial loss of performance at the same time. The minimal fraction of superfine pigment particles yielded by the process improves the application and coloristic properties. Scattering effects are reduced, and intense, pure interference colours are achievable through combination with the colourless aluminium oxide substrate. The pigments can be applied by the standard methods for conventional pearlescent and interference pigments.

Effect pigments based on alumina flakes are employed in all major application systems, i.e. in coatings, plastics, printing inks and cosmetic formulations. They can be given an additional surface treatment for use in solvent-borne and water-borne coating systems. The pigments can be used in combination with many conventional pigments as well as with other effect pigments. The outcome is virtually unlimited styling possibilities [79, 80].

### *4.1.3 Metal oxide pigments based on borosilicate flakes*

The latest development in transparent effect pigments is pigments based on borosilicate glass. Thin borosilicate glass flakes are completely colourless and transparent. These favourable properties can be exploited in the manufacture and application of special effect pigments. Coating the glass flakes with $TiO_2$ leads to very pure interference colours, without any mass tone colour [72, 79, 81].

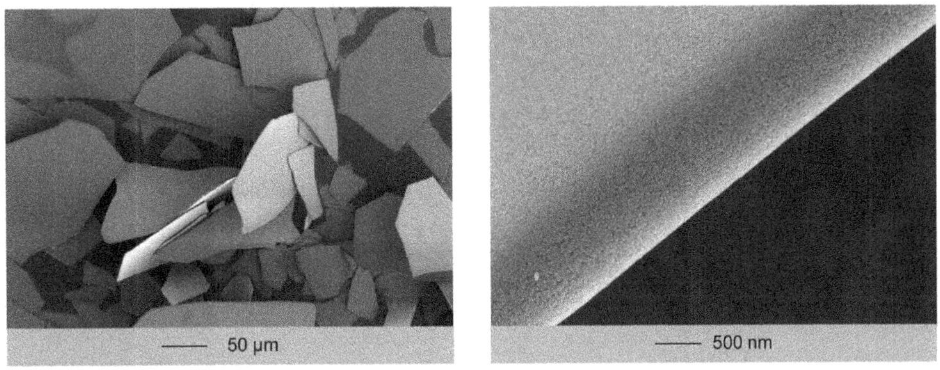


*Figure X.34: Scanning electron micrographs of pure (left) and $TiO_2$-coated (right) borosilicate flakes.*

Borosilicate pigments are made in three steps:

- Production of the raw flakes
- Grinding and classifying of the flakes
- Coating of the flakes with high-refractive-index metal oxides (titanium dioxide, iron(III) oxide)

The starting point in the synthesis of borosilicate flakes is the melt. The purity of the raw materials used guarantees that the flakes will have the desired colourlessness and transparency. The flakes are obtained directly by cooling the melt, and then grinding and classifying it. Coating of borosilicate flakes is similar to that for mica and aluminium oxide substrates:

- Coating with hydrous metal oxides in aqueous suspension
- Separation of the coated flakes by filtration
- Calcining of the particles to transform the hydrous metal oxides into high-refractive-index metal oxides.

This pigment consists of a borosilicate flake surrounded by a layer of silica, on top of which is a layer of high-refractive-index titanium dioxide. The $SiO_2$ encapsulates the substrate and acts as a protective layer; its surface is ideal for coating with $TiO_2$, which determines the colour. In addition, the $SiO_2$ intermediate layer imparts mechanical stability to the thin substrate flakes. Furthermore, at the end of the process it leads to interference pigments with improved lustre because the refractive index of the $SiO_2$ (1.42 to 1.47) is lower than that of borosilicate (1.50 to 1.58). The effect of this is create a large difference between the refractive indices of the $TiO_2$ layer and the adjacent layer.

The electron micrographs in Figure X.34 show the benefits of the coating. The edges of the pigment particles are perfectly covered and rounded. This promotes reflection of light while, at the same time, improving the feel on the skin for cosmetics applications [81].

Most borosilicate pigments possess particle diameters below 200 µm and are less than 1 µm thick. They provide application systems with a unique and striking multicolour sparkle effect. This optical phenomenon, which is particularly pronounced in the case of silvery-white pigments, is therefore also called the multicolour effect. These pigments serve in various systems to create manifold colour effects, ranging from simple lustre through to highly reflective formulations.

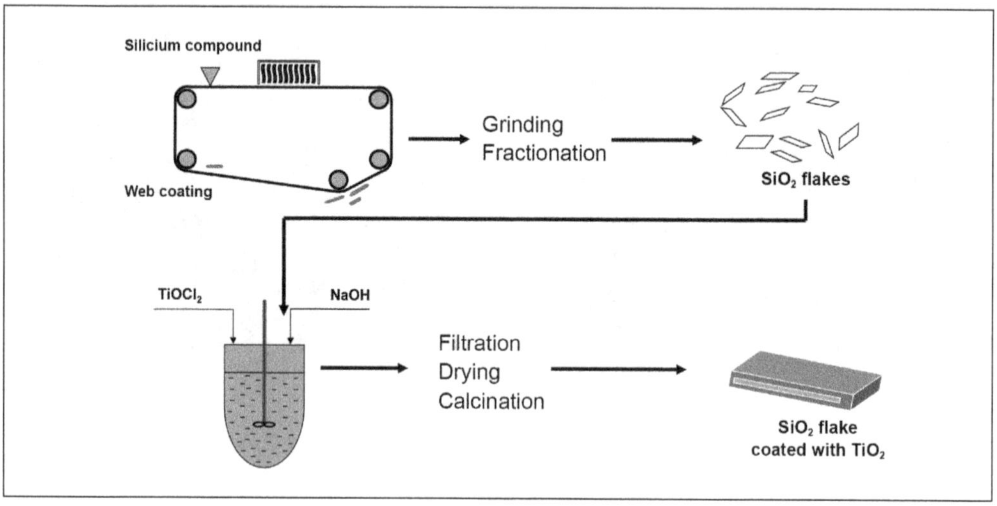


*Figure X.35: Production scheme for $SiO_2$ flake pigments.*

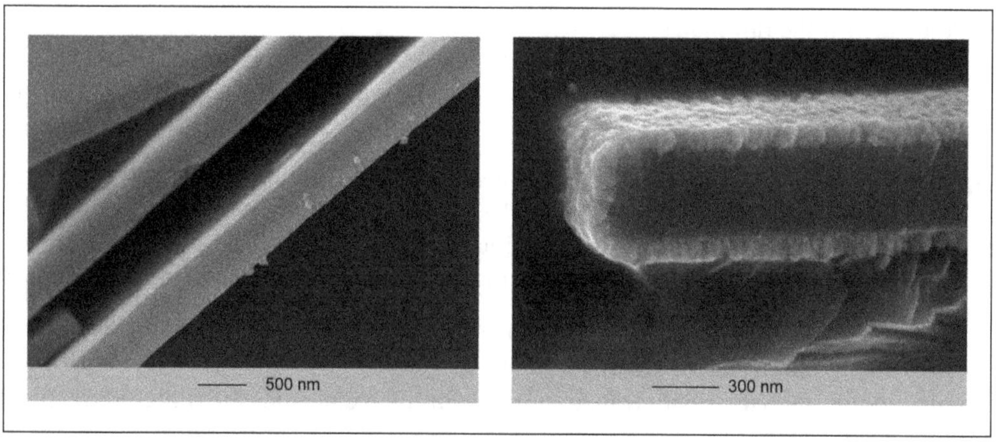


*Figure X.36: Scanning electron micrographs of pure (left) and $TiO_2$-coated (right) $SiO_2$ flakes.*

### 4.1.4 Metal oxide pigments based on silica flakes

Silica flakes (silicon dioxide flakes, $SiO_2$ flakes), together with their alumina and borosilicate counterparts, are innovative substrate materials that produce extraordinary colour effects in transparent and semi-transparent effect pigments. They are manufactured by a web-coating process, which yields thin flakes of completely uniform thickness. Precursors for the process are silicon compounds, such as sodium water glass. Figure X.35 shows the production of $SiO_2$ flakes as well as the downstream step in which the flakes are coated with high-refractive-index metal oxides, especially titanium dioxide or iron oxide [73, 75, 82, 83].

The structure of the pure $SiO_2$ flakes and the $TiO_2$-coated $SiO_2$ flakes can be seen in the electron micrographs in Figure X.36.

Pigments based on silica flakes are particularly remarkable for their multi-colour effects (angle-dependent colour, colour travel, colour flop). Precise adjustment of the thickness of the substrate and metal oxide coating ensures optimal development of the pigments' coloristic properties. At the same time, their gradual colour travel supports pronounced multi-colour changes in application systems. Combinations with transparent coloured pigments and black pigments yield extensive colour effects that offer colorists plenty of scope for creative styling.

Synthetic $SiO_2$ flakes differ from mica platelets in three main ways:

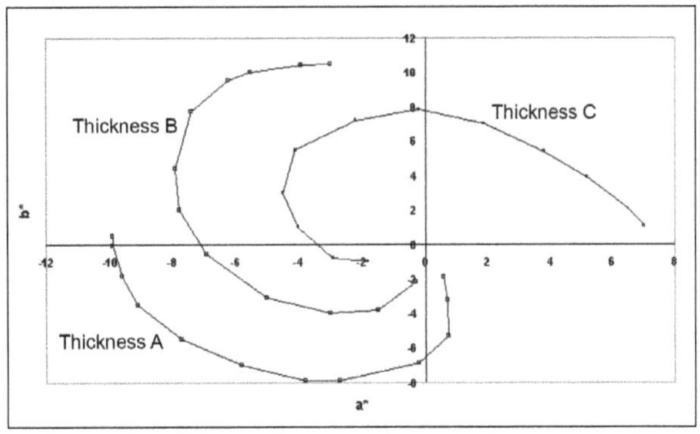


*Figure X.37: Angle- dependent colour travel of $Fe_2O_3$-coated $SiO_2$ flakes of different thickness.*

1. The thickness of the $SiO_2$ substrate can be controlled very precisely during manufacture. This is why the final pigments are genuine three-layer optical systems. The interference colour of these pigments is stronger than that of conventional mica pigments because mica possesses a broad thickness distribution as a result of its manufacturing process.
2. As synthetic substrates, $SiO_2$ flakes are free of iron impurities, which, in the case of natural mica, lead to a slightly yellow mass tone colour.
3. $SiO_2$ has a lower refractive index (1.46) than mica (approx. 1.58) and therefore gives rise to more intense interference effects [75]. The properties obtainable with $SiO_2$ flakes can be summarised as follows:
   - Improved substrate characteristics
     - Uniform, controllable substrate thickness
     - Smooth, uniform substrate surface
     - Transparent substrate with no mass tone colour
   - Improved reflection and refraction of light
   - Improved chromatic strength and purity
   - New colour travel properties

The thickness of the $SiO_2$ flakes produced in the web-coating process can be varied from 50 to 1000 nm. The flakes make an excellent substrate for coating with high-refractive-index metal oxides. Deposition of oxides, such as $TiO_2$ and a-$Fe_2O_3$, occurs by means of the wet-chemical coating process established for mica pigments. As is the case with mica substrates, direct coating yields $TiO_2$ in the anatase modification. Generation of rutile layers requires an $SnO_2$ intermediate layer to be applied before deposition of the $TiO_2$ layer. $SiO_2$ flake pigments with rutile layers demonstrate more brilliant colour intensity and exhibit superior light stability. Most silica-based pigments are therefore produced with rutile layers.

The colour travel described above can be adjusted over a broad colour range. Colour changes from golden silver to green and then from green to blue or even dark blue are possible. Figure X.37 shows angle-dependent colour travel in the CIELAB system for three pigments of constant $Fe_2O_3$ layer thickness on $SiO_2$ flakes of different thickness.

### 4.1.5 Metal oxide pigments based on iron oxide flakes

Iron oxide flakes can be used to make lustre pigments that exhibit more pronounced interference effects and particularly angle-dependent colour travel. This is achieved by applying additional metal oxide layers to the flakes [84]. Typically a layer structure of $Fe_2O_3/SiO_2/Fe_2O_3/SiO_2/Fe_2O_3$ is generated, in which the central $Fe_2O_3$ layer (core) consists of the iron oxide flake, while the two outer layers provide selective reflection and, together with the $SiO_2$ layers, create a pronounced colour change in violet-red-orange-gold colour space. This also imbues the pigments with good corrosion resistance.

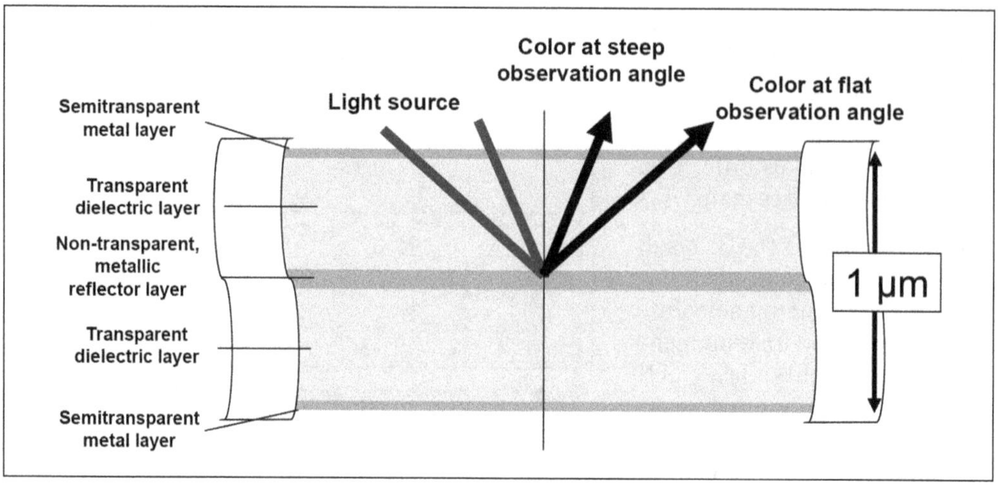


*Figure X.38: Schematic illustration of the layer structure of optically variable pigments (OVP) with Fabry-Perot structure.*

Such five-layer effect pigments are manufactured in a stirred vessel, where iron oxide flakes are coated with a precisely controlled $SiO_2$ layer by hydrolysis of tetraethyl orthosilicate in alcoholic suspension. After washing and drying stages, the outer $Fe_2O_3$ layer is deposited in a fluidised bed reactor by means of CVD. This entails making gaseous $Fe(CO)_5$ react with oxygen at elevated temperature in the presence of iron oxide flakes coated with $SiO_2$ to yield a pigment with a surface layer of a-$Fe_2O_3$ of defined thickness. Alternatively, the outer iron oxide layer can be deposited by wet-chemical co-precipitation.

### *4.1.6 Multi-layer pigments with a Fabry-Perot structure*

Besides the optically variable pigments just described, products which are made by **physical vapour deposition** (PVD) and which exhibit a strong angle-dependent colour effect have been on the market since 1990. These are mainly based on a metal-dielectric, multi-layer structure and have not only a distinctive colour travel, but also high colour intensity. Different colours are generated by matching the film thickness with the type of pigment used and carefully controlling the parameters. The film thickness must be kept within a tolerance of a few atoms in order to ensure minimum colour difference [56, 84].

Chromium (a semi-transparent absorber) and aluminium (a reflector impermeable to light) are often chosen for the metal layers. Magnesium fluoride or silicon dioxide usually serve as the dielectric layers. As shown in Figure X.38, symmetrical configurations are used for pigments, while asymmetrical layers are suitable for large-area, non-pigmented optical coatings. The chosen layer structures form the basis of an optical phenomenon known as the Fabry-Perot effect. Such multi-layer interference configurations can be constructed in such a way that some parts of the visible spectrum are extensively reflected, while neighbouring wavelengths undergo little reflection. This offers a way of generating very high colour intensity. Extremely precise film thicknesses are needed if the special spectral properties of the pigments are to be obtained.

For flake-shaped pigment particles that are ideal for heavily angle-dependent effects, the symmetrical layer configuration mentioned above is essential. At least five layers are necessary.

Fabry-Perot pigments are typically produced on a specific sequence of roll coaters located inside vacuum chambers. The first step involves depositing a release layer onto a moving transfer film. This layer consists of a soluble polymer or a salt and is soluble in solvents. This ensures that coatings deposited on the film at a later stage can be removed at the end of the process.

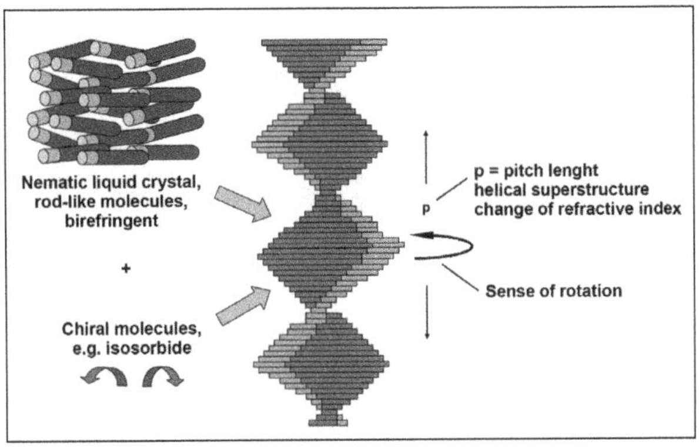


*Figure X.39: Formation of cholesteric liquid crystal using the reaction of a liquid-crystalline nematic phase with chiral molecules.*

The transfer film, complete with release layer, is then placed in the vacuum coater. The air is removed and the first layer of semi-transparent absorber metal (e.g. chromium) is vacuum-deposited onto the film. Then the first dielectric layer (e.g. $MgF_2$) and the non-transparent metal layer (e.g. aluminium) are applied. The second dielectric layer and the second semi-transparent metallic absorber layer are then applied in the same way. When the entire symmetric multi-layer configuration (e.g. Cr-$MgF_2$-Al-$MgF_2$-Cr) has been applied, the transfer film is removed from the vacuum equipment. The release layer is then dissolved to yield the multi-layer film.

In the following stage, the multi-layer film is ground and sorted to the required particle-size distribution. The resulting multi-layer pigment particles have a diameter of 1 to 100 μm and are from 0.2 to 2 μm thick. Pigments with a Fabry-Perot structure are used for anti-counterfeiting purposes, e.g. bank notes, share certificates, visas and passports [85].

### 4.1.7 Effect pigments based on liquid-crystal polymers (cholesteric effect pigments)

Angle-dependent interference colour effects can also be created with layers or particles of liquid crystal polymers (LCPs). They is done either with platelet-shaped particles, which themselves are made from LCP material, or with platelets that are uniformly coated with a cross-linked liquid crystal polymer in a chiral-nematic (cholesteric) arrangement [86–88].

Liquid crystals are organic compounds which are in a state intermediate between that of an isotropic liquid and an anisotropic, crystalline solid [89]. They consist of long, rod-shaped or disc-shaped molecules, the arrangement of which determines the various types of liquid crystal that can occur: nematic, smectic and cholesteric. In **nematic phases,** the molecules arrange themselves in a long-range order, and, on average, are aligned in the same direction. The molecules lie parallel to one another along the director. Nematic phases therefore behave like crystals that have one axis [89].

The **cholesteric phase** is the chiral opposite of the nematic phase. It is mostly used in displays. If chiral molecules are arranged in a nematic host phase, helical superstructures are formed. The resulting structure can be understood in the following way: the molecules arrange themselves from layer to layer at a slight angle to one another. One turn of the helix represents a rotation of 360°, i.e. the director is back in the original direction (Figure X.39). The thickness of the layer structure up to the complete turn of the director is the pitch (p) of the helix.

A continual change in refractive index within the helix causes interference effects that are combined with angle-dependent selective reflection. The reflected light is circularly polarised in the same rotational direction as the helix of the liquid crystalline phase. Circularly polarised light with the opposite rotational direction as well as all non-reflected

*Figure X.40: Scanning electron micrograph of the surface of a holographic pigment.*

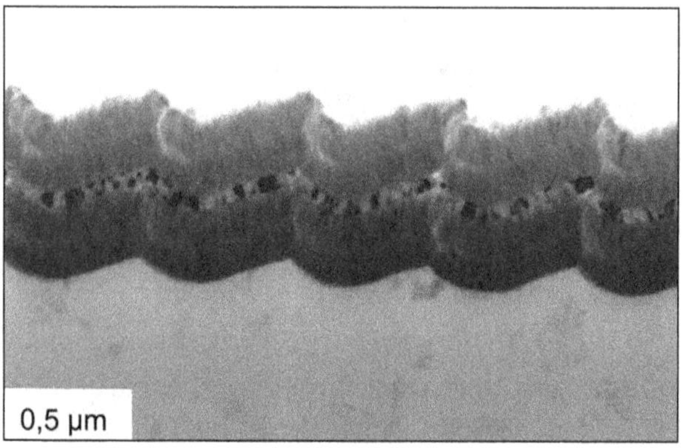


*Figure X.41: Transmission electron micrograph (cross-section) of a diffractive pigment with layer structure $MgF_2$-Al-$MgF_2$.*

wavelengths of light pass through the transparent liquid-crystalline material. Just one clockwise or anticlockwise rotating layer is enough to cause 50 % of the light striking the object to be reflected. If the two layers are combined, full reflection occurs. The wavelength of the reflected light depends on the pitch and the refractive index. The bandwidth $\Delta l$ of the reflected light follows the equation $\Delta l = p \cdot \Delta n$.

The intensity of the total reflection is strengthened by the reflection from the various layers of the helix. To obtain the maximum reflectivity of a polarised state, at least six helices or a thickness of approx. 3 µm are required [86]. The most effective reflection is achieved at film thicknesses of up to 10 µm.

Cholesteric materials are heat-sensitive and thermochromic. This is due to the temperature dependence of the pitch of the helix and the refractive index [86, 90, 91]. The three-dimensional crosslinking of the layers of liquid crystal fixes the existing state and the temperature dependence vanishes. At the same time, the high-viscosity state of the liquid crystals is transformed to a polymer network which permanently maintains its form.

Polymerised pigment platelets are often made from polysiloxanes. A thin film of the liquid crystal is photochemically polymerised and subsequently ground into platelets. The pigments are colourless and transparent. They demonstrate a strong angle-dependent interference effect in a paint film, especially against a dark background. When the substrate is white, only a nacreous sheen is seen whereas, on black areas, pure rainbow colours with strong colour travel become visible. Commercially available pigment particles are over 4 µm thick. Cholesteric pigments have found limited use so far because of the thickness of the platelets and their insufficient stability and workability for certain applications.

### 4.1.8 Structured effect pigments

#### Holographic pigments

Platelet-shaped pigment particles can be formed by crushing holographic films into small pieces [92, 93]. Figure X.40 shows the surface of a pigment particle with a holographic structure. Basically, any base material that has a symmetrical, holographic structure can be used to make holographic pigments. However, in practice, holographic mirrors are the most popular. The simplest of reflection holograms, these are produced by splitting a laser beam and then reuniting the divided beams at a photosensitive layer on a thermoplastic film (double-beam method). Alternatively, the hologram can be the result of projecting a laser beam through a photosensitive layer onto a mirror (single-beam method).

Pigment particles produced by crushing thermoplastic films have a minimum thickness of 10 μm and an average diameter of 10 to 300 μm. A thin metallic layer, preferably of aluminium vapour, is deposited onto the film prior to size reduction to ensure the necessary reflection [62]. The difficulty in producing thinner polymer films with a holographic structure has restricted their use to the printing and paint sectors.

***Diffractive pigments***

Diffractive pigments are produced by vacuum-depositing thin films onto a structured-polymer film template. Typical examples of the sequence of layering on the polymer film are $MgF_2$ Al $MgF_2$ and $SiO_2$-Al-$SiO_2$. There is a release layer between the film and the layers produced by the PVD process (see Section X.4.1.6). At the end of the deposition process, the release layer is removed. The resulting pieces of deposited material are then collected, ground into particles and fractionated [94, 95]. They are typically less than 1 μm thick. The structure has a periodicity of around 1 μm and a depth of several nanometers [95]. Figure X.41 shows a transmission electron micrograph of a diffractive pigment. $MgF_2$ or $SiO_2$ acts as mechanical support layers for the thin, structured aluminium film.

# 5 Literature

[1] Gunter Buxbaum (Hrsg.) Industrial inorganic pigments, VCH Verlagsgesellschaft, Weinheim, 1st Edition 1993

[2] [1] p. 43 ff

[3] [1] p. 141 ff

[4.] [1] p. 85 ff

[5.] [1] p. 120-121

[6.] [1] p. 118 ff

[7.] [1] p. 101 ff

[8] [1] p. 114 ff

[9] [1] p. 96

[10] [1] p. 103

[11] W. Herbst, K. Hunger, Industrial Organic Pigments - 3rd, completely revised edition 2004, WILEY-VCH Verlag GmbH & Co. KGaA, Weinheim

[12] [11] p. 474 ff

[13] [11] p. 504 ff

[14] [11] p. 328 ff

[15] [11] p. 303 ff

[16] [11] p. 480 ff

[17] [11] p. 490 ff

[18] [11] p. 295 ff

[19] [11] p. 307 ff

[20] [11] p. 276 ff

[21] [11] p. 574

[22] [11] p. 370-371

[23] [11] p. 415

[24] [11] p. 414

[25] [11] p. 225-226

[26] [11] p. 355

[27] [11] p. 353

[28] [11] p. 222-223

[29] [11] p. 451

[30] [11] p. 449

[31] [11] p. 448

[32] [11] p. 440 ff

[33] [11] p. 448

[34] [11] p. 533 ff

[35] [11] p. 465

[36] [11] p. 467

[37] [11] p. 468

[38] [11] p. 331

[39] [11] p. 465

[40] [11] p. 539

[41] [11] p. 518

[42] DIN 6175 - Teil 2 (03-2001): Farbtoleranzen von Automobillackierungen, Teil 2: Effektlackierungen

[43] A.B.J. Rodriguez, „Metallic Flop and its Measurement", J. Oil Colour Chem. Assoc., 75 (4), 150 (1992)

[44] A. Wooker et al., „Orientierung von Effektpigmenten", Farbe & Lack 11-2008, S. 34-39

[45] Pfaff, G., Spezielle Effektpigmente, Vincentz Verlag, Hannover, 2007

[46] Krüger, A., Perlen, Bibl. d. Unterhaltung und des Wissens, Union Deutsche Verlagsgesellschaft, Stuttgart, 1919

[47] DIN 55943:2001-10, Farbmittel (Begriffe)

[48] Glausch, R., Kieser, M., Maisch, R., Pfaff, G., Weitzel, J., Special Effect Pigments, Vincentz Verlag, Hannover, 1998

[49] Greenstein, L. M., Pigment Handbook, 2nd ed., Vol. 1, Wiley New York, 1988, 829

[50] Schmidt, C., Friz, M., Optical Physics of Synthetic Interference Pigments, Kontakte, Darmstadt, 2 (1992) 15

[51] Pfaff, G., Perlglanzpigmente, Vincentz Verlag Hannover, 1995, 13

[52] Benbow, B. J., de Jong, B., Adams, J. W., Ullmann's Encyclopedia of Industrial Chemistry, 5th ed., Vol. A 16, VCH Verlagsges. Weinheim, 1991, 551

[53] Gysau, D., Fillers for Paints, 2nd ed., Vincentz Network Hanover, 2011

[54] Atwood, F. C., US 2,278,970

[55] Linton, H. R., US 3,087,828

[56] Linton, H. R., US 3,087,829

[57] Kohlschütter, H. W., Rössler, H., Getrost, H., Hörl, W., Reich, W., US 3,553,001

[58] Quinn, C. A., Rieger, C. J., Bolomey, R. A., US 3,437,515

[59] Eskelinen, P., X-Ray Diffraction Study of $TiO_2$ Thin Films on Mica, J. Solid State Chem., **100** (1992) 356

[60] Hildenbrand, V. D., Doyle, S., Fuess, H., Pfaff, G., Reynders, P., Crystallization of Thin Anatase Coatings on Mica, Thin Solid Films, **304** (1997) 204

[61] Pfaff, G., Effektpigmente - Nano-Metalloxidschichten auf plättchenförmigen Partikeln, Praxis der Naturwissenschaften: Chemie, **55** (2006) 26

[62] Pfaff, G., Reynders, P., Angle-Dependent Optical Effects Deriving from Submicron Structures of Films and Pigments, Chem. Rev., **99** (1999) 1963

[63] Schmidt, C., Maurer, U., Interference for the New Millennium - Timiron Splendid Colors, Euro Cosmetics, **8** (1999) 28

[64] Huber, A., Pfaff, G., Rösler, M., A Novel System of Interference Pigments, Proc. Congress ECS 2009

[65] Schmidt, C., Bernhardt, K., Schoen, S., Heinz, D., US 6,599,355

[66] Ostertag, W., Effektpigmente, Nachr. Chem. Technol. Lab., **42** (1994) 849

[67] Ostertag, W., Bittler, K., Bock, G., EP 68.311

[68] Thurn-Müller, A., Hollenberg, J., Cosmetics and Toiletries Manufacture, Century Press Bushey Heath, 1 991, 72

[69] Thurn-Müller, A., Hollenberg, J., Liston, J., "Special Effect Pigments and Fillers for Cosmetics", Kontakte, Darmstadt, 2 (1992) 35

[70] Emmert, R., Weigand, M., EP 307 747

[71] Maisch, R., Pinture e Vernici, **66** (1990) (4) 9

[72] Pfaff, G., Industrial Inorganic Pigments, 3rd ed., Wiley-VCH, 2005, 230

[73] Pfaff, G., High Performance Pigments, ed. H. M. Smith, Wiley-VCH, 2002, 77

[74] Nitta, K., Shau, T. M., Sugahara, J., EP 763 573

[75] Teaney, S., Pfaff, G., Nitta, K., New Effect Pigments Using Innovative Substrates, Eur. Coat. J., 4 (1999) 90

[76] Pfaff, G., Innovationen bei Effektpigmenten, Welt der Farben, 1 (2000) 16

[77] Sharrock, S., Schül, N., New Effect Pigments Based on $SiO_2$ and $Al_2O_3$ Flakes, Eur. Coat. J., 1/2 (2000) 20

[78] Xirallic: eingetragener Warenname der Firma Merck KGaA, Darmstadt

[79] Huber, A., Pfaff, G., Spezielle Effektpigmente, Phänomen Farbe, 3 (2005) 34

[80] Pfaff, G., Huber, A., Spezielle Effektpigmente auf Basis von Aluminiumoxid-Plättchen, Welt der Farben, 9 (2005) 14

[81] Rüger, R., Oldenburg, N., Schulz, E., Thurn-Schneller, A., Tar Effects from Borosilicates, Cosmet. Toiletries, 5 (2004) 133

[82] Bauer, G., Osterried, K., Schmidt, C., Vogt, R., Knieß, H., Uhlig, M., Schül, N., Brenner, G., WO 93/08237

[83] Pfaff, G., Special Effect Pigments Based on Silica Flakes, Inorg. Mater., **39** (2003) 123

[84] Schmid, R., Mronga, N., Radtke, V., Seeger, O., Optisch variable Glanzpigmente, Farbe + Lack, **104** (1998) 44

[85] Phillips, R. W., Bleikolm, A. F., Optical Coatings for Document Security, Appl. Opt., **35** (1996) 5529

[86] Makow, D., Color Gamut of Liquid Crystal Polysiloxanes, Mol. Cryst. Liq. Cryst., **123** (1985) 347

[87] Robbie, K., Brett, M. J., Lakhtakia, A., Chiral sculptured thin films, Nature, **384** (1996) 616

[88] Haremza, S., Wenn Farben floppen - Effekte mit Flüssigkristallen, Welt der Farben, 3 (2000) 14

[89] Sage, I., Ullmann's Encyclopedia of Industrial Chemistry, 5th ed., Vol. A 15, VCH Verlagsges. Weinheim, 1990, 359

[90] Makow, D., Reflection and Transmission of Polymer Liquid-Crystal Coatings and Their Application to Decorative Arts and Stained Glass, Color Res. Appl., **11** (1986) 205

[91] Tamaoki, N., Parfenov, A. V., Masaki, A., Matsuda, H., Mater. Res. Soc. Symp. Proc., **488** (1998) 915

[92] Boyle, E., Fast & Flawless, Paper, Film & Foil Converter, May 1 (2004) 1

[93] Fitch, J. J., Fotlano, R. A., Josephy, K., Meikka, F. N., Meikka, R. G., EP 678 074

[94] Maile, F. J., Pfaff, G., Reynders, P., Effect Pigments - past, present and future, Progr. Org. Coat., 54 (2005) 150

[95] Argoitia, A., Witzmann, M., Society of Vacuum Coaters, 45th Annual Techn. Conf. Proc., 2002, 539

# XI Recipe prediction

## 1 Recipe prediction for solid colours

*Uwe Hempelmann*

A recipe in this context is a tabular list of the various colorants, especially the pigments, needed to obtain a certain colour, along with the exact masses required.

The objective is to determine the type and quantity of colorants in accordance with a template. Here, again, it is necessary to link the mass of the colorants used to the resulting reflectance spectrum. In the following, we look at a highly successful approach that is adopted for solid colours in practice and is based on the generalised Kubelka-Munk equation (see Section VII.4). The absorption and scattering coefficients of a mixture, $K_M$ and $S_M$, are given by the sum of the components ($K_1 + K_2 + ... + K_N$, and $S_1 + S_2 + ... + S_N$).

*Equation XI.1*

$$\frac{K_M}{S_M} = \frac{K_1 + K_2 + \ldots + K_n}{S_1 + S_2 + \ldots + S_n}$$

Using $c_1, c_2, ... c_n$ to denote the concentrations of the colorants or pigments:

*Equation XI.2*

$$\frac{K_M}{S_M} = \frac{c_1 K_1 + c_2 K_2 + \ldots + c_n K_n}{c_1 S_1 + c_2 S_2 + \ldots + c_n S_n}$$

Recipe prediction entails calculating concentrations $c_1, c_2, ... c_n$ of the pigments from R or $K_M/S_M$ values of the template and from calibration tables (see Section XI.2). The objective is to minimise the colour difference between the target and the calculated mixture.

There are two ways to find this optimum:

- Minimise either the colour difference $\Delta E^*_{ab}$ between measured and calculated colour values (L*, a*, b* or. X, Y, Z values, "XYZ prediction") or
- Minimise the difference between the measured and calculated reflectance spectra.

The latter case is mathematically equivalent to the minimum of the sum of the squares of the differences between the measured and calculated reflectance spectra. Sometimes this is also called "spectral matching".

The objective can be formulated as follows: Given reflectance values between 400 and 700 nm, calculate the concentration of the pigments in such a way that the following expression yields a minimum (see Equation XI.3).

*Equations XI.3 and XI.4*

$$\text{Min}\,(\Delta E^*_{ab}) = (L^*_{measured} - L^*_{calculated})^2 + (a^*_{measured} - a^*_{calculated})^2 + (b^*_{measured} - b^*_{calculated})^2 \quad \text{i.e.}$$

$$\text{Min}\left(\sum_{\lambda=400}^{700} (R_{measured}(\lambda) - R_{calculated}(\lambda))^2\right)$$

In the last variant, an equivalent term can be found via the Kubelka-Munk function. Here, "O" = original and "M" = calculated mixture.

*Equations XI.5 and XI.6*

$$\mathrm{Min}\left(\sum_{\lambda=400}^{700}\left(\left\{\frac{K_V(\lambda)}{S_V(\lambda)}\right\}_{\text{measured}} - \left\{\frac{K_M(\lambda)}{S_M(\lambda)}\right\}_{\text{calculated}}\right)^2\right) \quad \text{i.e.}$$

$$\mathrm{Min}\left(\sum_{\lambda=400}^{700}\left(\left\{\frac{K_V(\lambda)}{S_V(\lambda)}\right\}_{\text{measured}} - \left\{\frac{c_1 K_1(\lambda) + c_2 K_2(\lambda) + \ldots + c_n K_n(\lambda)}{c_1 S_1(\lambda) + c_2 S_2(\lambda) + \ldots + c_n S_n(\lambda)}\right\}_{\text{calculated}}\right)^2\right)$$

The solution leads to a system of equations which can be solved by the usual methods for iterative approximations and yields the concentrations $c_1, c_2, \ldots, c_n$. Often, the necessary methods are part of spreadsheet programs such as Microsoft Excel and Open Office Calc.

The XYZ approximation often leads to a very small colour difference ($\Delta E^*_{ab}$ = 0 is always true for the case of three pigments), but suffers from the major disadvantage that the results do not exclude highly metameric recipes. The solutions obtained with spectral matching are much more robust in this respect, but do not necessarily return the colorimetric minimum for the chosen specific standard illuminant.

There are several factors that must always be considered when the results are being evaluated:

1. The sensitivity or robustness of the recipe with respect to possible fluctuations in values of single components.
2. Correctability of the recipe
3. Compatibility with additives

Commercial programs for recipe prediction typically include the following factors:

1. Application aspects
2. Economic effects
3. Availability of components
4. Calculation of hiding power, i.e. the pigment volume concentration needed to obtain a hiding layer for a given coating thickness. Figure XI.1 summarises the various steps involved in recipe prediction.

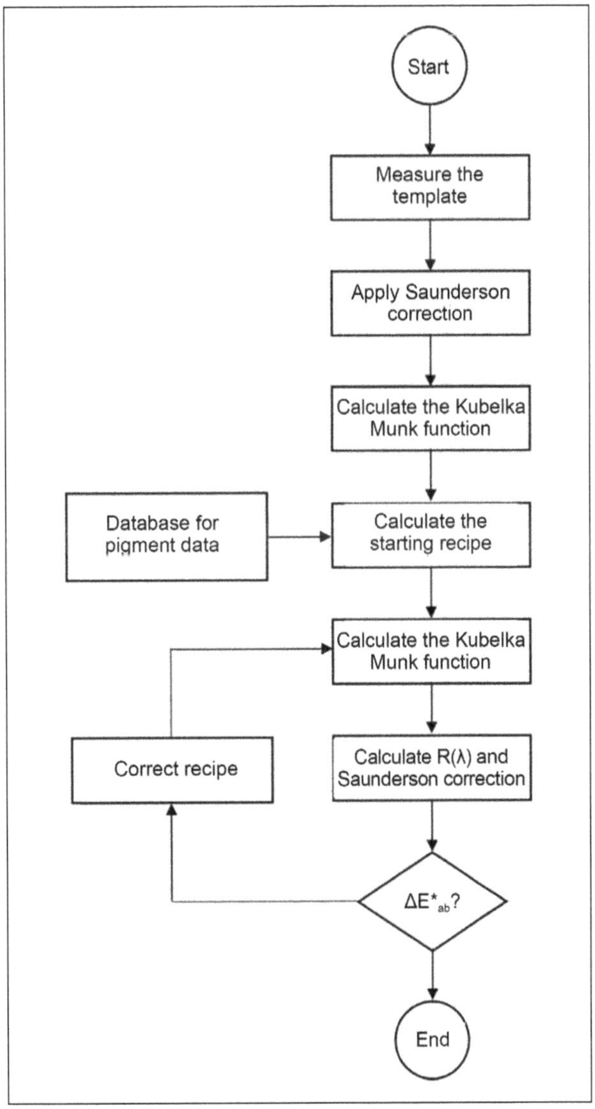


*Figure XI.1: Overview of the steps involved in recipe prediction.*

# 2 *Calibration of colourants*

*Uwe Hempelmann*

The preparation of **calibration samples** is essential for successful recipe prediction. As described in the preceding sections, calculation of absorption and scattering coefficients is difficult and prone to much higher experimental error than is the case for determining the quotient of the two values. With opaque calibration samples, this can be avoided by determining the absorption and scattering coefficient relative to a so-called white base. The downside is the impossibility of determining the hiding power and the necessary concentration for obtaining an opaque layer. The equations behind this approach are as follows.

*Equations XI.7 to XI.9*

$$\frac{K_M}{S_M} = \frac{c_1 K_1 + c_2 K_2 + \ldots + c_n K_n}{c_1 S_1 + c_2 S_2 + \ldots + c_n S_n}$$

$$= \frac{c_1 K_{1W} + c_2 K_{2W} + \ldots + c_n K_{nW}}{c_1 S_{1W} + c_2 S_{2W} + \ldots + c_n S_{nW}}$$

$$= \frac{c_1 \frac{K_1}{S_W} + c_2 \frac{K_2}{S_W} + \ldots + c_n \frac{K_2}{S_W}}{c_1 \frac{S_1}{S_W} + c_2 \frac{S_2}{S_W} + \ldots + c_n \frac{S_2}{S_W}}$$

Experimentally this is done by testing the respective pigments in different PVCs in blends with the chosen white base. The application of the Kubelka-Munk equation to yield these results is shown below. The equation systems need to be rearranged in a subtle way.

Thus, limiting ourselves to a mixture of just two components, the white pigment and pigment A, Equations XI.7 to XI.9 yield:

$$\frac{K_M}{S_M} = \frac{K_A + K_W}{S_A + S_W}$$

$$= \frac{\frac{K_A}{S_W} + \frac{K_W}{S_W}}{\frac{S_A}{S_W} + 1}$$

$$= \frac{\left(\frac{K_A}{S_A}\right)\left(\frac{S_A}{S_W}\right) + \left(\frac{K_W}{S_W}\right)}{\frac{S_A}{S_W} + 1}$$

The target variables are given by Equations XI.10 and XI.11:

*Equations XI.10 and XI.11:*

$$\frac{S_A}{S_W} = \frac{\frac{K_M}{S_M} - \frac{K_W}{S_W}}{\frac{K_A}{S_A} - \frac{K_M}{S_M}}$$

$$\frac{K_A}{S_W} = \frac{K_A}{S_A} \frac{\frac{K_M}{S_M} - \frac{K_W}{S_W}}{\frac{K_A}{S_A} - \frac{K_M}{S_M}}$$

Explicitly stating the concentrations of the pigments, we arrive at Equations XI.12 to XI.14:

*Equations XI.12 to XI.14:*

$$\frac{K_A}{C_W \hat{S}_W} = \frac{K_A}{S_A} \frac{\frac{K_M}{S_M} - \frac{K_W}{S_W}}{\frac{K_A}{S_A} - \frac{K_M}{S_M}}$$

$$K_{AW} = \frac{K_A}{\hat{S}_W} = C_W \frac{K_A}{S_A} \frac{\frac{K_M}{S_M} - \frac{K_W}{S_W}}{\frac{K_A}{S_A} - \frac{K_M}{S_M}} = C_A \hat{K}_{AW}$$

$$S_{AW} = K_{AW} \frac{K_A}{S_A}$$

Thus, the values for $K_M/S_M$ can be obtained from the slope of the straight lines obtained by plotting the values for $K_{AW}(\lambda)$ and $S_{AW}(\lambda)$ against the pigment volume concentration.

These calculations are done for each wavelength. Naturally, the concentrations used in the calibration series vary with the pigment used. Table XI.1 shows an example of this.

Figures XI.2 and XI.3 chart the K and S values for iron oxide pigments as a function of pigment volume concentration.

The results for the absorption and scattering coefficients relative to the chosen white base are entered into a database where they form the basis for all subsequent recipe predictions.

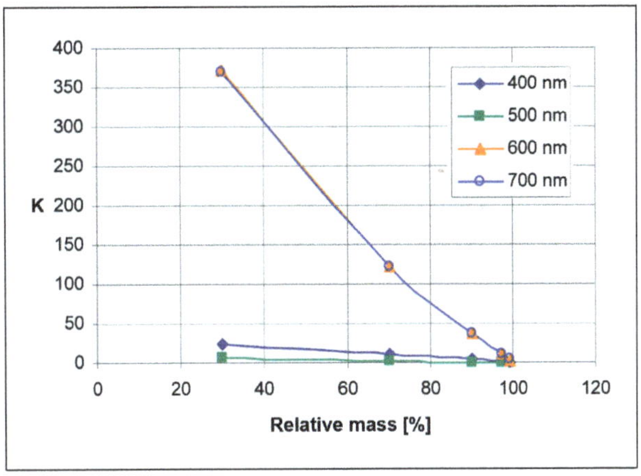


*Figure XI.2: Absorption coefficient as a function of mass of pigments.*

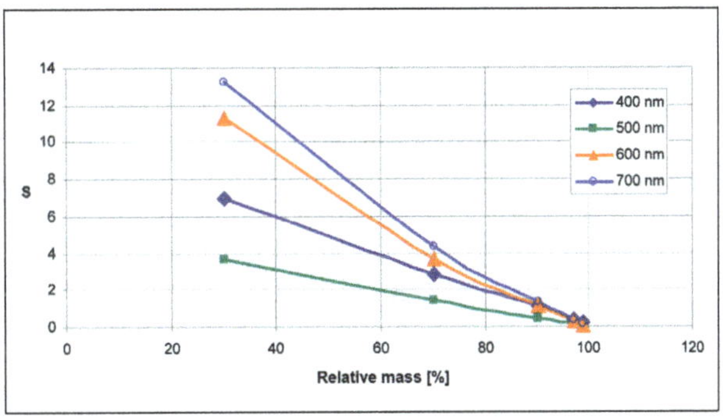


*Figure XI.3: Scattering coefficient as a function of mass of pigments.*

In the case of very pale pigments, especially yellows, blending with white pigments effects very little significant change in the reflectance spectra. This means that the calculated values for absorption and scattering coefficients are tainted with high experimental error. Here the necessary data for recipe prediction are obtained by blending with a black pigment.

The quality of recipe predictions is critically dependent on adherence to the following rules:

- Highest-possible precision when calibration samples are being prepared
- Very precise measurement of the coating thickness if transparent specimens are used
- Knowledge of the density of each component – important for calculating hiding power
- Calibration samples should be a close match for the actual application, i.e. should be of the correct thickness and always applied over a contrasting substrate.

*Table XI.1: Example of graduated concentrations in a calibration series.*

| (Colour) component [%] | Reference or standard [%] |
|---|---|
| 2.5 | 97.5 white |
| 5.0 | 95.0 white |
| 10.0 | 90.0 white |
| 20.0 | 80.0 white |
| 40.0 | 60.0 white |
| 70.0 | 30.0 white |
| 100.0 | 0.0 white |
| 99.0 | 1.0 black |

- If paper substrates are used, it is important to use only paper without optical brightener.

At least four different concentrations must be used per component. That allows the calibration samples to be checked.

# 3 Computer-aided correction of colour recipes

*Wilhelm Kettler*

In the ideal scenario of an error-free recipe prediction system, perfectly standardised paint materials and negligible process errors, computer-aided recipe calculation should directly yield a colour match with no residual colour difference to the given standard. However, experience shows that the limitations of the radiative transfer model employed, the quality of the optical material parameters, conditionally standardised tinting components, variance in sample preparation and measurement error will in any event give rise to a finite residual colour difference between the standard and the predicted formula which represents the total error of the process. This list of expected shortcomings implies that the total error of the process is composed of statistical and systematic contributions. For this reason, a correction algorithm based on the radiative transfer model cannot be accurate, especially since not even a single sample provides enough information to support a perfect **single-step correction procedure**. Hence, the objective must be to develop a **multi-step correction procedure** for recipe correction that has good convergence properties. In the literature, various methods based on different objective functions have been proposed [1-7]. As in the case of recipe prediction, there is a need in all cases to solve a constrained non-linear fitting problem. The only way to linearise the problem and solve it directly is to employ appropriately small concentration increments.

With regard to the target function, recipe correction can be carried out in different spaces of differing dimensionality, namely in reflectance space and in colour space. A third variant operating in concentration space [3] is numerically unstable and should be avoided. For solid colours, the space of optimisation in the first case would typically have 31 dimensions (spectral match), and in the second case correction would be carried out in 3-dimensional (XYZ-colour match) space. Effect colours are measured at $m > 1$ geometries and so in their case the dimensionality of the space needs to be multiplied by the $m$ number of geometries.

According to Graßmann's third law, transitions in colour space are continuous everywhere, so that an $m$-dimensional colour vector $\mathbf{f}$ can always be developed in a Taylor expansion about a reference point $\mathbf{f}_r$. The colour vector $\mathbf{f}$ on one hand can subsume the $m = 3$ components of a colour position or more generally the $m = 31$ elements of a discrete radiometric function (reflectance spectrum). Where colorimetric quantities are used, the $m = 3$ components of the mixing function are given by the function triple $\mathbf{f(c)} = (X(\mathbf{c}), Y(\mathbf{c}), Z(\mathbf{c}))^T$ and $\mathbf{f(c)} = (L^*(\mathbf{c}), a^*(\mathbf{c}), b^*(\mathbf{c}))^T$, respectively. For terms up to the first order, the colour vector $\mathbf{f} = (f_1, \ldots, f_m)^T$ in component notation can be written

*Equation XI.15*

$$f_1(\mathbf{c}) \cong f_{r,1}(\mathbf{c}_r) + \sum_{i=1}^{n} \frac{\partial f_{r,1}(\mathbf{c}_r)}{\partial c_i} \Delta c_i$$

$$f_2(\mathbf{c}) \cong f_{r,2}(\mathbf{c}_r) + \sum_{i=1}^{n} \frac{\partial f_{r,2}(\mathbf{c}_r)}{\partial c_i} \Delta c_i$$

$$\vdots$$

$$f_m(\mathbf{c}) \cong f_{r,m}(\mathbf{c}_r) + \sum_{i=1}^{n} \frac{\partial f_{r,m}(\mathbf{c}_r)}{\partial c_i} \Delta c_i$$

for small deviations about the reference point $\mathbf{f}_r = (f_{r,1}, \ldots, f_{r,m})^T$, where the number of colouring ingredients of a recipe is equal to the upper running index $n$. When spectral differences are being corrected, the number of degrees of freedom typically is $m = 31$. In vector notation, these equations can be expressed by the relation

*Equation XI.16* $$\Delta\mathbf{f} = \mathbf{f}(\mathbf{c}) - \mathbf{f}(\mathbf{c}_r) \equiv \mathbf{W} \cdot \Delta\mathbf{c} \quad \text{where}$$

*Equation XI.17* $$\mathbf{W} = \begin{pmatrix} \frac{\partial f_{r,1}(\mathbf{c}_r)}{\partial c_1} & \frac{\partial f_{r,1}(\mathbf{c}_r)}{\partial c_2} & \cdots & \frac{\partial f_{r,1}(\mathbf{c}_r)}{\partial c_n} \\ \frac{\partial f_{r,2}(\mathbf{c}_r)}{\partial c_1} & \frac{\partial f_{r,2}(\mathbf{c}_r)}{\partial c_2} & \cdots & \frac{\partial f_{r,2}(\mathbf{c}_r)}{\partial c_n} \\ \vdots & \vdots & \ddots & \vdots \\ \frac{\partial f_{r,m}(\mathbf{c}_r)}{\partial c_1} & \frac{\partial f_{r,m}(\mathbf{c}_r)}{\partial c_2} & \cdots & \frac{\partial f_{r,m}(\mathbf{c}_r)}{\partial c_n} \end{pmatrix}$$

represents the so-called *m × n action* matrix and $\Delta\mathbf{c} = \mathbf{c} - \mathbf{c}_r = (\Delta c_1, \ldots, \Delta c_n)^T$ the vector of concentration increments about the reference point $\mathbf{c}_r$. The column-wise-combined components of the action matrix can be termed the **tinting vector** of the respective colorant. All components of the action matrix $\mathbf{W}$ for any given recipe $\mathbf{c}$ are either synthesised numerically from the solution for the adopted radiative transfer model with the aid of the known optical material parameters of a mixing system or are derived experimentally from appropriate mixing series. The concentration increments $\Delta\mathbf{c}$, which should lead to a decrease in the residual colour difference between sample and colour standard, can be estimated by inverting Equation XI.16:

*Equation XI.18* $$\Delta\mathbf{c} = (\mathbf{W}^T \cdot \mathbf{W})^{-1} \cdot \mathbf{W}^T \cdot \Delta\mathbf{f}$$

This functional form of the solution is a consequence of the fact that, where $m > n$ (i.e. there are more equations than unknowns), only one solution for the set of equations can be obtained (least squares). The calculated concentration increments $\Delta\mathbf{c}$ are added to the concentration vector $\mathbf{c}_r$ from the previous step to determine the corrected recipe

*Equation XI.19* $$\mathbf{c}_{corr} = \mathbf{c}_r + \Delta\mathbf{c}$$

Repeated application of this approach creates an iterative process which eventually leads to an adjustment, where $\Delta\mathbf{f} = 0$ or $\Delta\mathbf{f}$ is at least negligible. This mathematical formalism can be readily generalised to goniochromatic colours, but is left to the reader, due to a lack of space.

For a tristimulus match, the dimensionality of the correction space is crucial. A perfect match requires a formulation to possess three independent degrees of freedom. Due to the constraint arising from the normalisation of the concentrations $\sum_{i=1}^{n} c_i = 1$, opaque paints therefore require four independent pigments to ensure the requisite three degrees of freedom and hence a complete solution to the set of the three simultaneous equations

*Equations XI.20 to XI.22*

$$\Delta L^* = \frac{\partial L^*(\mathbf{c}_r)}{\partial c_1}\Delta c_1 + \frac{\partial L^*(\mathbf{c}_r)}{\partial c_2}\Delta c_2 + \frac{\partial L^*(\mathbf{c}_r)}{\partial c_3}\Delta c_3$$

$$\Delta a^* = \frac{\partial a^*(\mathbf{c}_r)}{\partial c_1}\Delta c_1 + \frac{\partial a^*(\mathbf{c}_r)}{\partial c_2}\Delta c_2 + \frac{\partial a^*(\mathbf{c}_r)}{\partial c_3}\Delta c_3$$

$$\Delta b^* = \frac{\partial b^*(\mathbf{c}_r)}{\partial c_1}\Delta c_1 + \frac{\partial b^*(\mathbf{c}_r)}{\partial c_2}\Delta c_2 + \frac{\partial b^*(\mathbf{c}_r)}{\partial c_3}\Delta c_3$$

*Table XI.2: Table XI.2: Example for the correction of a large paint batch of a four-component colour recipe for a mixing system. The maximum factor $f_{max}$ for scaling up the batch is obtained for colorant 2. Where pigment powders are incorporated, the pigment-to-binder ratio must additionally be factored into the calculation.*

| | Before correction $c_r$ [%] | After correction $c_{corr}$ [%] | f | $f_{max} \cdot c_{corr}$ [%] | $h_i$ [%] | $\Delta m_i$ [kg] |
|---|---|---|---|---|---|---|
| Colorant 1 | 5 | 6 | 5/6 | 7.5 | 2.5 | 50 |
| Colorant 2 | 15 | 12 | 15/12 | 15 | 0 | 0 |
| Colorant 3 | 20 | 23 | 20/23 | 28.75 | 8.75 | 175 |
| White | 60 | 59 | 60/59 | 73.75 | 13.75 | 275 |
| Σ | 100 % | 100 % | | | | 500 kg |
| Batch size | 2000 kg | | | | | |
| Admixture | 500 kg | | | | | |

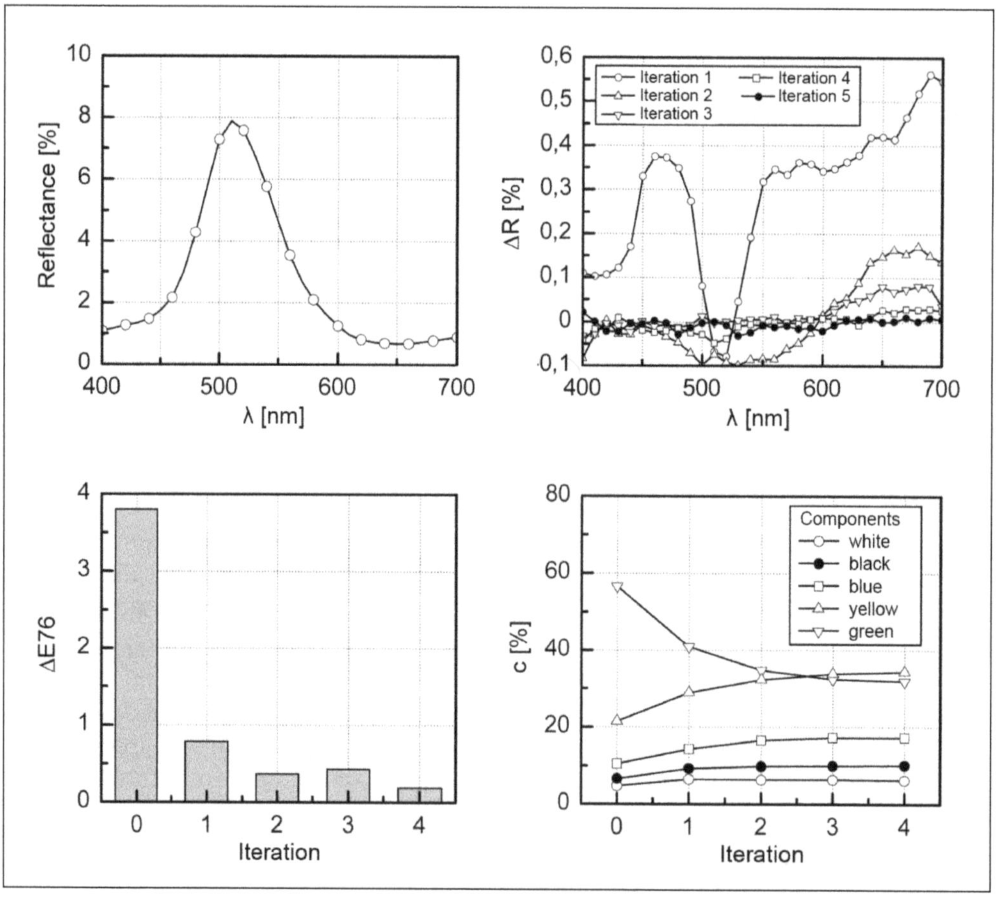


*Figure XI.4: Example of computer-aided matching of a green solid colour. The diagram at top left shows the reflectance spectrum of the colour standard to be matched. The diagram at top right displays the spectral differences between the colour match and the colour standard at the various correction stages. Both lower diagrams show the corresponding residual colour differences of each correction step from the colour standard and the change in amounts of tinting components used. An acceptable match is achieved after just two correction steps, and after four corrections, a quasi-isomeric match with negligible residual colour difference to the standard.*

If a recipe consists of more than four chromatic ingredients, further equations need to be generated by considering yet another illuminant and be added to the set of equations.

Figure XI.4 shows a typical example of the colour development of a dark green solid colour standard, whose reflectance spectrum is shown at top left. The correction results shown are based on the optimisation of a spectral objective function. The diagram at top right shows the spectral differences at each tinting step to the colour standard. The two lower diagrams display the corresponding residual colour difference of all iterations (left) and the variation in amounts of all recipe ingredients (right). All diagrams confirm the convergence of the correction method which, from a coloristic point of view, could have been stopped after just the second step.

The concentration changes $\Delta c_i$ and the admixture amounts returned by the correction algorithm can be positive or negative. In a colour laboratory, this does not create a problem, as a recipe is always mixed from scratch, because the batch size is small (e.g., 100 g). For large production batches of up to several tons, the mixer content must be tinted to the colour standard in the knowledge that once added quantities cannot be removed. Therefore, a correc tion proposal must be produced such that only positive admixture amounts $\Delta c_i \geq \forall\, i$ have to be dealt with. For this purpose, the first step is to determine the maximum factor $f_{max} = \max\,(c_{r,\,i}/c_{corr,\,i})$, $i=1,\ldots, n$, from the ratio of concentrations of all colorants before $(\mathbf{c}_r = (c_{r,1}, \ldots, c_{r,n})^T)$ and after $(\mathbf{c}_{corr} = (c_{corr,1}, \ldots, c_{corr,n})^T)$ the correction. The relative fraction $h_i$ of the tints to be added to a batch is then obtained by subtracting the starting concentration from the product $f_{max} \cdot c_{i,corr}$, i.e., $h_i = f_{max} \cdot c_{i,corr} - c_{r,i}$; for the calculation of the corresponding absolute quantities to be admixed, the total quantity $M$ already in the mixer must be taken into account: $\Delta m_i = h_i\, M = (f_{max} \cdot c_{i,corr} - c_{r,i})M$. The total admixture amount for the correction step is then given by $\Delta\mathbf{M} = \sum_i \Delta m_i = (f_{max} - 1)M$. Table XI.2 shows a simple arithmetic example for correcting a large batch of tints consisting of a 4-component recipe. Where pigment powders are added, the pigment-to-binder ratio must also be accounted for. For a good colour match between production batch and colour standard, considerable amounts of admixture may need to be incorporated in certain circumstances. In that event, the aim must be to find a tolerable admixture amount by accepting a minor residual colour difference.

In batch tinting, both the limited capacity of the container and the goal of creating a manufacturing process that yields a given total batch size require that a second boundary condition be taken into consideration: for reasons of economy, the batch size and the delivery quantity should be identical. This necessity requires the adoption of special strategies for recipe correction that prevent the overshooting of the corrected batch and ensure a convergent and efficient correction process. To minimise risks, only those components in a correction step which will in any event shift the colour position of the batch towards the target colour position should be varied. It is also possible to incorporate a risk factor that limits the step size of a corrective action.

# 4 Practical colour-recipe prediction of gonioapparent colours

*Wilhelm Kettler*

The basic approach to recipe prediction for effect colour shades is the same as that for solid colours[1, 2, 10, 11]. It has two parts, namely pigment calibration, in which the optical material parameters for a colorant assortment are determined from a calibration echelon, and the actual recipe simulation. Both problems are highly non-linear in nature and require sophisticated numerical algorithms for the purpose of non-linear data-fitting with constraints.

### *Material parameters and degrees of freedom*

The first step in determining the optical material parameters or characterisation data of a mixing system for effect coatings is to create a calibration echelon and evaluate the measured reflectance spectra against the adopted radiative transfer model. The minimum number of blends not only depends on coloristic criteria, but also on the parameterisation of the scattering properties of the pigments to be modelled. Besides the scattering and absorption coefficients, $\sigma$ and $\kappa$, the parameters of the model for the scattering function, $p(\theta)$, must be derived from the calibration echelon. In the model function for $p(\theta)$ which was introduced by *Henyey* and *Greenstein* [34] for simulation purposes, only a single additional (asymmetry) parameter, g, needs to be determined, bringing the total number of material parameters to be determined to three. The minimum number of calibration panels here is also two if their reflectance properties are measured at more than one measuring geometry, since all material parameters are angle independent. The angle travel of the reflectance function affords further degrees of freedom for determining the model parameters. The number of available degrees of freedom per wavelength, $f$, results from the product of the number of calibration panels $N$ and the number of measurement geometries $M$, i.e., $f = N \cdot M$. For $N = 2$ and $M = 5$, there are $f = 10$ degrees of freedom, so that models with at most 10 parameters can theoretically be solved. For more complex models of the scattering function, the number of parameters to be determined can be much higher than in the case of the simple structured scattering function of *Henyey* and *Greenstein*. However, from a statistical point of view, an overdetermined set of equations should always be worked with in practice.

From a coloristic point of view, at least five to eight blends should be prepared in order that a balanced and powerful set of characterisation data may be obtained. As in the case of isotropic scattering solid-colour pigments, the mixing partner for an effect pigment in a calibration echelon should preferably have a neutral colour, i.e., exhibit low dispersion* of spectral and optical properties, and also be transparent in order that the angle travel of the effect pigment to be characterised may not be contaminated by additional structures. Carbon black pigments are suitable here, because they meet these criteria to a satisfactory extent. As in the case of calibration of solid-colour pigments, the selected ladder of blends in the space of reflectance should lead to a uniform distribution between the extreme points.

### *Mixtures with white pigments*

Strongly scattering white pigments make poor mixing partners when data for characterising effect pigments need to be determined because, even when they are added in small amounts, they markedly reduce the angle travel of the effect pigments, i.e., they "destroy" the effect to be characterised. An exception to this rule is the micronised variant of $TiO_2$ which, in blends with colour-neutral aluminium pigments, gives rise to angle-dependent colour travel [26] and hence behaves like an independent effect pigment. In the case of effect pigments with object colours, such as iron oxide-coated interference pigments, whose mass tone exhibits comparatively low brightness close to the specular angle, colour-neutral aluminium pigments can also serve as a mixing partner to reinforce the effect to be characterised.

### *Approximation method*

For hiding layers, unit-less relative scattering and absorption coefficients are determined which are expressed in terms of the scattering coefficient of the white base, which is arbitrarily taken to be unity. The optical material parameters are determined by fitting the adopted radiative transfer model to the experimentally determined reflectance surfaces of

---

* *dispersion in this context means wavelength dependence*

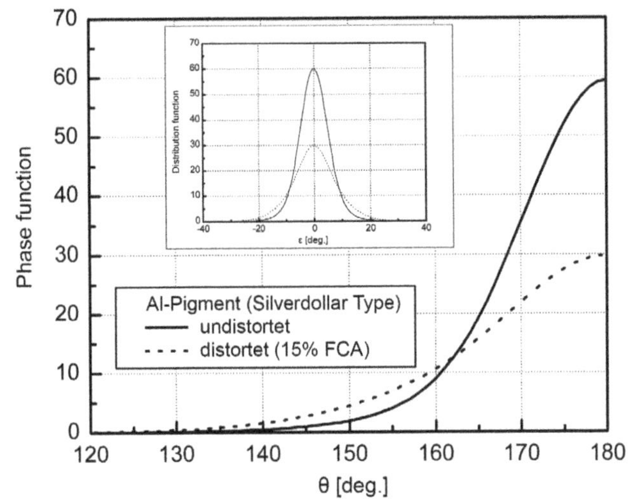


*Figure XI.5: Experimentally determined scattering function p(Θ) for a mass tone of a conventional aluminium pigment (solid line) and blended with a coloristically inert organic flop control agent (FCA; dotted line), which only influences the topology of the platelet-like effect pigments within the coating. The inset shows the distribution functions derived from the scattering function for the polar angle.*

the calibration echelon, whereby, in addition to the scattering and absorption coefficients, the model parameters of the scattering function need to be derived. Since enough reflectance data are available to over-determine the set of equations for the radiative transfer model, an approximation approach in the sense of the Euclidean or $L_2$ norm appears obvious. This non-linear optimisation problem entails minimising a quadratic function while making allowance for constraints and non-negativity bounds for the parameters. To determine the optical material parameters, we first minimise the quadratic function[10]

*Equation XI.23* $$\chi^2(\lambda) = \sum_{j=1}^{N_\varepsilon} w_j \{ L_{exp}(0,-\mu_j,\lambda) - L_{theor}(0,-\mu_j,\lambda) \}^2 = \min$$

while making allowance for the constraints of the scattering-function parameters and the non-negativity of the scattering and absorption coefficients. $L_{exp}(0,-\mu_j, \lambda)$ and $L_{theor}(0,-\mu_j, \lambda)$ denote the measured and theoretical radiances at wavelength $\lambda$, $w_j$ is a weighting function, and $\mu = \cos\theta$ is the cosine of the polar angle $\theta$. The summation index $j$ runs over all experimentally available viewing angles $N_\varepsilon$. As is the case for solid-colour pigments, effect pigments in the calibration echelon are mixed with a blending partner (generally carbon black), which does not contribute any structures to the spectra and hence does not introduce distortions into the derivation of the optical properties.

### *Scattering by aluminium pigments*

Figure XI.5 shows experimental results, derived from calibration echelons, for the scattering function of a conventional aluminium pigment exhibiting optimum (solid line) and strongly distorted orientation relative to the paint surface[12]. As can be seen from Figure XI.5, the contributions from both scattering functions are significant only in the backscattering region ($\theta$ = 180°) and, moreover, diminish rapidly with decrease in angle. The scattering function for the aluminium pigment markedly distorted in topology compared to the undistorted reference state is considerably broadened and less intense at $\theta$ = 180°. Apart from that, no wavelength-dependent differences can be observed. Within the scope of the azimuth-independent formulation of the radiative transfer problem, the scattering function is symmetric with regard to the azimuth angle and only contains information about the distribution of

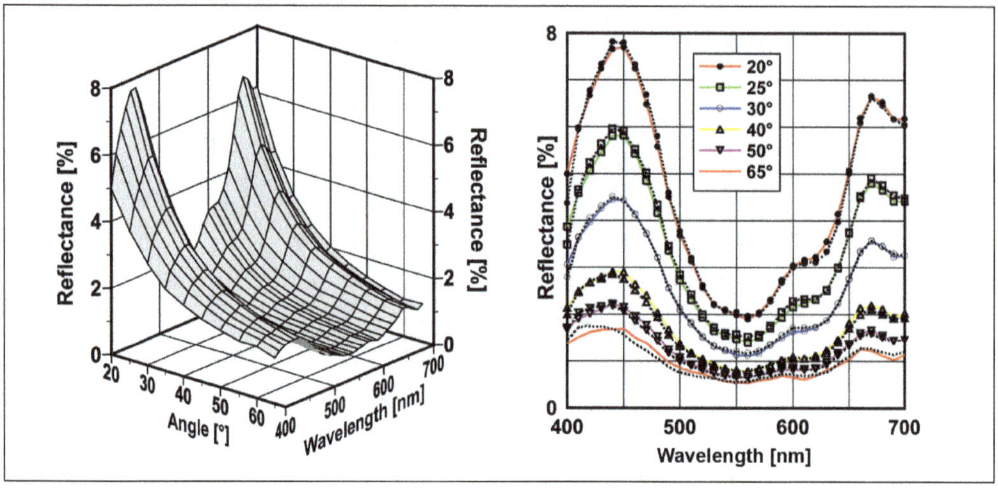


*Figure XI.6: Results of a colour recipe prediction for Byzantium metallic, an effect colour shade. The left diagram shows the reflectance factor of the colour to be matched as a function of wavelength and viewing angle. The right diagram shows the spectrum predicted by the simulation (dotted line with symbols) and the measured colour standard (solid line with symbols). The pigmentation employed consists of an aluminium pigment, a red and a yellow interference pigment, two solid-colour pigments (violet, magenta), and a transparent black pigment.*

the polar angle; in any event, the scattering function is a sensitive probe for the topology of the pigments in a surface coating.

Knowledge of the optical material parameters is essential for the actual simulation calculation for determining the concentrations of all pigment species of a given recipe. In the quadratic function to be minimised [10],

*Equation XI.24* $$\chi^2 = \sum_{i=1}^{N_\lambda} \sum_{j=1}^{N_\varepsilon} w_{ij} \{ L_{exp}(0,-\mu_j,\lambda_i) - L_{theor}(0,-\mu_j,\lambda_i) \}^2 = \min$$

summation is now carried out over all angles (index *j*) and all wavelengths (index *i*).

### *Massive processing power makes it possible*

Massive computing power is needed for solving the numerical problem, since the function values (intensities) as well as their derivatives have to be computed numerically. The physical model for recipe calculation of gonioapparent colours can be employed for all pigment classes, because it has been formulated for universal use. The simulation determines if an effect colour shade can be matched with a given pigment palette, along with the type and quantity of components needed. This can be exercised interactively or automatically. Modern recipe prediction systems also provide recipe correction functionality with which the colour difference between a mixed and sprayed-out recipe and the colour standard can be reduced significantly in a subsequent step. The quality of the results rivals that of a solid-colour recipe prediction based on the *SKM* model.

Figure XI.6 shows an example of a computer-aided recipe prediction for Byzantium metallic, an effect colour shade. To simulate the reflectance indicatrix of the colour standard, the following pigmentation was used: an aluminium pigment, two interference pigments (red, gold), two solid-colour pigments (violet, magenta), and a transparent black pigment. The results of the recipe prediction show that all spectral characteristics of the given reflectance indicatrix of the effect colour shade can be readily matched with the pigmentation selected from a given pigment palette.

## 4.1 Topology of effect pigments in surface coatings

The coloristic properties of platelet-like effect pigments not only depend on the type of interaction with electromagnetic radiation and on their particle size, but to a great extent also on their topology within a surface coating [12].

Platelet-like aluminium pigments behave like microscopically small mirrors, which reflect incident light directionally. If all aluminium pigments could be aligned parallel to the paint surface, the reflection indicatrix would only have a value greater than zero at the angle of regular reflection, while no measurable intensity would occur at any other angle. This ideal case can never be met in practice. The pigment particles are only approximately aligned parallel to the paint surface and, depending on the application conditions, exhibit a distribution of finite deorientation angles. Proceeding from an assumed maximum-possible parallel orientation, in which, at the smallest viewing angle (in terms of the gloss component), the highest reflectance signal and also the highest dynamic range for the angle variation occurs, it is observed that an increase in deorientation of the pigment particles is accompanied by a significant reduction in reflectance at the smallest viewing angle and that there is an associated diminution in lightness variation (Figure XI.8). The scattering function for aluminium pigments used in the radiative transfer equation must account for this. The scattering function of an ideally specular pigment particle can formally be described by a δ-function (mathematically more correct: distribution) which, for a given angle of incident light, assumes a finite value only for the direction of the light regularly reflected from the surface and stays zero everywhere else. For an ensemble of pigment particles, the range of angles over which the light is reflected widens with increase in deorientation, and this is accompanied by a reduction in intensity at the specular (gloss) angle. Figure XI.5 illustrates this fact for two experimentally determined scattering functions of a well-oriented and a highly deoriented ensemble of aluminium pigments. The same phenomena can be observed when a coarse aluminium pigment is replaced by a finer variant, which exhibits more pronounced scattering at the particle edges due to the poor surface/circumference ratio. The two curves exhibit the highest intensity at a reflectance angle of $\theta_r = 180°$, and rapidly converge towards negligibly small values with increase in deviation from $\theta_r$. Almost all scattered contributions are concentrated within the angle range $\Delta\theta < 30°$. Both the angle variation and the width of the scattering function in the backscattering region correspond well with the experimental results obtained by Billmeyer *et al.* [13] for aluminium pigments.

The topology of the platelet-like effect pigment particles is a sensitive function of the binder system concerned. To achieve a high-quality match in colour development, the particle size distribution of the effect pigments must generally be adapted to that of the colour standard to be matched. If the orientation of these effect pigments in the binder system is worse than in the colour standard, there is basically no possibility of an exact effect

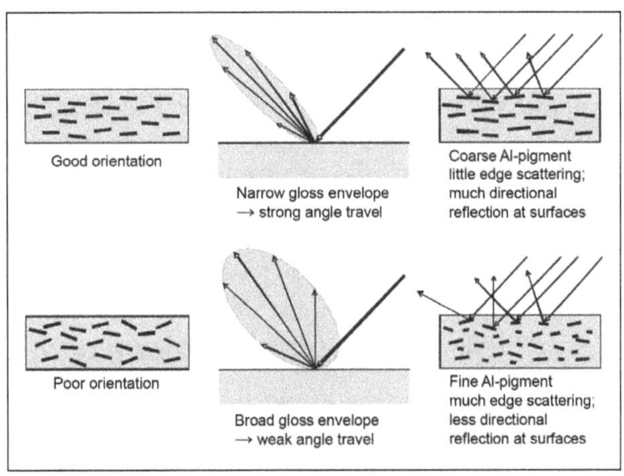


*Figure XI.7: Qualitative characteristics of the reflectance indicatrix of a platelet-like aluminium pigment with optimum (upper left figure) and distorted (lower left figure) orientation of the particles and for coarse (upper right figure) and fine (lower right figure) particle sizes.*

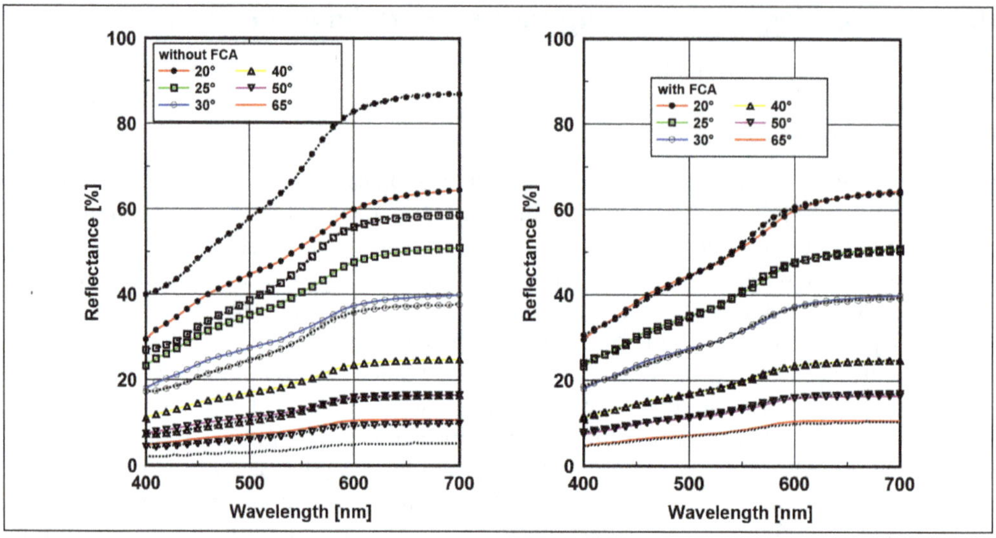


*Figure XI.8: Results of a colour recipe prediction for Toyota Beige, an effect colour shade. The left diagram shows the reflectance factor of the colour to be matched as a function of wavelength and viewing angle (solid lines with symbols) as well as the predicted curves without allowance for a deorientation agent (dotted curves with symbols). The right diagram shows the simulation with allowance for a deorientation agent (flop control agent, FCA) in comparison to the colour standard. The pigmentation employed consists of an aluminium pigment, a red and a black solid-colour pigment, and a deorientation agent.*

match, as the orientation cannot be improved. In the reverse situation, namely better orientation of the effect pigments in the selected binder system than in the colour standard, the colorist can distort their alignment and destroy the effect by systematically adding a coloristically neutral "spacer" material of appropriate particle size. Figure XI.8 illustrates this situation with the example of a yellow effect colour shade, which can be produced with a comparatively simple formulation comprising a silver-dollar-type aluminium pigment and a yellow and a black absorption pigment. As can be seen from Figure XI.8, the orientation of the aluminium pigment in the chosen binder system is better than in the colour standard. Thus, the colour match exhibits a markedly higher reflectance level close to the specular, and is therefore brighter than the colour standard, but a much lower reflectance level (i.e., is darker) far from the specular. The resulting greater angle travel can only be destroyed by adding a flop control agent (FCA) and can be adjusted such that the spectra of the colour match and the colour standards are more or less congruent (see Figure XI.8, right). The spacer particles lower the reflectance level of the pigmentation in the near-specular region and increase the reflectance level far from the specular. This effect can be simulated by mixing the scattering functions of ideally oriented and strongly deoriented aluminium pigments, which will also provide the quantity of flop control agent.

## 4.2 The limitations of colour recipe prediction

The assumption that the particle concentration is sufficiently low to allow the interaction between adjacent particles to be ignored cannot always hold in real paint systems. Above a certain particle density, the state of a light wave incident on a particle deviates substantially from that of a plane wave; consequently, the frequency dependence of the scattered radiation is programmed to deviate from Mie theory computations (see Section VII.3). If the inter-particle distance falls below a minimum level, the intensity scattered by an ensemble of particles can no longer be modelled as the sum of the radiation scattered by the individual particles. Finite phase differences between the radiation scattered by different particles give rise to interference effects (coherent scattering) which cause significant deviations from the incoherent scattering contributions. Rigorous theoretical treatment of the scattering problem of two dielectric spherical particles [14] has shown that significant deviations from Mie theory occur if the inter-particle distance is less than two to three particle diameters.

***Accounting for dependent scattering events***

If **dependent scattering** is expected, the additive law of mixing for the optical material parameters, which only applies to sufficiently dilute systems, loses its validity and precision. Above a critical pigment-to-volume concentration (P/B; pigment to binder), laws of mixing that consider only individual pigment particles are no longer useful. The concentration at which dependent scattering starts to set in varies with the specific properties of the system. One well-studied system is that of $TiO_2$, the primary scattering pigment in the majority of surface coatings. Above a P/B ratio of 20%, which is easily reached in highly packed (high-solids) paint systems, dependent scattering can exert a marked influence on coating opacity.

***High-solids systems and polarisation***

*Goedecke* [15] has shown that the continuum hypothesis is generally invalid. He has derived a new differential equation for radiative transfer that accounts for the angle travel of the radiation field and the particle-scattering properties, without the need to invoke the continuum assumption. Scattering by particles in high-solids systems is treated by means of suitably defined differential cross-sections for the near-field range. Preliminary analyses of the new model have revealed that marked deviations from the prognoses of the classical approach must be expected in high-solids systems or highly absorbing particulate media; however, a quantitative comparative test of both transport models has yet to be performed.

Likewise, the impact of light polarisation on the colour-measurement and recipe-prediction results has not yet been conclusively investigated. In view of the aforementioned theoretical limitations, the precision, which can be achieved with the standard radiative transfer models used in computer-aided recipe prediction is impressive. The user benefits from the fact that the error sources mentioned, which will have an impact on both the determination of optical material parameters and the prediction of colour recipes, compensate each other to a first approximation, extending the theoretical model limits enough to support efficient industrial application.

# 5 The profitability of colour recipe calculation

*Walter Franz*

## 5.1 Review

The basic principles of colour recipe calculation emerged primarily in the applied laboratories of the pigment manufacturers. There, over 40 years ago, algorithms were developed that made it possible for the first time to calculate recipes in advance. Besides natural scientific curiosity, there were strong economic interests for this development. Increasing numbers of colour matches had to be produced for customers in ever shorter time periods. This created a huge manual workload for experienced colour specialists and so it made sense to let a computer system take over. As PCs were not yet available back then, expensive mini-computers were used. The first commercially available systems, complete with high-performance spectrophotometer, cost more than DM 200,000 (EUR 100,000). They were simply beyond the financial reach of medium-sized paint factories in those days.

The situation has changed dramatically in the meantime. Even small paint manufacturers now need to have colour formulation systems if they want to stay competitive. At the same time, the requirements imposed by the markets on such systems have grown enormously, and just calculating the right combination of colorants is no longer enough.

In the following section, we look at the general savings potential afforded by colour recipe calculation and at the cost reductions made feasible by the latest computing algorithms.

It must be borne in mind, however, that it is difficult to produce clear figures that prove profitability. Often, profitability is a combination of raw material costs and time, which must be evaluated separately in each case.

## 5.2 General savings potential afforded by colour recipe calculation

### 5.2.1 Swift feasibility analysis

A resolute NO is often a good answer. Where the sample, e.g. a colour from a graphic art fandeck, is unsuitable, colour matches are frequently not possible or, if so, only at great detriment to quality (metamerism, brilliance). This can be demonstrated quickly and accurately with a formulation system. At the same time, the customer can be shown precisely what is feasible.

Expensive, time-consuming trial and error is thus eliminated and the personnel that would be needed for this task can be put to more efficient use.

### 5.2.2 Low-cost metamerism-free recipes

For many years, these were the laboratories' only objective – and little has changed in this respect. With the aid of spectral matching, the metameric properties of a recipe can be predicted with great precision for all known illuminants. Not even the best colorist can do this.

Once suitable colorants have been preselected, the colour formulation system shows all the recipes which can be produced with the available pigments and which will yield the desired effect within set tolerances. This ensures that the lowest-cost combination is obtained.

Many paint manufacturers have indicated in surveys that this approach produces pigment savings of up to 10% compared to manual mixing procedures. The correction rate is halved, with the result that 1.5 to 1.8 adjustments (corrections) can be expected across all applications.

A further savings effect is a general reduction in the number of colorants. It is also found that "traditional" pigments will yield suitable recipes, which further reduces the use of expensive "specialty pigments".

## 5.2.3 Accurate corrections in production

Initially, colour recipe calculation was the sole preserve of the colour laboratory. With the advent of special correction programs, it shifted into production, where it offered the following crucial advantages:

- Accurate calculation of the minimum tinting quantity, after allowance for metamerism
- Virtual total elimination of excess quantities
- Drastic reduction in production lead time and more reliable planning.
- All these combine to yield huge savings, particularly on production resources.

### 5.2.4 *Fewer complaints*

In the worst case, complaints about colour give rise to returns. The cause is frequently poor coordination of different illuminants (metamerism), and cannot always be avoided in manual procedures, even when there is a vast pool of experience to draw on.

In contrast, laboratories frequently do not have time to redevelop a basic recipe that has proved unsuitable. In such cases, compromises are frequently necessary if delivery deadlines are to be met. A colour formulation system could possibly avoid this or at least indicate whether the basic recipe is suitable or not (see also Section XI.5.2.1).

### 5.2.5 *Computer algorithms for saving costs in specific applications*

#### *Industrial coatings – pigment load calculation*

The quality of recipe prediction systems was often assessed in the past by what was known as the "first hit rate"; in other words, recipes that were well calculated in advance did not require any correction – not in the laboratory, at any rate. Naturally, low metamerism and the pigment costs of the colorant combination were taken into consideration. Calculation of the correct pigment quantity in a given binder proved inadequate, especially for brilliant saturated colours, owing to the limitations of the Kubelka-Munk theory here (see Section VII.7.1).

More recent formulation programs use mathematical models (multi-flux theory; see Section VII.7) which are highly precise at calculating the absorption and scattering properties of pigments, and especially those of transparent and translucent binders. This is a basic condition for accurate pigment load calculations.

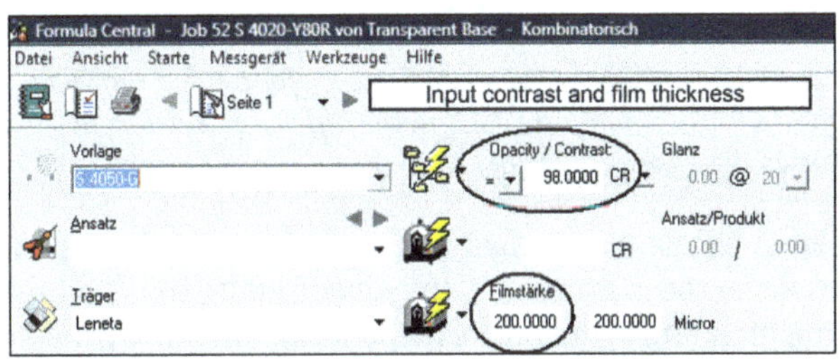


*Figure XI.9: Input box for parameters needed for pigment load calculation, such as contrast ratio and film thickness.*

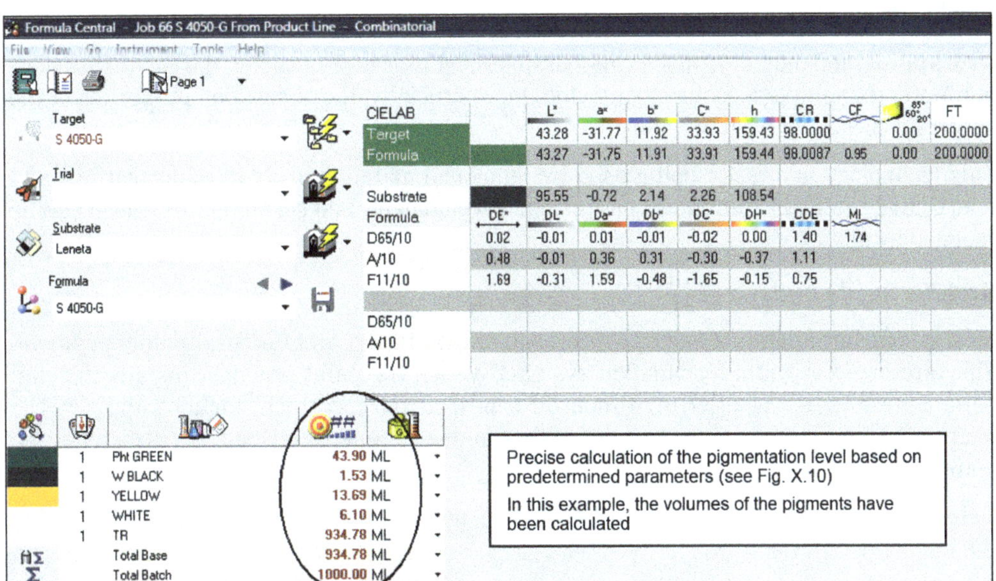


| CIELAB | L* | a* | b* | C* | h | CR | CF | 85° 60° 20° | FT |
|---|---|---|---|---|---|---|---|---|---|
| Target | 43.28 | -31.77 | 11.92 | 33.93 | 159.43 | 98.0000 | | 0.00 | 200.0000 |
| Formula | 43.27 | -31.75 | 11.91 | 33.91 | 159.44 | 98.0087 | 0.95 | 0.00 | 200.0000 |
| Substrate | 95.55 | -0.72 | 2.14 | 2.26 | 108.54 | | | | |

| Formula | DE* | DL* | Da* | Db* | DC* | DH* | CDE | MI |
|---|---|---|---|---|---|---|---|---|
| D65/10 | 0.02 | -0.01 | 0.01 | -0.01 | -0.02 | 0.00 | 1.40 | 1.74 |
| A/10 | 0.48 | -0.01 | 0.36 | 0.31 | -0.30 | -0.37 | 1.11 | |
| F11/10 | 1.69 | -0.31 | 1.59 | -0.48 | -1.65 | -0.15 | 0.75 | |

*Figure XI.10: Result of a colour recipe calculation: The precise ratio of colorant to binder for a certain film thickness and desired contrast ratio was achieved.*

| Match | | | Best Colorimetric | | | | | |
|---|---|---|---|---|---|---|---|---|
| Recipe Nr. | | | 1 | 2 | 3 | 4 | 5 | 6 |
| CieLab Default[D65], Weig | | 1.0 | 0.000 | 0.000 | 0.000 | 0.000 | 0.011 | 0.017 |
| Metam.. "A" | | 1.0 | 0.585 | 0.570 | 0.584 | 1.476 | 0.521 | 0.543 |
| Metam.. "F11" | | 1.0 | 0.177 | 0.153 | 0.176 | 1.390 | 0.326 | 0.288 |
| Priority | | 0.0 | 4.000 | 4.000 | 5.000 | 13.000 | 14.000 | 14.000 |
| Price | | 0.0 | 1.08 | 1.12 | 1.09 | 1.28 | 1.19 | 1.09 |
| Contrast on dE | | | 21.835 | 22.150 | 21.846 | 7.602 | 18.072 | 20.461 |
| Thickness | | | 1.0000 | 1.0000 | 1.0000 | 1.0000 | 1.0000 | 1.0000 |
| Color of standard | | | | | | | | |
| Color of recipe | | | | | | | | |
| Color of recipe/Black | | | | | | | | |
| SM reference [Size] | | | 529 a[15] | 529 a[14] | 529 a[19] | 510 a[15] | 529 a[19] | 529 a[20] |
| Colorants [g] | Nr | | ☑ | ☐ | ☐ | ☐ | ☐ | ☐ |
| Total | | | 1010.99 | 1016.64 | 1011.17 | 1030.49 | 1023.72 | 1011.26 |
| Base TC | 1 | | 1000.00 | 1000.00 | 1000.00 | 1000.00 | 1000.00 | 1000.00 |
| White_Tio2 | 2 | | | | | 12.29 | 2.19 | 0.80 |
| Black | 3 | | 5.93 | 6.82 | 5.96 | | 5.05 | 4.78 |
| Yellow | 4 | | | 7.53 | 0.24 | | 14.80 | |
| Red | 5 | | 4.71 | 2.29 | 4.64 | 16.63 | 1.69 | 5.26 |
| Blue | 6 | | 0.34 | | 0.33 | 1.58 | | 0.43 |

*Figure XI.11: Table of calculated recipes for a paint on wood; scattering pigments were also included, e.g. $TiO_2$. This manifests itself in the differing translucency of the recipe. The desired effect can be varied to suit requirements.*

When pigment manufacturers who have used recipe calculation for many years conducted trials on these programs, they made additional savings on pigment costs of approx. 5 to 7 %, and on hours spent on corresponding laboratory tests.

### *Staining and glazing –recipe calculation for wood coatings*

For a long time, it was believed that recipes could not be calculated for transparent or translucent wood coatings. The introduction of a special technology (see [35]) and the use of special programs has now made this possible. The procedure utilises the intrinsic colour of the wood (multiple measurements) and, more crucially, the absorption properties of the substrate chosen for each pigment.

This technology is already being used by renowned manufacturers of wood coatings and offers huge savings potential for new colour development. The correction workload can be reduced by up to 70 %.

### *Architectural paints*

Most colour formulation systems are probably used in the architectural paints segment. They are low-cost "mini-formulation systems" which the paint producer installs in retail paint stores. As well as offering standard recipes, search & correct functions and new formulations, their success hinges on an ability to use volumetric dispensers directly at the point of sale.

Here, the system's task is to find the best recipe, not colorimetrically, but rather in terms of the following criteria:

- Access, where possible, to stored and tested standard recipes, with possible subsequent correction to the required colour
- Preference to iron oxide pigments for new formulations, where possible

- Lowest price
- Ease of dispensing, i.e. small dosing quantities must be avoided
- Calculation and output in units of volume to avoid rounding errors, e.g. when grams are converted into shots.

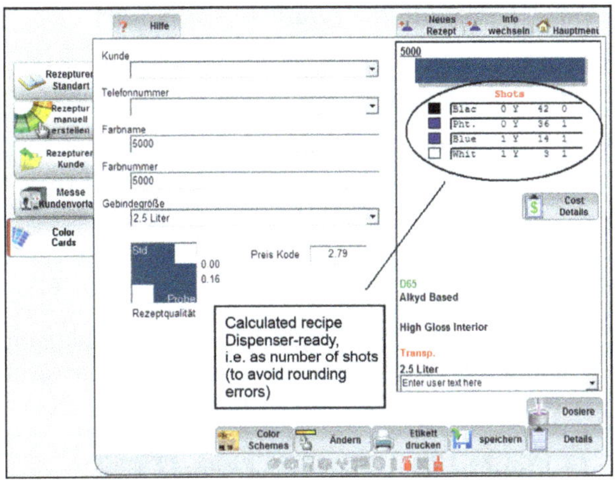


*Figure XI.12: Results of recipe finding, with output in "shots" for direct transfer to the dispenser. Only the best recipe match for the described criteria is displayed.*

### *Use in car repair workshops*

Professional repair workshops are under huge pressure to be efficient, because they operate under fixed cost and time constraints when repairing damaged vehicles. Naturally, this also affects the paint shops, where there is often great uncertainty about producing the right colour. This is made all the harder by the widespread use of effect coatings.

Furthermore, it is necessary to be accepting of pre-existing colour deviations in the OEM finish as a result of different film thicknesses, minor recipe changes and differing production processes (at different production sites). Natural aging of the coating also plays a prominent role.

As a result, the colour number for a vehicle usually stands for a number of recipe variations.

The traditional way to find the right recipe has been for the manufacturer to provide each customer with accurately produced colour standards, which are valid for one colour number. This approach is extremely time-consuming, expensive and uncertain and can be further compounded by changes in the colour standard, e.g. due to aging and surface scratches.

The paint shop worker has to find the correct variant in the physical standards archive and then adjust this visually to the colour of the vehicle being repaired (manual searching and correcting).

For several years, major manufacturers of car repair paints have been offering their customers a new colorimetric solution based on mobile multi-angle spectrophotometers with at least three different viewing angles.

- The colour standards produced by the paint makers are measured colorimetrically and stored in a searchable data archive under various criteria:
  - Colour number
  - At least three R% curves for the corresponding viewing angles
  - Graininess, i.e. properties of the aluminium pigments used (size and distribution)
  - Vehicle manufacturer, type, model year, etc.
- The archive is made available to the repair work shop, along with the spectrophotometer and "search & correct" software.
- The undamaged paintwork of the vehicle being repaired is cleaned and then measured with the mobile multi-angle spectrophotometer.
- This is followed by the colorimetric "search & correct" procedure.

Tens of thousands of businesses already employ this method, benefiting from the greater profitability which ensues from a much shorter, predictable process. Deadlines are met because uncertainties in the paint shop are eradicated.

### 5.2.6 *Profitability analysis – summary*

In the highly unlikely event that there are paint or ink makers which still do not use a colour formulation system, the advantages are as follows:

- Up to 10% reduction in pigment costs
- At least 50% reduction in laboratory correction cycles
- Max. one to two corrections in production
- Practically no complaints about colour

The use of the very latest technology brings about the following efficiency gains:

- Optimum pigment load calculation reduces pigment costs by a further 5 to 7%.
- Accurate series recipe calculation for manufacturers of architectural paints. Recipes for new colour charts with thousands of standards or for new products can be produced in just a few hours.
- Up to 70% reduction in the correction cycles for glazes and stains.

These advantages need to be balanced by the costs of purchasing a new system or converting an existing one:

- Procurement costs for an efficient colour formulation system, including spectrophotometer, software, computer hardware, installation and training: EUR 30,000
- Annual upkeep (maintenance etc.) approx. EUR 2,000
- Depreciation over fife years = EUR 6,000 p.a.
- Total costs = EUR 8,000 p.a.
- Upgrade costs for the latest software, including training: approx. EUR 10,000

The profitability can be easily estimated by the reader on the basis of these data. A complete system will often pay for itself by avoiding just a single complaint.

## 6 *Guidelines for formulating and matching object colours*

*Wilhelm Kettler*

When colours are being developed in industrial practice, there are a number of different requirements which need to be considered. Some these are conflicting and so present an obstacle to the ideal solution. Faced with having to satisfy the conflicting fields of **coloristic criteria**, **processing aspects**, and **economic** and **environmental demands**, the colorist must arrive at a compromise formulation by carefully weighting and balancing the various factors to arrive at a solution acceptable to all parties [16–19].

In colour development, coloristic aspects are particularly important. Here, good hiding power, negligible metamerism, tolerable residual colour difference, weather-resistant pigmentation and paint formulation make high demands on the colorist's store of knowledge. The colorant range employed must not only lend itself to the development of the largest-possible colour gamut (see Section XI.10), but also be environmentally compatible and satisfy financial concerns in terms of cost structure.

One aspect of growing importance is that of process reliability during the manufacture of colours and application by end users. For both of these processing phases, it is economic constraints nowadays that determine the model – not basic feasibility. Designers of (effect) colour shades must contend with process engineering restrictions and focus some of their creativity on striking the right compromise between colour design concept and technical feasibility.

Many of these criteria can be formulated in colour prediction software packages as a set of rules for controlling the end quality. Due to lack of space, we can only present the most important coloristic guidelines here.

## 6.1 Rules for mixing pigments

Colour mixtures can be produced by **additive** and **subtractive** principles[1]. The theoretical basis for additive and subtractive colour mixing is presented in Section I.3. Only those facts relevant to the overall subject will therefore be addressed below.

In colour space, the colour position of a pigment is represented by a single point. If a second pigment is added, mixing of both components can create a curved line in colour space (see Figure XI.13). The colour positions of mixtures of three pigments are located on a curved surface in colour space. It takes four or more pigments mixed together to create a continuous volume in colour space. The optimum mixing system maximises the continuous volume in colour space (colour gamut) and hence the number of feasible colours.

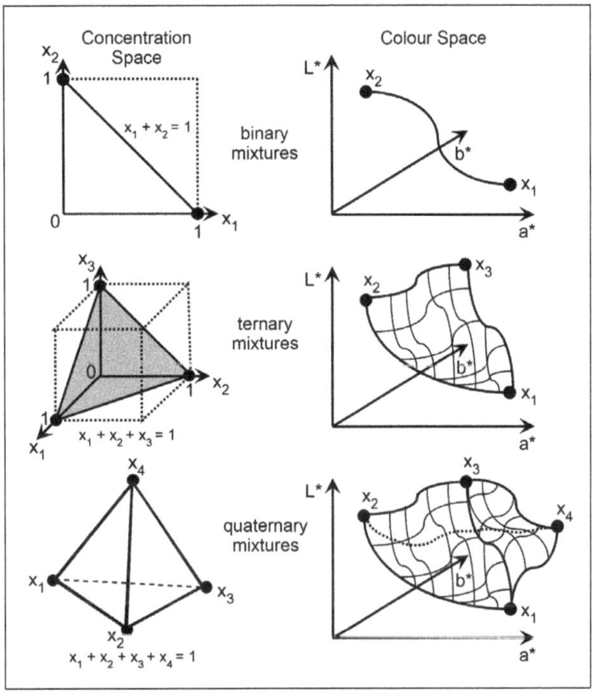


*Figure XI.13: Colour mixing for solid-colour pigments in CIELAB76 colour space (right column of diagrams) and in the corresponding concentration space (left column of diagrams).*

### Subtractive mixing for coloured pigments

In mixtures, coloured pigments tend to behave subtractively in the manner of consecutive colour filters. By means of absorption, each component of a mixture extinguishes its own specific range of the electromagnetic spectrum, i.e., the reflectance curve of a pigment mixture is shaped by selective absorption by each of the constituent pigments. Figure XI.14 illustrates this for a

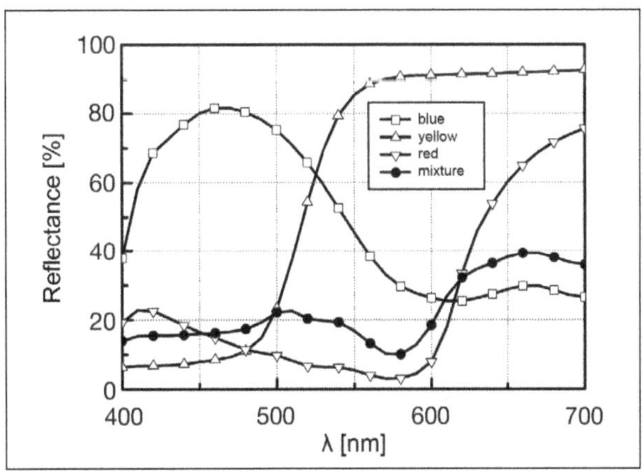


*Figure XI.14: Colour mixing for a blue, a yellow, and a red solid pigment. The mixing process, which can only be considered quasi-subtractive on account of the scattering properties of the pigments, can be described by the Schuster-Kubelka-Munk theory[13, 14].*

[1] *Combinations of both mixing principles as in case of the* **partitive** *or* **autotypic** *colour mixture is also possible[11, 12]. If tiny colour spots are set side by side and can no longer be distinguished visually, they mix additively and coalesce into a single colour. The spots themselves can yet consist of subtractive colour mixtures.*

colour mixed from a blue, a yellow, and a red pigment. Whereas the reflectance minimum of the yellow pigment is in the short-wavelength spectral range and significantly impacts on the mixing curve between 400 and 480 nm, the red pigment shapes the mixing curve in the middle of the spectrum between 500 and 580 nm. The blue pigment is responsible for the variation of the mixing curve in the long-wavelength range of the spectrum above 600 nm.

***Colour mixing for gonioapparent pigments***

For gonioapparent colours, i.e., colours that vary with the viewing angle, the situation is more complex. In three-dimensional colour space, each sequence of measuring geometries for the mass tone of an effect pigment already yields a curve, the extent of which depends on the measuring geometry and the optical properties of the respective pigment. The two end-points of the curve can be assigned to the configurations closest to and furthest from the specular angle. In a combination of two pigments, of which at least one should belong to the group of effect-imparting pigments, mixtures create a curved surface in colour space, whose precise location and structure again depend on the actual measuring geometries and the coloristic properties of the pigments involved.

The basic model for quantitatively describing radiative transfer in particulate media loaded with isotropic scatterers is the Schuster-Kubelka-Munk theory [19, 24] (SKM). For the purpose of describing the angle-dependent optical properties of effect colour shades, the general radiative transfer equation must be evaluated [19, 25]. An analytical equivalent to the SKM model cannot be derived for the radiative transfer equation and so it can only be approximated by numerical methods.

## *6.2 Solid colours*

A **basic rule** when developing solid colours is to keep the pigmentation as simple as possible . Complex mixtures of more than seven to eight pigments are difficult to tint and often more expensive than necessary and less weather resistant. Nor does it make any sense to dull expensive high-chroma pigments by blends with several additional pigments; this can be achieved by choosing cheaper pigments with chroma values close to the target colour.

For reasons of tintability, solid colour shades require a minimum of **four pigments**: two chromatic pigments for the hue and two achromatic pigments (white and black) for adjustment of the lightness and chroma level. The prime pigment of a colour formulation should always be as close as possible to the hue of the desired target colour position. An example of the pigmentation of a medium blue colour would be a mixture of a copper phthalocyanine blue pigment for determining the base colour, titanium dioxide (white) to adjust the lightness, dioxazine violet to shift the shade, and carbon black to adjust the chroma. The function of carbon black can generally be performed by other desaturating pigments instead, e.g., iron oxide yellow, iron oxide red, chromium titanium yellow, or by pigments that have light absorption comparable to that of carbon black pigments, e.g., phthalocyanine blue, phthalocyanine green, indanthrone blue, or dioxazine violet.

***Prime and secondary pigments***

It is standard practice to extend the scope for producing shades by using three chromatic pigments in a formulation, with two secondary pigments coloristically bracketing the prime pigment. However, this may have a slight adverse effect on the brilliance of the colour. For example, a red prime pigment P.R. 254 can be mixed with a combination of an orange pigment P.O. 36 (yellow side) and a bluish quinacridone pigment P.R. 122 (blue side). Compared to binary mixtures of chromatic pigments, this extended 3-component variant of the coloristic structure described offers additional scope for shading.

### *Colour differences due to different illuminants: metamerism*

One phenomenon that cannot always be avoided during actual colour development is that of metamerism (see Section I.9). Of its various manifestations, we are concerned here with illuminant metamerism, which becomes apparent when the light source is changed. To an observer, a metameric pair of object colours will only look the same if illuminated by a specific illuminant. Switching to a different illuminant (e.g., from $D_{65}$ to A or F11) will give rise to a more or less pronounced residual colour difference, which must be eliminated or minimised to an acceptable level. Metrics have been developed for colorimetric characterisation of metamerism which provide clear hints about inadequate pigmentation.

Spectrophotometrically, it is easy to identify a formulation which is metameric to a colour standard, as the spectral reflectance curves for the respective pair of object colours have different spectral structures and several points of intersection at various wavelengths. The problem of a metameric colour match can only be mitigated by changing the selected pigmentation. However, a colorist might not have access to all pigments needed for metamerism-free colour match. In such cases, the only remedy is to attempt to minimise metamerism via the choice of pigments.

This problem frequently occurs in practice when, e.g., print colour charts, such as Pantone standards, have to be matched with an automotive paint system. As these transparent (i.e., non-hiding) colour charts are produced with pigments and/or dyes which, due to their poor coating properties, are not used by the automotive industry, there is a high probability that only a metameric colour match can be achieved in certain colour regions. The problem is even more complex in some colour regions, where only parameric matches are possible, i.e., residual colour differences to the standard can also occur under the primary illuminant if the colour position of the colour standard is located outside the colour gamut of the available pigment assortment. Other colour systems, such as the RAL register, are less critical in regard to pigmentation, as the selection of available pigments for producing the colour standards concerning all fastness properties is oriented towards the criteria adopted in the automotive industry.

### *Subtractive colour mixing*

Compensative mixtures of pigments should be avoided wherever possible. Given a suitable mixing ratio, the chroma of the two mixing partners cancels each other out (compensate for their respective effects), giving rise to a dark achromatic (grey) colour which, in a complex formulation, can compete with the greyness created by white and black pigments, thereby rendering instrumental colour matching and correction unstable. An achromatic colour can always be achieved at lower cost and more efficiently with a mixture of white and black pigments. A pigment combination typical of the situation just described is the subtractive mixing of green and red pigments, where the absorption range of one pigment coincides with the scattering range of the other and gives rise to a poorly structured reflectance spectrum of achromatic character.

### *Tri-coats*

In the case of non-hiding colour shades, the background also contributes to the colouring. In practice such three-layer coatings are more critical than hiding coatings, because thickness fluctuations in the latter will not affect the appearance as long as the hiding criterion is fulfilled. Three-layer coatings require an extremely stable application process. They also take more time to repair, because both the recipes for the coloured background and the base-coat layer must be correct.

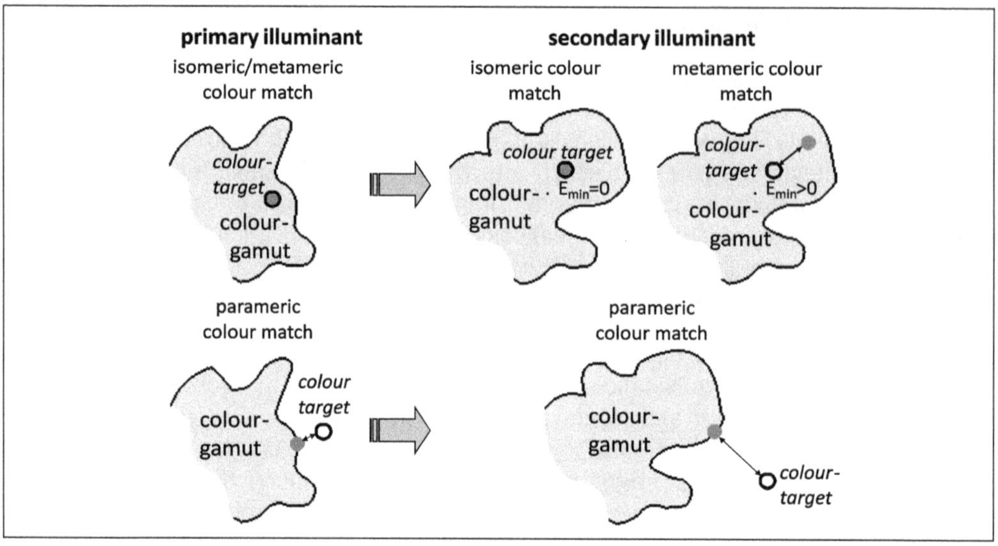


*Figure XI.15: Two-dimensional sketch showing the matching of solid colours under two illuminants. The target colour is denoted by an open circle and the developed colour match by a solid circle.*

Just as critical, because of its greater complexity in practice, is a pigmented base coat that is covered by a coloured clearcoat. Here, too, fluctuations in film thickness quickly impart the coating an inhomogeneous appearance.

### *Colour matching after exposure*

In automotive applications, surface coatings are subject to mechanical and chemical stress and exposed to moisture and radiation. Particularly critical is the high-energy ultraviolet region of the spectrum, which has the greatest potential for damage. UV radiation is energetic enough to cleave many chemical bonds in polymers and pigments. The consequences range from colour shifts through to total pigment fading, matting of the surface of a coating and in extreme cases cracking and delamination of the finish. From a coloristic point of view, it is inadvisable to mix pigments of high weathering fastness with pigments of poor weathering fastness and markedly different hue. Differential bleaching of the pigments inevitably gives rise to unacceptable shifts in hue. Complementary mixtures are particularly sensitive, e.g., such as those involving yellow and blue pigments. The overall problem of weathering fastness can be mitigated by "blending" weak pigments with strong pigments of similar hue.

### *Isomeric – metameric – parameric colour matches*

Developed solid colours can be divided into three quality categories (see Figure XI.15). The first category, namely an **isomeric** (identical) colour match, is considered the best solution, because the reflectance spectra of the colour standard and the colour match and thus their colour positions are congruent. Furthermore, this state is independent of the illuminant employed. The second category is a **metameric** (conditional) colour match, in which the reflectance spectra of the colour standard and the colour match are not congruent, but the residual colour difference of the sample pair can still be adjusted to zero under the primary illuminant. However, a change of illuminant alters this state and gives rise to a more or less excessive residual colour difference. The third, and least desirable, category is a **parameric** (deviating) colour match which, even under the primary illuminant, exhibits a finite residual colour difference, generally accompanied by illuminant metamerism. The colour position of the standard is located outside the colour gamut of the mixing system employed. Changing the illuminant will yield a different residual colour difference.

## 6.3 *Gonioapparent colours*

Generally, effect colour shades are more complex than solid colours and as a rule contain **five to six pigments**, with up to eight in the case of colour matches. It is much more difficult to match effect colour shades than solid colours, as typical mixing systems comprise 30 to 40 solid-colour pigments and a comparable number of effect pigments, all of which can be combined in any ratio. Moreover, effect pigments are also available in different particle sizes. Furthermore, additives need to be incorporated to control the orientation of platelet-like effect pigments within the coating. Consequently, the number of permutations of five, six and even more pigments of a mixing system is astronomical. The use of optical microscopy to identify all effect-imparting components (see Section XII) can and must reduce the number of pigment permutations to a manageable level.

The basic model of an effect colour shade comprises just four components. Besides the two chromatic pigments responsible for the hue and a black pigment for controlling chroma and lightness, an aluminium pigment is added to adjust the lightness level and the angle travel, and serves as a replacement for white. Unlike the case for solid colours, it is normal practice to use only transparent coloured pigments that will not overly diminish the **light-dark variation**. In more complex pigmentations, three chromatic pigments, one of which determines the hue while the other two serve as tinting components bracketing the prime pigment for better tintability in colour space, are mixed with carbon black and two effect pigments. Aluminium pigments essentially give rise to a light-dark variation, whereas the use of interference pigments as a recipe ingredient can additionally lead to **iridescence**. It may be necessary to adjust the angle variation in a formulation by incorporating additives for controlling the orientation of the effect-imparting platelet-like pigments.

To influence the hue of an effect colour shade at an oblique viewing angle, two different aspects need to be considered. A **dark appearance** at an oblique viewing angle can only be achieved with transparent solid pigments and coarse types of effect pigment. The coarser the effect pigment, the less the appearance is influenced by scattering at the edges of pigment particles. This optical gain comes at the expense of an inferior topcoat gloss and poorer hiding power. In the case of aluminium pigments, it is advisable to use silver-dollar types that have well-defined particle edges.

The opposite effect, a **bright appearance** at an oblique viewing angle, can be obtained by adding small quantities of strongly scattering hiding pigments (titanium dioxide, iron oxide red, iron oxide yellow, chromium titanium yellow, nickel titanium yellow) to a formula. Organic hiding pigments also belong in this category. With effect-imparting components, it is more common to use fine-particle grades in which edge scattering predominates.

Designers also have various options when it comes to adjusting a specific **colour appearance** of a pigmentation at an oblique viewing angle. To illustrate the interrelations, consider blue colours, whose oblique viewing angle can be adjusted to a greenish or reddish hue. For the greenish hue, highly chlorinated α- and β-copper phthalocyanine pigments and also the copper-free phthalocyanine blue P.B. 16 are used. The reddish hue can be achieved with α-copper phthalocyanine and in particular ε-copper phthalocyanine P.B. 15:6.

A small number of commercial effect colour shades are formulated so as to be **non-hiding**. Here, designers can create special **iridescent effects** by taking advantage of the background colour. One consequence of the necessity to have a translucent base-coat layer of constant thickness is that, as a general rule, this tri-coat is more difficult to process than a simpler coating structure with hiding coloured layers. It also takes much more effort to obtain an optical match for such paint coat structures in the refinish shop. Another variant

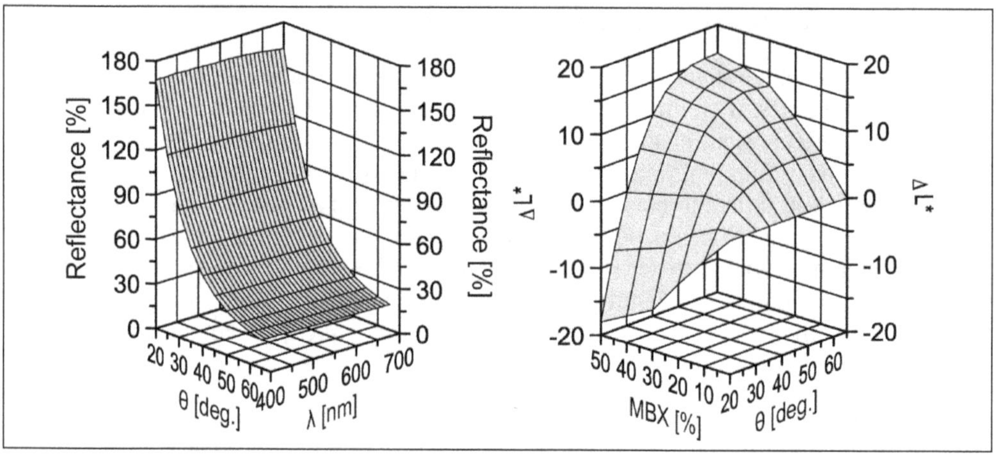


*Figure XI.16: Reflectance indicatrix of an undisturbed aluminium base coat (left diagram), and lightness variation in the system whose topology was disturbed by adding a deorientation agent (MBX)[3] (right diagram).*

is a hiding, two-layer design with a slightly tinted clearcoat. This variant affords a way of achieving highly chromatic, brilliant colours, which a conventional paint structure comprising a single pigmented layer struggles to match.

In mixtures of coarse and fine aluminium pigments, the coarse pigments will lack **brilliance**, and the mixture will appear more greyish than the straight formulation with coarse pigment. Spectrally, this means that the reflectance level is lowered in the specular region and raised in the far specular region. However, the fine aluminium pigment improves the hiding properties.

Where goniochromatic colours of unknown pigmentation are being matched, three distinct situations may arise, of which only two are amenable to coloristic treatment. As the topology of effect pigments within a coating depends not only on the **conditions during application**, but also on the paint system employed, the orientation of the effect-imparting pigments within a coating of the chosen paint system can be better or worse than in the paint system of the standard. Ideally it will be equivalent. The second case, namely worse orientation, is more problematic, as these circumstances cannot be controlled and generally cannot be improved. Even if the pigmentation is correct, the match will be darker than the colour standard in the near-specular region and too bright in the far-specular region. The converse problem can be mitigated coloristically by blending with a **deorientation agent** to disrupt the alignment of the pigment platelets. The third case, namely equivalent orientation, is not critical and requires no additional action.

The foregoing can be illustrated by considering a pure aluminium base-coat system, which has been disturbed in systematic fashion by adding a deorientation agent. The latter acts merely as a spacer between the pigment particles and, because its matched refractive index, is virtually invisible in the pigmented layer and so makes no coloristic contribution. The results of the goniospectrophotometric studies are shown in Figure XI.16. The reflectance surface of the undisturbed aluminium system exhibits hardly any wavelength dispersion. The addition of deorientation agent close to the specular ($\theta < 25°$) angle leads to a decrease in lightness ($L^*$). At higher angles ($\theta > 25°$), this situation is reversed; increases in the quantity of deorientation agent cause a substantial increase in lightness. The deorientation agent "destroys" the optical effect of the aluminium pigment.

The total **lightness variation** (flop) of the undisturbed sample over the measured range of angles can be reduced considerably by the deorientation component. The influence of the "flop killer" is such that the regular reflectance component of the aluminium pigments is distributed over a wider solid angle in the region close to the specular, whereas, far from the specular, scattering and reflection at edges of the deoriented aluminium pigments lead to in an increase in the diffuse reflectance component.

The use of **opaque pigments** in goniochromatic colours also destroys the effect of the anisotropically reflected component. In a bright metallic coating, even small quantities of $TiO_2$ will considerably reduce angle travel. Reflection and thus lightness are substantially lowered in the specular region and increased and brightened in the far-specular region. The situation is different when the **transparent** variant of $\mathbf{TiO_2}$ is used; its particle diameter is roughly one-tenth the size of its opaque counterpart, which has a particle size of about 250 nm. This $\mu$-$TiO_2$ hardly destroys any of the effect of the angle travel caused by the aluminium pigments, but gives rise to a special type of colour travel, ranging from yellowish close to the specular to bluish far from the specular (**frost effect** [26]). To maximise this effect, only transparent coloured pigments should be used in gonioapparent surface coatings.

For certain **translucent effect pigments**, which per se do not possess hiding power, the colour of the background can be incorporated into the colouring. This category includes **liquid crystal pigments** [27], which selectively reflect and transmit the incident light. As both reflected and transmitted components are complementary and would combine to form the achromatic spectrum of the illuminant, marked colour travel can only be achieved by using a black background to absorb the transmitted part. This affords a way of realising colour changes between highly chromatic green and blue reflectance spectra.

Whereas solid-colour standards can generally be matched isomerically with a typical pigment assortment of a **refinish mixing system**, such a situation virtually never arises when it comes to the matching of effect-colour standards. As a result of the macroscopic anisotropy of the reflected light, a complex curve in 3N-dimensional space has to be matched which is approximated by 3N discrete degrees of freedom (N = number of measuring geometries). This problem essentially only permits parameric solutions, i.e., matches with tolerable or merely acceptable residual colour difference and metamerism.

## 6.4 *Hiding power and pigmentation level*

A crucial property of pigmented media in practical work is their hiding power [28]. By this is meant the ability of a coating to obliterate the colour of the background. The hiding power (or opacity) of pigments is achieved by selective absorption and scattering of light. Both of these elementary optical properties are determined by the ratio of the complex refractive index of the pigments to that of the embedding medium and its particle size and particle shape. The scattering power increases with increase in refractive index ratio of pigment to embedding medium and assumes particularly high values at particle diameters comparable to the wavelength of the incident light (~400 to 700 nm).

**Inorganic coloured pigments** are characterised by very high refractive indices, coupled with large particle diameters, and generally possess a higher opacity than coloristically comparable organic pigments. Due to their broader absorption range, red pigments have higher opacity than yellow pigments. The hiding power of greenish pigments is usually inferior to that of corresponding reddish variants (e.g. copper phthalocyanine pigments). Two end-points of the scale are formed by aselective scattering (and barely absorbing) **white pigments** and a selective absorbing (and barely scattering) **black pigments**. As a result, the opacity of white pigments is low whereas that of fine black pigments is particularly high.

The coloristic properties of black pigments are also impacted by their particle size (e.g., lamp black ~100 nm, gas black ~10 nm). A small particle diameter gives rise to a bluish hue in the mass tone region and to a brownish hue in blends with white; the opposite applies to large particle diameters. This observation also applies to metallic coatings, in which the fine carbon black in the far-specular region also gives rise to a brownish hue combined with a high degree of transparency, while the coarser black exhibits a bluish hue. At the same time, the fine variant possesses a higher tinting strength than the coarse material and is darker due to less scattering.

Closely related to the hiding power is the pigmentation level of a formulation, which also determines its price. It is imperative, while making allowance for fluctuations of a given film build, to adjust the pigment-to-volume ratio of a formulation in a way that the hiding criterion is always fulfilled. A further increase in the pigmentation level increases the cost of a formulation without providing any extra benefit and moreover can have a negative impact on the technological properties of paint. The **hiding limit** should not be undershot, as fluctuations in the film thickness will then be accompanied by fluctuations in colour and the background will be exposed to radiation which not all chemical bonds of the material are capable of resisting. The inevitable consequence is delamination. Moreover, colour fluctuations related to film thickness also cause problems in colour matching or repair.

The price of a pigment is not indicative of its contribution to the total cost of a formulation. Evan a high-price pigment can lower the cost of pigmentation of a paint by improving the hiding power and/or chroma (allows for a higher proportion of a cheaper inorganic pigment or carbon black in a formulation). As a rule of thumb: to achieve a dry hiding film thickness of about 50 µm in the plain-yellow region, generally at least 60 % inorganic and 40 % organic pigments are required. The more brilliant and pure the desired colour, the less the amount of inorganic pigments that can generally be used and the lower the hiding power of the formulation.

# 7 *Recipe dosability*

*Walter Franz*

It is almost inconceivable to think of a paint manufacturer without a dosing system. Gravimetric and volumetric systems have been in use for many years, whether in the laboratory, in production or at the point of sale in paint retail stores. Many are directly linked with a colour formulation system (see paint retail store); in other words, once a calculated recipe has been given approval, it is transferred directly to the recipe management module in the dosing system. This necessitates compliance with the following points:

- As a basic rule, the minimum quantity of any individual component in a recipe should be at least 10-times (preferably 20-times) the accuracy of the dosing system.
- This can be stipulated as a parameter for recipe calculation, i.e. the calculation takes account of the quantities and ensures that they never fall short. Where necessary, the formulation system reverts to another recipe component.
- Gravimetric dosing systems **must use gravimetric** calculation; volumetric dosing systems **must use volumetric** calculation.
- Straight conversion from gravimetric to volumetric variables is only permitted where large quantities are involved. Where small dosing quantities are concerned, rounding errors, e.g. when grams are converted to shots, can lead to major errors. Modern formulation systems can make allowance for this in advance, too.

The economic benefits are obvious: fewer faulty batches, accompanied by savings on time and costs. Manual weighing systems depend on how carefully the operator works and should only be used for small batches in the laboratory.

# 8 Structure of colour mixing systems

*Wilhelm Kettler*

## 8.1 Introduction

Colour recipe prediction can only be used efficiently if a properly standardised and characterised mixing system is available. In this context the term "colour mixing system" means an assortment of standardised finished coatings and pastes of high colour consistency which is capable of producing a multiplicity of different colour shades by combining several of the colorants in different proportions.

The number of colorants employed in colour mixing systems can vary substantially. As a rule, however, the greater the number of carefully chosen colorants, the more colours generally can be matched. Coloristic aspects aside, an assortment of ten colorants can generate up to 252 distinguishable permutations of five of these colorants. Increasing the number to 40 colorants increases the number of permutations for any five combinations to 658,005. However, not all of these combinations can be distinguished visually. Under the conservative assumption that each of the 40 colorants can be adjusted to some 30 different lightness and 30 different chroma levels and two different hues, roughly 72,000 visually distinguishable colours can be created.

**Mixing paint** is supplied, for example, by paint manufacturers to automotive repair shops, thereby providing the user with a high degree of mixing convenience. This includes the provision of mixing formulae for producing a variety of different colours in different quantities, agitators to homogenise the base colorants, micro-fiche systems or computer programs to store and read recipe information, as well as dispensing and weighing devices. Moreover, there are established systems on the market which are equipped with colour-measuring technology and which automatically search for and correct mixing repair recipes efficiently. Unlike finished coatings for colour mixing systems, pigment pastes act as intermediates for use within the paint production process.

## 8.2 Alternative methods of paint production

### 8.2.1 Production of OEM paint material

The classic method of paint production consists in dispersing all the pigments of a formulation in a portion of the binder and, after the formulation has been completed, to adjust the final colour position by tinting (see Figure XI.17). This procedure still makes sense today where, for example, a certain colour has to be produced in large quantities or the delivery programme comprises only a very limited number of colours. However, as a result of the large variety of colours required today, another production method involving standardised **intermediate products** has become increasingly popular. The intermediate products in this case are binder-containing pigment pastes, each containing just a single pigment. As the pigment has already been dispersed in these pastes, the desired colour can be adjusted by simply mixing the pastes in a given ratio.

### 8.2.2 Advantages of colour mixing systems

The introduction of a mixing system into production makes considerable preparatory demands on the part of the coloristic laboratory as regards pigment selection, paste formulation, and the standardisation process (see Figure XI.17). Nevertheless, given the attainable advantages, this outlay is justified in most cases.

In paint production, the concept of **pigment pastes** benefits the technically complex and cost-intensive dispersion step by allowing millbase formulation and dispersion conditions

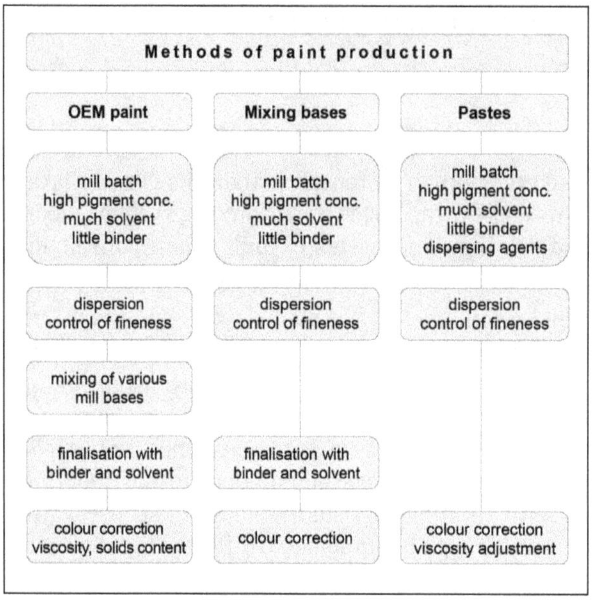


*Figure XI.17: Production of automotive OEM paint (left column), mixing paint (centre column) and pastes (right column).*

to be optimised to the pigment concerned. In addition, production of the pigment pastes can be based on the planned consumption in the medium term and independently of the day-to-day business. This makes the most efficient use of existing dispersing capacity, and keeps cleaning times and related product losses to a minimum. As the time required for dispersion is already eliminated when pigment pastes are used, production output can to a large extent be adapted to suit daily incoming orders. This adds up to improved delivery capabilities (for small runs, too) and shorter delivery times.

As the coloristic properties are **standardised**, the desired target colour is generally created directly by mixing all the pastes in the given ratio for the formulation. Where necessary, colorimetry affords a way of efficiently making colour corrections and virtually avoiding excess quantities. Computer-aided recipe prediction also facilitates the production of colours beyond the standard programme quickly and cost effectively. The standardisation of the rheological properties also supports the use of modern automated dispensing equipment within a certain range of tolerances.

The **performance** of a colour mixing system depends on further, closely interwoven criteria, such as the coloristic properties, logistics, and process and environmental safety. The colorants must cover the broadest possible colour gamut in colour space and provide a balanced hiding-power profile in conjunction with the required weathering fastness. The optimised coloristic behaviour, however, must support the processing properties required for paint production. Rapid, flexible availability of the full production assortment in the required volume, paired with minimal internal and external complexity, contributes to cost-efficient production. Both the colorants and the paint technology itself must ensure that the end product is environmentally compatible.

### *8.2.3 Paint mixing systems*

The mixing system concept can be realised in a simple fashion by using the main binder in the respective product line as the binder for the pastes (see Figure XI.17). The chosen pigment concentration must reflect the different binder demands of the individual pigments. While the highest-possible pigment concentration may be advantageous, there must be enough binder to safeguard the flow properties and storage stability of the pastes.

Pastes formulated in this way can be mixed in any ratio according to the recipe specification. The mixture is then completed to the final paint by adding further binders and binder components. The product then has the same composition and hence the same coating properties as an equivalent paint prepared by the traditional production process. Paint mixing systems can therefore also be used with advantage to meet the strictest or most specific requirements imposed on paint quality.

### 8.2.4 Universal mixing systems

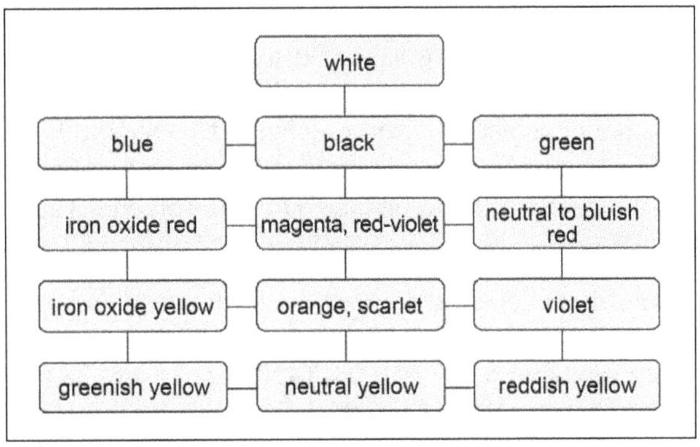


*Figure XI.18: Colour regions which a pigment assortment must cover in a mixing system for solid colours.*

A universal mixing system further streamlines the concept of a paint mixing system. The idea here is to use a single set of pastes to pigment all the paints in a production programme. The key to a universal tinting system is therefore the particular binder employed for the pigment pastes. This **grinding resin** must be compatible with all the binder systems in a production programme and should not exert an influence on their coating properties.

This maximum requirement needs compromises to be made. Therefore, it is accepted from the outset that the pigment pastes may only be added to each binder system in limited proportions. This reduces compatibility problems and any influencing of properties. However, it also creates a new problem. If limited quantities of pastes are added, it will be difficult in many cases to achieve the required pigment-to-binder ratio. This problem can be solved in two ways:

- try to achieve the **highest-possible pigment concentration** in the pigment pastes - if necessary, by incorporating appropriate additives, and
- use so-called base colours - in addition to the universal pigment pastes - which already contain a substantial proportion of the required pigments.

These base colours must be available for all product lines, i.e., all binder systems in the production programme. One example would be to work with several white base colours of different titanium dioxide content as well as with chromatic base colours, especially in the yellow-orange-red colour region. The chromatic base colours are crucial to achieving acceptable hiding power in the mass tone region.

Tinting the base colours with the universal pastes allows a huge number of colours to be matched, but also exerts an influence on the coating properties that must be accepted. Universal mixing systems are only of limited suitability for paints that must meet the strictest quality specifications.

In a diversified production programme, it is often impossible to develop a universal mixing system that will meet all demands. In such cases, operating a **limited universal mixing system** that covers not the entire production programme, but instead only a limited number of product lines of related properties constitutes a compromise that is sustainable and economical in technical terms.

In any event, the pigments chosen for the pastes (see Figure XI.18) must have a high fastness level because they are intended for universal use. In contrast, the pigments for the base colours can be modified to the requirements profile of the product line concerned.

## 8.3 Coloristic demands on colour mixing systems

#### Standard and tolerance framework

Colour mixing systems impose very high demands on the coloristic constancy of the colorants. Various ways have evolved to ensure the coloristic stability of mixing systems. These are outlined below.

The pigment supplier and paint manufacturer agree on a **standard** which, in all future deliveries, is to be reproduced as far as possible without residual colour difference or metamerism. Furthermore, a tolerance framework is specified for these deliveries. For the sake of objectivity and precision, this should be based on chromaticity coordinates. To avoid communication problems between pigment supplier and paint manufacturer, release procedures should not be based on a purely visual assessment.

On account of the **three-dimensionality** of colour space, the tolerance solid, too, has a three-dimensional structure which reduces to a sphere only in the case of a perfectly uniform colour space. This ideal case is only approximately attained in commonly employed colour systems (CMC, CIE94, DIN99, etc.; see Section V). Generally, real tolerance solids deviate substantially from the highly symmetrical shape of a sphere and are approximated for practical use by a cuboid or an ellipsoid. Accordingly, a delivery coloristically conforms to type if its colour position is located well within the tolerance ellipsoid. To avoid exaggerated judgements and to come to a better agreement on visual assessments, it can be helpful to define three different categories:

- conforms to type with no perceptible difference → accepted;
- conforms to type despite small perceptible differences → (conditionally) accepted;
- does not conform to type due to large perceptible differences → rejected.

The first two are "pass" categories, whereas the third is a "fail" or rejection. In order to avoid erroneous judgements, special attention must be paid to the colour measurement process.

Not only the standard and the tolerance framework, but also the **test methods** need to be specified. The reproducibility of the test method concerned should be considerably better than the agreed specific tolerances. A confidence interval of 95 % has proved to be an expedient measure of test-method reproducibility. Statistical methods such as the **F-test**[29-31] should also be adopted to confirm whether the results obtained for a given confidence interval (e.g., 95 %) deviate significantly from each other or whether the differences are more of a random nature. Particularly in the case of chromaticity coordinates, it is necessary to remember that distributions of colour positions in colour space may exhibit marked asymmetries or anisotropies and require the application of multivariate statistics[31].

Where pigments depart from the agreed standard, the **options for coloristic correction** are rather limited and will meet the goal of adjustment with varying degrees of success. The goal of correction is to adjust a colour characteristic of the non-conforming pigment (e.g., in a binary mixture with white), whose deviation from the colour characteristic of the standard is located within a coloristically acceptable tolerance framework. To a certain extent, different particle size distributions between standard and batch can be compensated by adjusting the tinting strength. For this purpose, various criteria, such as the **lightness adjustment** (tristimulus value Y), the weighted **sum of K/S values**, or the **colour depth adjustment** have been developed[28] (see Chapter V). This **tinting strength adjustment** is always associated with a strengthening or weakening of a colour paste. In many cases, the residual colour difference after tinting strength adjustment is located within the agreed tolerance solid for the standard. As a rule of thumb, differences in relative tinting strength of less than 2 % are acceptable and are barely visible in complex formulations.

Tint strength adjustment on its own will not always produce the desired result. The correction of **differences in colour position** requires the availability of tinting pigments, which in often cases is not the case. Correction of a mixing paint by tinting with another chromatic pigment inevitably leads to a change in colour characteristics. The delivery of an overly pure pigment is less critical; it can be blended with a black pigment to render it less chromatic. The reverse case can rarely be corrected in practice.

With effect pigments, further degrees of freedom need to be considered. The appearance of an effect coating, which is determined by its colour and texture, heavily depends on the orientation of the platelet-like pigments within the paint film. The orientation of the pigment particles can be substantially influenced by the application conditions, and this imposes high demands on the reproducibility and constancy of the application process.

# 9 Optimisation of colour mixing systems

*Wilhelm Kettler*

If the paint manufacturer is forced to substitute a pigment in a mixing system or to accept a change of type for an existing pigment (type mutation), he must develop a new mixing paint and new formulations for all recipes containing the original pigment. Recipe substitution can be very costly and will only be done by the paint manufacturer if there are compelling reasons to do so. The assortment of colorants in a mixing system as well as the raw materials supplier need to be chosen carefully to avoid supply bottlenecks and reformulation costs. The colorimetric methods described below will support the selection process.

## 9.1 CIELAB colour maps of colour mixing systems

When the practicability of colorants for a mixing system is being assessed, it can be helpful to construct a "map" of all the colouring components of the mixing system in the form of an $a^*$-$b^*$ diagram, which provides information about their distribution in colour space. To avoid having to create numerous colour samples, it is advisable to prepare a single calibration echelon for both the colour recipe prediction and the determination of the optical material parameters of all colorants. This information can be used to simulate the spectral and colorimetric behaviour of arbitrary mixtures and to evaluate them with, e.g., the SKM model for solid colours.

Figure XI.19 shows an $a^*$-$b^*$ diagram for a refinish system generated as described from a 1 : 1 mixing ratio of respective coloured component and white base. Closer examination of Figure XI.19 reveals that all the components of this paint system are unevenly distributed across the $a^*$-$b^*$ plane. Particularly evident is the high accumulation of data points in the positive $a^*$ range, which contains many red, violet, and yellow components. By contrast, there are large gaps in the negative $a^*$ range. No pigments are available for the regions of strongly greenish yellow and blue colours in the **coloristic map** of this refinish system. Colours in these sub-regions can only be produced by means of pigment

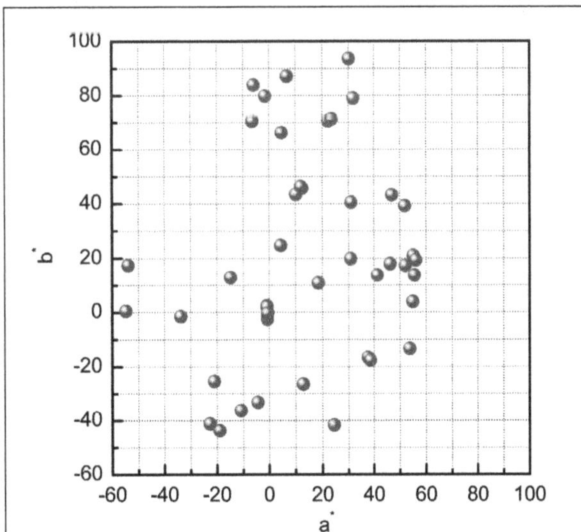


*Figure XI.19: Map showing the colour positions of all mixing components of a refinish paint system in CIELAB76 colour space for 1 : 1 blends of all chromatic components with white. The colour positions of the calibration echelon are located at different lightness levels, which is of no interest in the present context.*

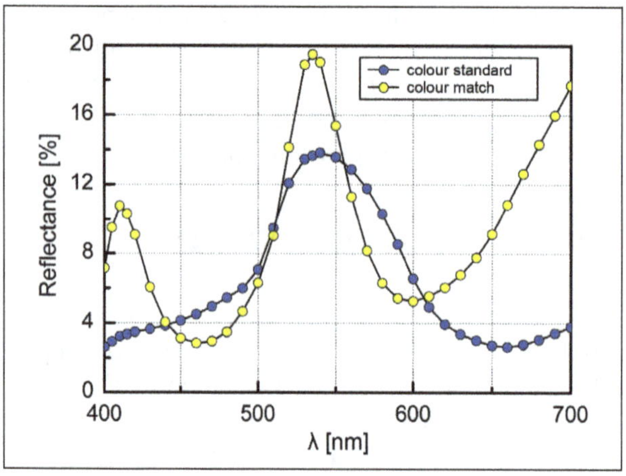


*Figure XI.20: Reflectance spectra of two metameric green solid colours. The colour standard formulated with the chromium-containing pigment P.G. 17 cannot be matched with the chromium-free pigment P.G. 7 available for the colour match.*

mixtures, although only comparatively low chroma values can be achieved. The pigments possessing the highest chroma in **CIELAB76** colour space are all located in the yellow region. All other colour regions covered by the pigment palette only comprise colorants of medium chroma.

The question arises as to whether all components depicted in Figure XI.19 are absolutely necessary for recipe prediction and whether parts of the assortment might be redundant. For instance, the transparent iron oxide yellow possesses almost the same hue as the opaque iron oxide yellow (P.Y. 42). From a coloristic point of view, opaque iron oxide yellow is superior to the transparent variant because it is more chromatic at the same concentration ratio. Consequently, all opaque colours formulated with the transparent iron oxide yellow can also be mixed with the opaque pigment. If the transparent iron oxide yellow were not needed for matching effect colour shades, it would unnecessarily burden the database and stock-keeping. The same considerations apply to the red colour family, which contains a wide variety of transparent and opaque pigments.

Before components are eliminated from a toner or paste assortment, it needs to be verified that the characteristic features of the reflectance spectrum of the replacement colorant do not deviate too much from those of the component to be replaced. This problem can be illustrated by considering green pigments. As can be seen in Figure XI.19, it would be possible for the foregoing reasons to replace the green colorant P.G. 17 (chromium oxide green) with component P.G. 7 or P.G. 36 (phthalocyanine green types). However, the w-shape of the reflectance spectrum for chromium oxide green in Figure XI.20 is so characteristic that there is limited scope for replacing it with either P.G. 7 or P.G. 36, and any such replacement bears the potential risk of metamerism.

## *9.2 Colouring characteristics of pigments*

Figure XI.21 shows a further application for an $L^*$-$a^*$-$b^*$ diagram, namely the colouring characteristics of different types of pigments. This is the curve in **CIELAB76** colour space, which results from all mixtures of a chromatic pigment with **white or black**. Figure XI.21 not only shows the three-dimensional curves of pigments P.B. 15:2 (copper phthalocyanine blue, α-modification), P.G. 7 (copper phthalocyanine green), P.Y. 34 (chromium yellow) and P.R. 254 (diketo-pyrrolo-pyrrole red), but also their projections onto the $a^*$-$b^*$ principal plane of CIELAB76 colour space. As can be seen from Figure XI.21, pigments that already have **high chroma** values when applied as a mass tone, e.g., the chromium yellow (P.Y. 34) or the red (P.R. 254), gradually become less chromatic in mixtures with white or black. With increase in dilution, chroma $C^*$ continuously approaches the achromatic point of the $a^*$-$b^*$ plane.

With regard to pigments that are comparatively **less chromatic** in mass tone as, e.g., phthalocyanine blue (P.B. 15:2) or phthalocyanine green (P.G. 7), only in mixtures with black the chroma decreases monotonically with increasing dilution. In mixtures with white, by contrast, $C^*$ initially increases with increasing dilution, reaching a maximum value at medium brightness. Only above this extreme point $C^*$ continuously decreases with increasing dilution, finally ending in the achromatic point.

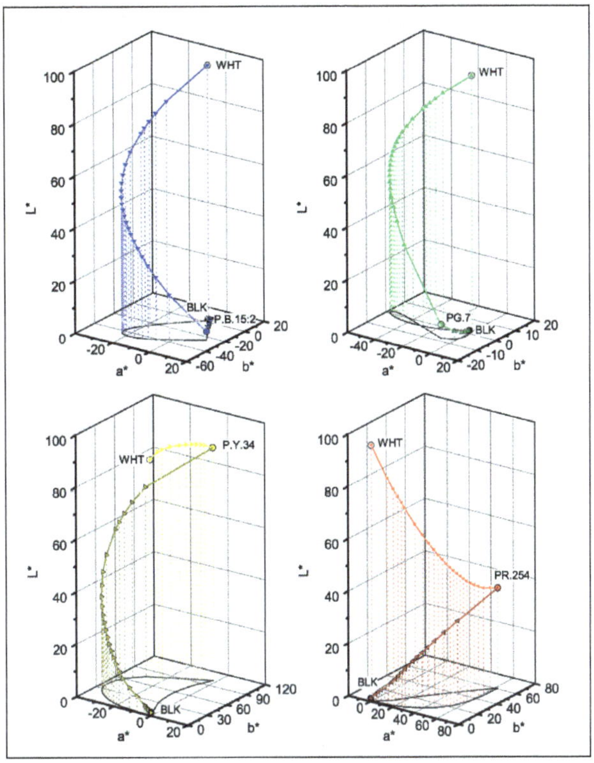


*Figure XI.21: Colouring characteristics of various solid-colour pigments as 3-dimensional mixing curves in CIELAB76 colour space. Mixtures with black are depicted as black-bordered symbols and shaded colours, while mixtures with white are shown in brilliant colours.*

Moreover, as can be seen from Figure XI.21, the chromium yellow pigment undergoes a monotonic decrease in chroma with increasing dilution with white, whereas the mixing series with black has a marked green tinge. For all other pigments, the positions of the two mixing curves are totally different. Only the mixing series with black takes the direct path towards the achromatic point. The mixing series with white is characterised by substantial variation in hue angle. Thus, the dilution function for the red pigment (P.R. 254) reflects the tendency towards a bluish hue as the white content increases; this is also observed in practice.

The curves of the two mixing series for the phthalocyanine blue (P.B. 15:2) pigment behave similarly. Note that the tendency of the mass tone to assume a reddish hue diminishes with increase in brightening, giving way to an almost pure blue perception.

The most extraordinary curve is that for the phthalocyanine green pigment (P.G. 7) in mixtures with white, whose loop-shaped structure in projection onto the $a^*$-$b^*$ plane has an **intersection point** which can be assigned to two different mixing ratios due to its ambiguity. Since the variation in the hue angle is relatively small, the hue character is maintained for the entire curve. The curve of the binary black mixtures does not contain an intersection point and approaches the achromatic point monotonically.

A pigment's colouring characteristics can also be exploited to optimise the set of **calibration panels** of a paint system used for determining the optical material parameters (absorption and scattering coefficients) needed for recipe prediction. From a coloristic point of view, it is expedient to use mixtures of the highest chroma and medium chroma values within the calibration echelon. A close scrutiny of Figure XI.21 reveals that mass tones are preferable for brilliant red and yellow pigments, as these possess the highest chroma. Medium chroma values are achieved in mixtures containing about 20 % black or 80 % white. For the two other depicted pigments (blue, green), Figure XI.21 suggests that only mass tones and mixtures with about 50 % white content, which yield the highest chromaticity, should be used in a calibration echelon.

# 10 Colour gamuts and the limits of colour matching

*Wilhelm Kettler*

The term colour gamut refers to the tonal range of colours. In other words, it is the colour solid in colour space which can be physically realised with a given colour reproduction medium. In the paints and coatings sector, the colour reproduction medium is a mixing system with a carefully chosen pigment palette. The latter should be as small as possible, while making it feasible to match the largest-possible volume in colour space.

The envelope of all theoretically realisable object colours is defined by the colour positions of the **optimal colours**. The enclosed volume is considerably larger than the colour gamuts of commercially available mixing systems. Only in the regions of brilliant yellows and reds do the colour gamut limits of the pigments approach the theoretical limit; there are no pigments with the character of optimal colours among the blues and greens.

If the colour gamut of a colorant assortment is known and a colour standard has been colorimetrically characterised, from its corresponding colour position known prior to recipe prediction it will be obvious, whether an isomeric, metameric, or parameric match can be achieved with the available pigment assortment as, e. g., in matching of colour standards of the printing industry (for instance, Pantone colours) with an automotive mixing system, where it is frequently only possible to achieve parameric matches with, in some cases, considerable residual colour difference and substantial metamerism.

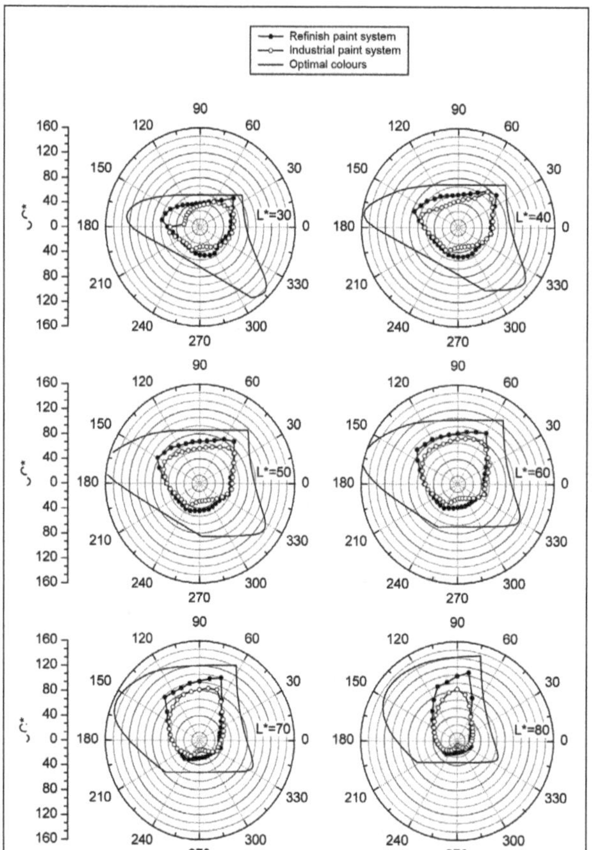


*Figure XI.22: Cross-sectional views of the envelope of the colour gamut for a typical refinish paint system (black circles) and a commercial industrial paint system (open circles) in CIELAB76 colour space at various lightness levels. The analogous colour positions of optimal colours are also shown (solid lines) for comparison.*

The goal of colour development is to find a pigmentation having the fewest-possible number of components, whose individual colour gamut encloses the colour position of the standard. The number of theoretically possible formulations can be dramatically reduced by applying a further independent criterion of minimal metamerism.

Pigment assortments of commercial mixing systems for automotive applications are geared to achieve the largest-possible volume in colour space with a minimum number of pigments. For solid colours, the vast majority of colour matches have the status of isomeric or low-metamerism colour development. Metameric colour matches only occur in comparatively few exceptional cases, such as when colour standards containing lead (yellow, red) or chromium oxide (green) are matched with a lead-free or chromium oxide-free pigmentation. Typically, parameric matches may be expected only when printed colours are being matched with automotive paint. These matches have a finite residual colour difference and metamerism potential relative to the colour standard, and neither can be reduced to zero.

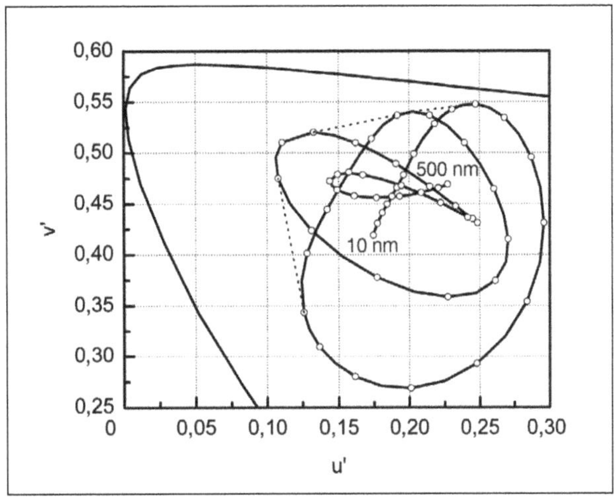


*Figure XI.23: Simulation results for the ideal case of a mica substrate having a geometrical thickness of 1 µm, which is covered with a progressively increasing $TiO_2$ layer and embedded in a binder, and depicted as colour position in a u'-v' diagram. The dotted line indicates the convex envelope of the curve.*

Figure XI.22 depicts how the colour gamut computed for a commercial refinish mixing system (21 pigments) derived with the aid of the **SKM** model compares with a smaller industrial colorant assortment (12 pigments). The computation is based on maximising the chroma $C^*$ of the pigment assortment in colour space for a given lightness level L* and the radius direction in this plane defined by the hue angle $h_{ab}$ [32]. For better clarity, only 2-dimensional cross-sections through the colour gamut at different lightness levels between $L^* = 30$ and $L^* = 80$ are shown. In addition, Figure XI.22 demonstrates that the colour gamut of pigment assortments is not necessarily convex and so cannot be accurately displayed by standard algorithms employed in 3 dimensional computer graphics, which presume convexity for the object to be displayed. By contrast, the cross-sections through the respective colour gamuts in Figure XI.22 are an accurate representation of the existing conditions. For reference, analogous cross-sections through the envelope of all optimal colours are included in the diagram. If the volume of the colour gamut of optimal colours is set to 100%, the refinish mixing system still covers about 30% of this volume, whereas the industrial pigment palette reaches only about 20 %.

The situation regarding the colour gamut of effect pigments is much more complex than the example of computing and interpreting the colour gamut of solid-colour-pigment assortments just discussed. Due to the additional degree of freedom of the macroscopically anisotropic reflectance, isomeric colour matches tend to be the exception for effect colours as compared with solid colours. The colour gamuts at the various measuring geometries are unable to reach their maximum dimensions simultaneously. Their maximum dimension needs to be computed separately for each measuring geometry. This inevitably means that only suboptimal colour gamuts are available for all other measuring geometries.

Usually the pigment processor optimises the pigment palette of his mixing systems to suit the properties of the respective application. A leading manufacturer has taken on this task for interference pigments, extending the range of products with a colour-gamut-optimised pigment palette and offering user-specific solutions [33].

The most elementary structure of an interference pigment consists of a sequence of three layers, with the central, transparent substrate layer (e.g., muscovite) sandwiched between

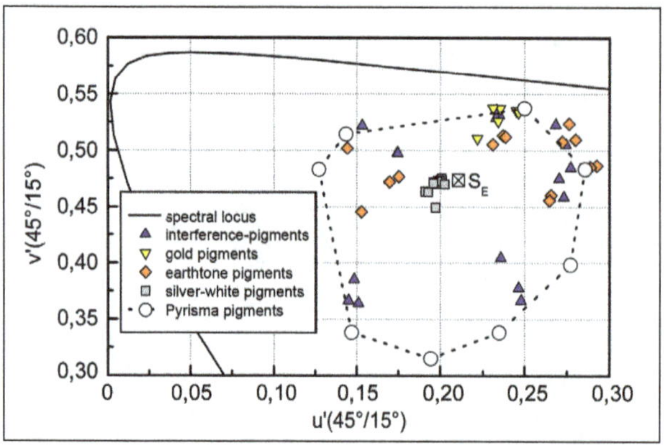


*Figure XI.24: Colour positions of commercial interference pigments in a u'-v' diagram (10° observer, illuminant $D_{65}$). An assortment of eight interference pigments optimised with respect to their colour gamut is depicted as circles, whose connecting polygonal line defines the colour gamut. The visible part of the enclosing curve represents the positions of the spectral colours; $S_E$ denotes the achromatic point.*

two layers of highly refractive transparent metal oxides (e.g., $TiO_2$), whose film thickness and refractive power determine the resulting interference and object colours. Since in blends colours of interference pigments mix additively like light sources, it is obvious to display the colour positions in a coordinate system, in which all mixtures of two interference pigments are located on the straight line connecting their colour positions. Additive mixtures of colours are usually displayed in the **CIE** chromaticity diagram, which meets this requirement, but which is still far away from being a uniform colour space. Transformation of chromaticity coordinates into the **CIELAB76** colour space mitigates the problem of lack of uniformity, however, due to the non-linearity of the mapping, the straight mixing line is transfomed to a strongly bended curve. The only solution is the **CIELUV76** model, which was developed by the CIE at the same time as **CIELAB76 colour space** and which, due to the projective transformation equations, converts straight lines in chromaticity space into straight lines in the target space, while additionally exhibiting improved perceptual uniformity. The optimisation process can be carried out in the *u'-v'* plane, as the lightness of a formulation is also determined by the choice of further mixing components.

Continuous variation in the thickness of the optically active layers of interference pigments causes the corresponding colour positions in the *u'-v'* diagram to cycle in a helical path about the achromatic point [33]. Figure XI.23 shows simulation results in the form of a contInous curve in the *u'-v'* diagram for the ideal case of a mica substrate of 1 $\mu m$ thickness, which is coated with a progressively increasing $TiO_2$ layer and embedded in a binder. The convex envelope of this curve defines the maximum colour gamut that can be achieved with this set of interference pigments. To define an optimum assortment of $N \geq 3$ effect pigments, their colour positions need to be distributed on the envelope in such a way that the connecting polygon encloses the largest possible area. However, the number $N$ of optimally distributed pigments should be kept as low as possible. The contradictory nature of these two requirements means that the solution to the optimisation problem is less than optimal. Compared to the ideal colour gamut, the variant involving eight mixing components (colour areas: blue, indigo, violet, purple, red, yellow, turquoise, green) chosen by the manufacturer has a total area in the *u'-v'* diagram that is just 7 % smaller than the optimum and represents a good economic and coloristic compromise (see Figure XI.24).

The substitution of pigments in existing mixing systems that have a comprehensive stock of mixing recipes is not without difficulties. If an existing interference pigment has to be replaced in a formulation by a mixture of other interference pigments, the quantities of solid-colour pigments usually need to be readjusted. There need to be good reasons for replacing pigments in established mixing systems, before the steps involved in reworking formulas in a coloristic laboratory are taken. However, if there are plans to introduce a new mixing system, it is imperative to consider optimisation of the coloristic profiles of all the mixing components.

# 11 Literature

## 11.1 General literature

The literature has been selected with the aim of providing an overview and is suitable for gaining further insights into the field of colorimetry. The references are provided in alphabetical order and make no claim to completeness.


Barber, P. W.; Hill, S. C.; Light Scattering by Particles: Computational Methods, World Scientific, Singapore; 1990

Bohren, C. F., Huffman, D. R.; Absorption and Scattering of Light by Small Particles, John Wiley & Sons, New York; 1983

Bohren, C. F., Multiple Scattering at the Beach, Weatherwise 1983, 36, 197-200

Bohren, C. F., Multiple Scattering of Light and Some of its Observable Consequences, Am. J. Phys. 1987, 55(6), 524-533

Bohren, C. F., Salzman, G. C., Palmer, B. R. und Stamatakis, P., A Multiple-Scattering Model for Opacifying Particles in Polymer Films, J. Coat. Technol. 1989, 61(779), 41-47

Brockes, A., Der Einfluss glänzender Oberflächen auf Remissionsmessungen, Die Farbe 1960, 1/3, 53-62

Brockes, A., Der Zusammenhang von Farbstärke und Teilchengröße von Buntpigmenten nach der Mie-Theorie, Optik 1964, 21, 550-566

Chandrasekhar, S., Radiative Transfer, Clarendon Press, 1950

van de Hulst, H. C.; Light Scattering by Small Particles, Dover Publications, New York; 1957

van de Hulst, H. C.; Multiple Light Scattering - Tables, Formulas, and Applications, Volume 1, Academic Press, New York; 1980

van de Hulst, H. C.; Multiple Light Scattering - Tables, Formulas, and Applications, Volume 2, Academic Press, New York; 1980

Ishimaru, Akira; Wave Propagation and Scattering in Random Media, IEEE, 1997

Kerker, Milton; The Scattering of Light and other Electromagnetic Radiation, Academic Press, New York; 1969

Kerker M., Scheiner, Cooke und Kratohvil, J. Coll. Interface Sci. 1979, 71(1), 176-187

Klein, Georg A.; Industrial Colour Physics, Springer, New York Dordrecht Heidelberg London, 2010

Kortüm, Gustav; Reflexionsspektroskopie, Springer-Verlag, Heidelberg, 1969

Mehta, K. T. und Shah. H. S.; Simplified Equations to Calculate Mie-Theory parameters for Use in Many-Flux Calculations for Predicting the Reflectance of Paint Films, Colour Research and Application 1987, 12(3), 147-153

Mudgett, P. S. und Richards, L. W.; Multiple Scattering Calculations for Technology, Appl. Opt. 1971, 10 (7), 1485-1502

Mudgett, P. S. und Richards, L. W.; Multiple Scattering Calculations for Technology II, J. Coll. Interface Sci. 1972, 39 (3), 551-567

Mudgett, P. S. und Richards, L. W.; Kubelka-Munk Scattering and Absorption Coefficients For Use with Glossy, Opaque Objects, J. Paint Techn. 1973, 45 (586), 44-53

Nassau, Kurt; The Physics and Chemistry of Colour - The Fifteen Causes of Colour, John Wiley & Sons, New York, 1983

Palik, Edward D. (Hrsg.); Handbook of Optical Constants of Solids, Academic Press, Boston 1985

Querry, Marvin R.; Optical Constants, Govern. Rep., CRDC-CR-85034, Kansas City, Missouri, MO 64110

Quinten, Michael; Optical Properties of Nanoparticle System, Wiley, VCH, 2011

Rybicki, George B.; Radiative Transfer, J. Astrophys. Astr. 1996, 17, 95-112

Ullmann's Encyclopedia of Industrial Chemistry, Dyes, General Survey, A9, VCH Verlagsgesellschaft, Weinheim, 1987

Ullmann's Encyclopedia of Industrial Chemistry, Pigments Inorganic, A20, VCH Verlagsgesellschaft, Weinheim, 1992

Völz, Hans G.; Industrielle Farbprüfung, Grundlagen und Methoden, VCH Verlagsgesellschaft, Weinheim, 2001

Winter, Gerhard; Anorganische Pigmente: Disperse Festkörper mit technisch verwertbaren optischen und magnetischen Eigenschaften, Fortschr. Miner. 1979, 57 (2), 172-202

## *11.2 References*


[1] Allen, E., Basic equations in computer colour matching, II. Tristimulus matching, two-constant theory, J. Opt. Soc. Am. 64, 991 (1974)

[2] Allen, E., Colorant formulation and shading, in Optical Radiation Measurements, Bd. 2, Colour Measurement, Herausg. F. Grum und C. J. Bartleson, Academic Press, NewYork, USA (1980)

[3] Kühni, R. G., Computer colorant formulation, Lexington Books, Lexington, Massachusetts, USA (1975)

[4] Colour Physics, Herausg. R. McDonald, Society of Dyers and Colourists, Bradford, England (1987)

[5] Berns, R. S., Principles of Colour Technology, John Wiley & Sons, Inc., New York, USA (2000)

[6] Klein, G. A., Farbenphysik für industrielle Anwendungen, Springer, Berlin (2004)

[7] Unterforsthuber, K., Theoretische Grundlagen der Farbrezeptberechnung, published privately by the author, Frankfurt (2005)

[8] Richter, M., Einführung in die Farbmetrik, de Gruyter, Berlin, Deutschland (1981)

[9] Wyszecki, G., Stiles, W. S., Colour Science: Concepts and Methods, Quantitative Data and Formulae, John Wiley & Sons, New York, USA (1982)

[10] Kettler, W. H., Kolb, M., Numerical evaluation of optical single-scattering properties of reflective pigments using multiple-scattering inverse transport methods, Die Farbe, 43, S. 167, 1997

[11] Kettler, W. H., Kolb, M., Inverse multiple scattering calculations for plane-parallel turbid media: application to colour recipe formulation, Proceedings of the international workshop „Electromagnetic and Light Scattering: Theory and Applications", S. 19, (Moscow, Russia), 1997

[12] Kettler, W. H., Richter, G., Investigation on topology of platelet-like effect pigments in automotive surface coatings, Progress in Organic Coatings 31, S. 297, 1997

[13] Billmeyer, F. W., Carter, E. C., Colour and appearance of metallised paint films, J. Coat. Techn. 48, S. 53, 1976

[14] Trinks, W., Zur Vielfachstreuung an kleinen Kugeln, Ann. d. Phys. 22, S. 561, 1935

[15] Goedecke, G. H., Radiative transfer in closely packed media, J. Opt. Soc. Am. 67, S. 1339, 1977

[16] Sadowski, F., Schwab, H. M., Theorie und Praxis des Farbtönens (1), Industrie-Lackier-Betrieb 35, S. 193, 1967; ibid. S. 229 (Teil 2); ibid. S. 285 (Teil 3); ibid. S. 335 (Teil 4)

[17] Sadowski, F., Gesetze und Methoden des Farb- und Effektangleiches, erläutert am Beispiel der Autoreparaturlackierung, Die Farbe, 39, S. 187, 1993

[18] Sadowski, F., Basiswissen Autoreparaturlackierung, Vogel Buchverlag, Würzburg, Germany, 2000

[19] Klein, G. A., Farbenphysik für industrielle Anwendungen, Springer-Verlag, Berlin, Germany, 2004

[20] Richter, M., Einführung in die Farbmetrik, de Gruyter, Berlin, Germany, 1981

[21] Lang, K., Farbwiedergabe in den Medien, Muster-Schmidt Verlag, Göttingen, Germany, 1995

[22] Schuster, A., Radiation through a foggy atmosphere, Astrophys. J. 21, S. 1, 1905

[23] Kubelka, P., Munk, F., Ein Beitrag zur Optik der Farbanstriche, Z. Tech. Phys. 12, S. 593, 1931

[24] McDonald, R., Herausg., Colour Physics for Industry, Society of Dyers and Colourists, Bradford, UK, 1987

[25] Chandrasekhar, S., Radiative Transfer, Dover Publications, New York, USA, 1960

[26] Kettler, W. H., Richter, G., Einsatz von ultrafeinem Titandioxid in Metalleffekt-Lackierungen, Farbe + Lack 98, S 3, 1992

[27] Pfaff, G., Spezielle Effektpigmente, Vincentz Network GmbH & Co. KG, Hannover, Germany, 2007

[28] Völz, H. G., Industrial Colour Testing. Fundamentals and Techniques (2nd Edition), Wiley-VCH, New York, USA, 2001

[29] Stange, K., Angewandte Statistik. Eindimensionale Probleme, Bd. 1, Springer-Verlag, Berlin, Germany, 1970

[30] Stange, K., Angewandte Statistik. Mehrdimensionale Probleme, Bd. 2, Springer-Verlag, Berlin, Germany, 1971

[31] Kreyszig, E., Statistische Methoden und ihre Anwendungen, Vandenhoeck und Ruprecht, Göttingen, Germany, 1974

[32] Unterforsthuber, K., Theoretische Grundlagen der Farbrezeptberechnung, veröffentl. durch K. Unterforsthuber, Dannstadt-Schauerheim, Germany, 2005

[33] Roesler, M., Huber, A., System von Interferenzpigmenten, Internationale Patentanmeldung WO 2007/054144 A1 (2007)

[34] Henyey, L. C. Greenstein, J. L., Diffuse radiation in the galaxy, Astrophys. J. 93, 70, 1941

[35] Datacolor Case Study "Hesse-Lignal"

# XII Microscopic analysis of effect pigments

*Gerhard Wilker*

## 1 Matching of effect colour shades

***Metal effect pigments***

The introduction of metallic finishes in the automotive industry in the early 1970s presented totally new problems when it came to colour matching. Initially, such metallic effects were achieved with aluminium flakes (non-leafing types [1]) of different particle size distribution. For a more pronounced or dark flop, coarser flakes were used, whereas fine aluminium flakes were employed for a less pronounced or light flop. At the end of the 1980s, the new silver-dollar flakes augmented the cornflake types (see Section X.3.1.3).

The major problem that arises in the colour matching of metallics is the angle dependence of the colour, which varies with the type of aluminium flakes and the transparency of the chromatic pigments used. Both have a pronounced influence on

- light/dark flop and/or
- colour flop.

In other words, for a non-metameric match, the reflectance curve of colours must be replicated in the same visible wavelength region as for solid colours and, additionally, at all different viewing angles. In most cases, this can only be achieved by using aluminium flakes, which have identical

- particle shape
- particle size, or
- particle size distribution.

The use of a light microscope to determine these properties has been state of the art for many years. Combining the performance of colour matching systems with a light microscope has made it possible to solve nearly all the problems that arose during colour matching.

***Pearlescent pigments***

The advent of weather-fast pearlescent pigments brought additional problems for the colour matching process due to the very intense colour flop of these new effect pigments in automotive colours. As with aluminium flakes, the particle size of the pigments is responsible for the intensity of their effect.

### 1.1 Application of light microscopy

As they have a geometric structure, all effect pigments can be easily identified with the aid of a light microscope because their particle size normally ranges from 8 µm to 45 µm. Consequently, at magnifications of 100-times to 500-times, effect pigments can be determined

Wilhelm Kettler et al.: Colour Technology of Coatings

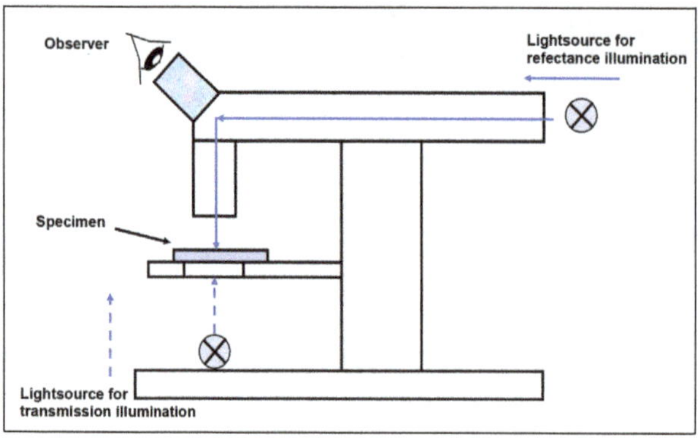


*Figure XII.1: Type of light sources: Reflectance/transmission illumination.*

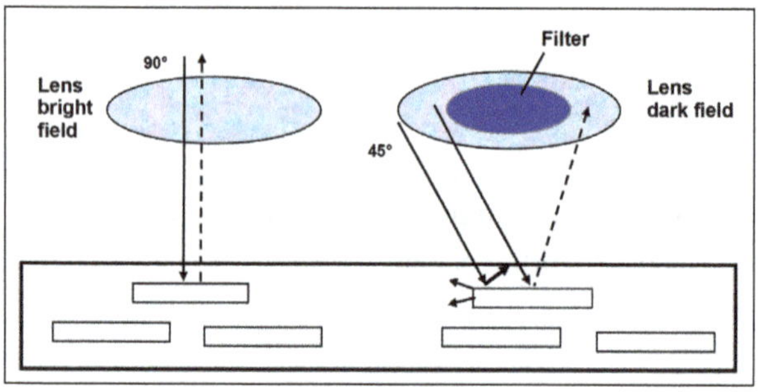


*Figure XII.2: Light path bright field/dark field illumination.*

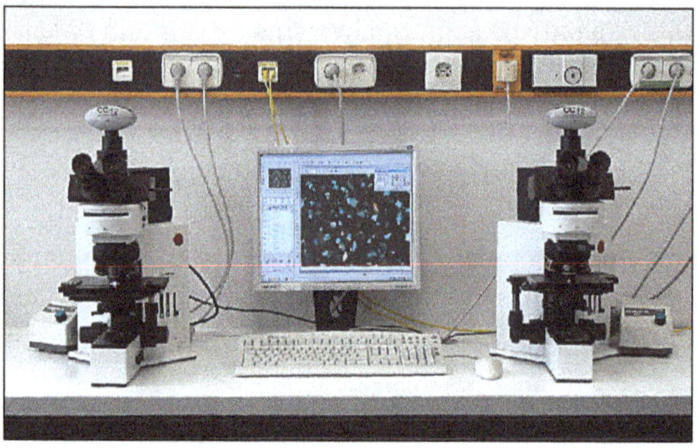

*Figure XII.3: Workplace microscope.*

both quantitatively and qualitatively. This applies to both solid and liquid coatings.

Well-dispersed organic or inorganic pigments cannot normally be studied with a light microscope. The particle size of properly dispersed pigments is below 1 μm (resolution at 500-times magnification is 1 μm). Flocculated pigments appear as coloured dots.

## 1.2 Accessories for the light microscope

The light microscope should be equipped with two types of light source:

- reflectance illumination
- transmission illumination.

Under reflectance illumination, the incident light is reflected by the specimen and generates the microscopic image.

Under transmission illumination, the incident light passes through sample, e.g. liquid paint, to generate the microscopic image (Figure XII.1).

Consequently, reflectance illumination is used for analysing painted substrates whereas transmission illumination is used for liquid paints and pigment powders.

Besides these light sources, the microscope should be equipped with the following types of illumination:

- bright field illumination
- dark field illumination.

Under bright field illumination, the light beam is incident at an angle of 90°. This contrasts with dark field illumination, where the light beam comes in at an oblique angle of 45° via a circular filter system or a cone placed in the light path [3].

Bright field illumination reveals the colour and the structure of the particle surface whereas, in dark field illumination, the combination of indirect incidence and reduced light intensity, yields information about edge scattering (and the colour of edge scattering), the fine structure of the surface, and the transparency or opacity of the analysed effect pigment.

The last of these properties provides important information about the metal oxide layers applied on top of the base substrates. Also detectable are de-orientation agents.

The use an analyser/polariser system to produce polarised light can prove very helpful (see Section XII.3.5).

The optimum magnification is 200-times. This is usually sufficient to permit proper identification of the effect pigments. Where there is a doubt, 500-times magnification can be used. Focusing may prove to be a problem, due to the fairly thick layer of clearcoat (40 to 45 µm) used with automotive paints. For quantifying the average amount (statistically) of particles on a given surface, 100-times magnification proves very useful, especially for mixtures of effect pigments.

The recommended laboratory equipment for analysing all kinds of effect pigments is as follows:

- two microscopes equipped with reflectance/transmission light sources, bright/dark field illumination, polariser/analyser, magnification: 100-times, 200-times, 500-times
- TFT screen (19" or larger)
- CCD colour cameras of high resolution (2x)
- particle-size measuring equipment
- PC for analysis and archiving system, and plotter

Where an analysis and archiving system is available, a second microscope and CCD camera are unnecessary; this will significantly lower the investment costs.

# 2 Effect pigments

## 2.1 Overview

A systematic approach to the analysis of effect pigments or mixtures of such pigments is strongly recommended.

***Systematic approach***
Systematic analysis under a light microscope is best carried out under bright field/dark field illumination. Determination of

- particle size of effect pigments
- mixtures of effect pigments
- shape of effect pigments
- colour of effect pigments
- opacity/transparency of effect pigments

is the standard approach. Special cases will be covered in Section XII.3.

Determination of particle size and particle size distribution of the pearlescent pigment is very important because pearlescent pigments are produced in the following manner [4]

- fine: below 11 µm
- medium: 15 to 35 µm
- coarse: 25 to 45 µm

Aluminium flakes have even more particle size gradations and particle size distributions. For our purpose, therefore, it is essential to create a library of the most commonly employed effect pigments (aluminium flakes, pearlescent pigments etc.) on the market to serve as reference panels.

***Application to the reference panel***
It is recommended that effect pigments be applied to a black substrate to avoid any interference, e.g. reflected light from the metal substrate.

The pigment concentration should be quite low to minimise overlapping of particles, as this can influence the determination of opacity/transparency of the individual pigment during the analysis. The finer the grade of effect pigment, the lower the pigment concentration must be to prevent overlapping (higher surface area concentration). For a cardboard version, black/white contrast cards are suitable.

Panels for building up such reference libraries should be composed as follows:

- effect pigment layer (0.5 to 1 %) applied over a black primer, clearcoat on top (PE/CAB-TSA). Substrate: steel panel
- effect pigment layer (0.5 to 1 %) applied over a black/white cardboard, clearcoat on top (PE/CAB-TSA).

## *2.2 Aluminium flakes*

Particle size aside, aluminium flakes can be categorized as

- cornflake types
- silver-dollar types
- PVD flakes
- anodised flakes

Cornflake and silver-dollar types can be subdivided additionally into anodised types and the "Paliocrom" range [5] (see Section XII.2.4).

***Cornflake types***
Cornflake types can be easily distinguished under the microscope by their rough and irregular edges and the pronounced surface structure.

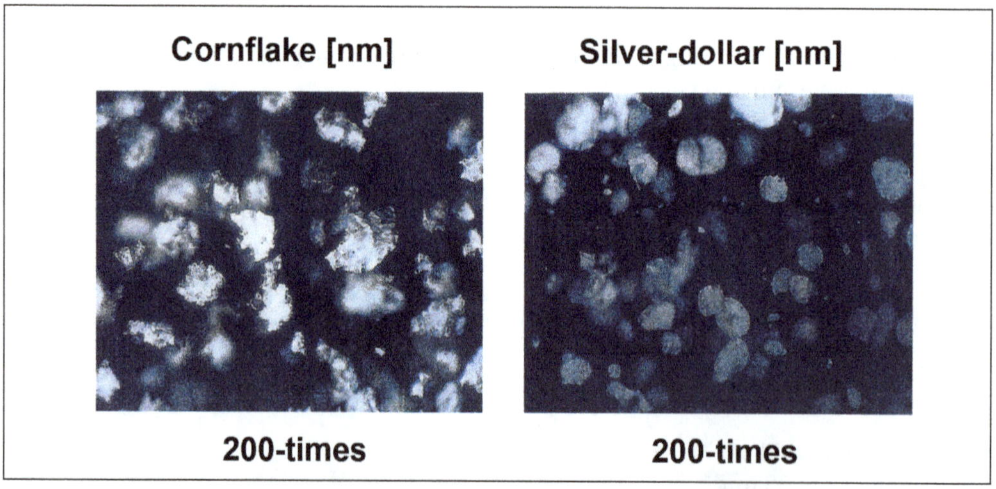


*Figure XII.4: Examples of aluminium flakes, left: cornflake, right: silver-dollar.*

*Figure XII.5: PVD aluminium flakes, left: 200-times, right: 200-times magnification.*

***Silver-dollar types***
In contrast, silver-dollar types are usually circular or elliptical, with a mostly smooth/planar surface. This imbues the particles with less edge scattering (less greyish colour) and greater surface reflectance (darker flop).

***Anodised flakes***
Under the microscope, anodised flakes have a characteristic champagne colour, which makes proper identification very difficult in mixtures with organic/inorganic yellow pigments.

***PVD flakes***
PVD flakes possess very sharp edges, sometimes are triangular or rhombohedral in shape and have an extremely smooth and even surface (due to the production method). Applied on a substrate, PVD flakes exhibit highly pronounced mirror-like and/or chrome effect.

They also have a characteristic (angle-dependent) rainbow effect (see Section XII.4).

## 2.3 Mica-based pearlescent pigments

Section X.4 discusses the structure of pearlescent pigments as well as other effect pigments covered in Section XII.2.4 in detail. Responsible for the colour and flop of these effect pigments are the base substrate, the thickness of the oxide layer and the chemistry of the oxide layer (refractive index). See also Sections X.4 and XI.6.3.

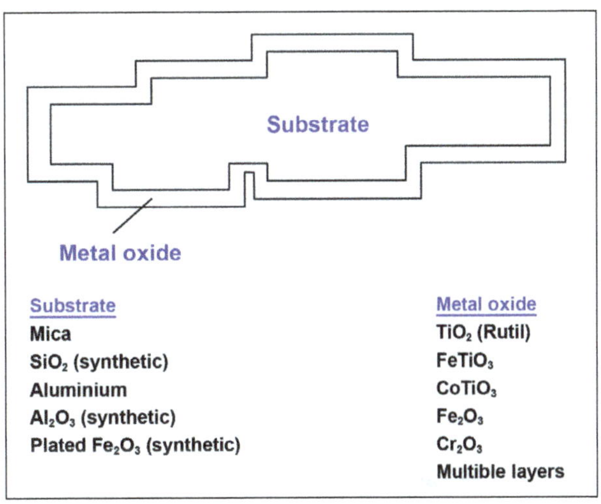


*Figure XII.6: Structure of effect pigments (see also Section X.4 and XI.6.3).*

For all the pearlescent pigments described in the following section, so-called "pigment profiles" are provided which consist of a microscopic bright field/dark field image taken at 200-times

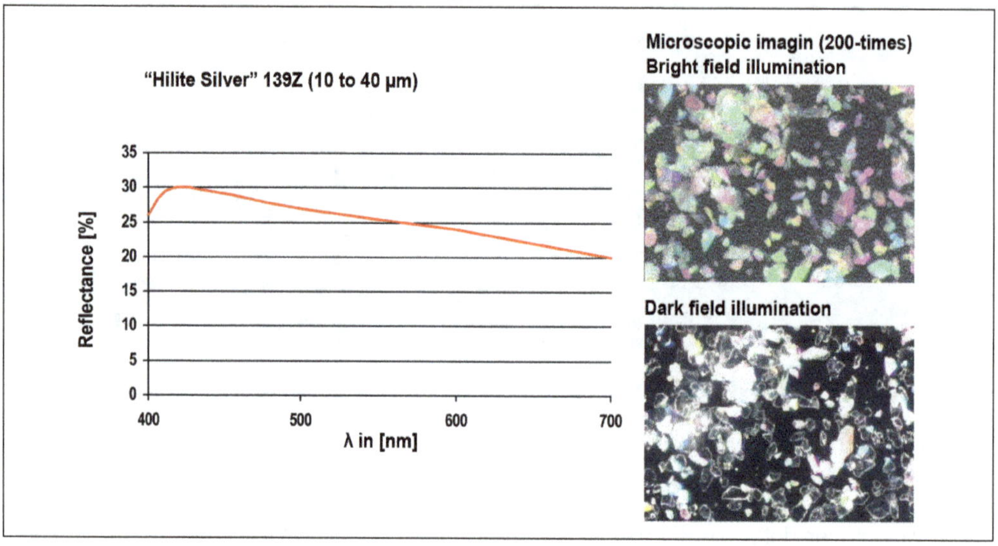


*Figure XII.7: Reflectance graph and microscopic image of a silver interference pearlescent pigment.*

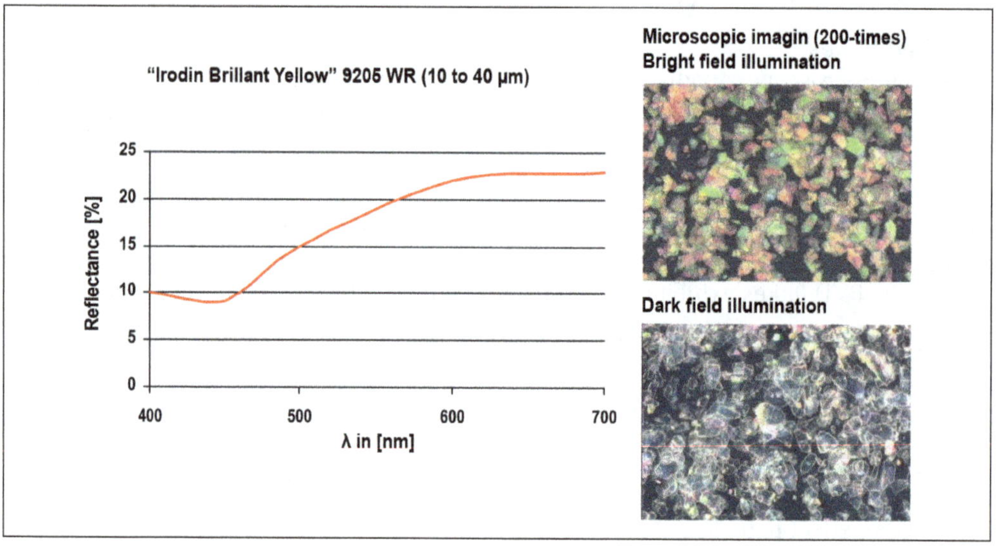


*Figure XII.8: Reflectance graph and microscopic image of a yellow interference pearlescent pigment.*

magnification and a reflectance curve [d:8°] against a black background. These reflectance curves, together with those of the most important organic/inorganic coloured pigments, can be collected to form a library that can be helpful when it comes to systematically colour-matching effect shades (see Section XII.4). In this Section XII.3, we will look at examples of selected pearlescent pigments with the following coatings:

- titanium dioxide
- iron oxide
- combinations or multiple layers

***Interference/pearlescent pigments***
The pearlescent silver pigment in the following diagram is rainbow coloured under bright field illumination. The main colour is silver grey, in accordance with the rule of additive colour mixing (see Section XI.6). Under dark field illumination, parallel particles are highly

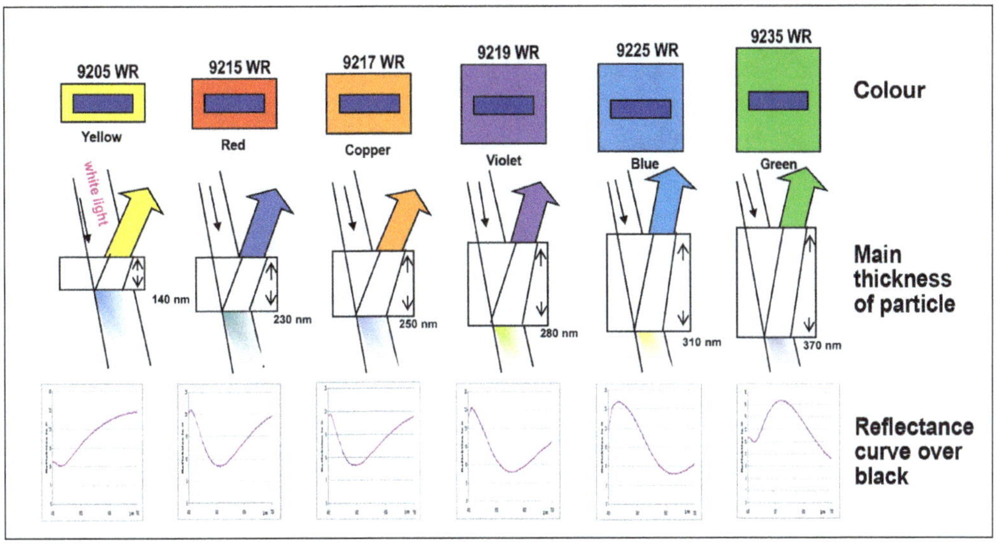


*Figure XII.9: Interference pearlescent pigments.*

transparent whereas angularly oriented particles exhibit the above-mentioned colour under bright field illumination. The white edge scattering by the individual particles is striking.

Similar considerations apply to colour development of pearlescent pigments (titanium dioxide coating of increasing thickness). Starting with interference yellow (yellow is the main colour in bright field illumination), some red, blue and green particles can be observed, whose colour results from a thick coating of titanium dioxide.

Under dark field illumination, the particles are still transparent. Angularly oriented particles have a yellow colour, with pronounced edge scattering in yellow. An important mark here that distinguishes them from yellow combinations of pigments is the content of blue particles under bright field illumination. The additional coating with iron oxide yields a more intense yellow/gold colour and the blue colour disappears. The reflectance curves of the combination yellows have a steeper absorption edge.

The subsequent red, violet, blue and green interference pigments exhibit similar behaviour. This means that the majority of the particles exist in the mass tone colour, with some in the colours with the thinner or thicker coating, e.g. yellow interference and copper pearlescent pigments.

***Iron oxide-coated pearlescent pigments***
As the thickness of the iron oxide coating increases, the colour changes from bronze to copper, red, red- violet and interference red/green under bright field illumination.

Under dark field illumination, the particles are opaque because the iron oxide is much more opaque than titanium dioxide.

Interference red/green is unmistakable under both bright field and dark field illumination on account of the mixture of red and green particles and their high opacity. Particles, which are red under bright field conditions change to green in dark fields and vice versa. When this "Scarabeus" pigment is used in a paint, it can be easily identified by its characteristic contribution (d:8° geometry) to the reflectance curve. A multi-angle spectrophotometer shows that the reflectance maximum is in the green region at measuring angles of as15°

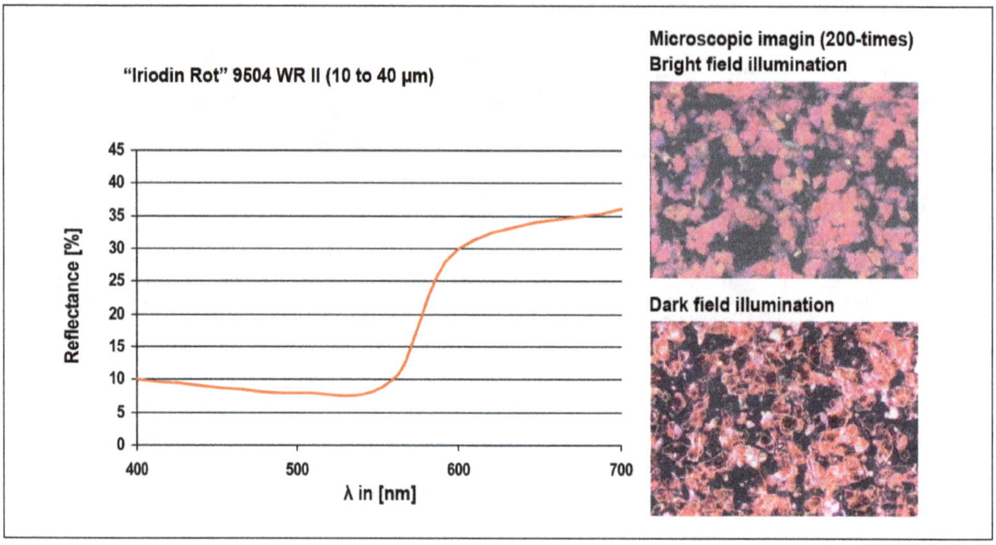


*Figure XII.10: Reflectance graph and microscopic image of a red pearlescent pigment.*

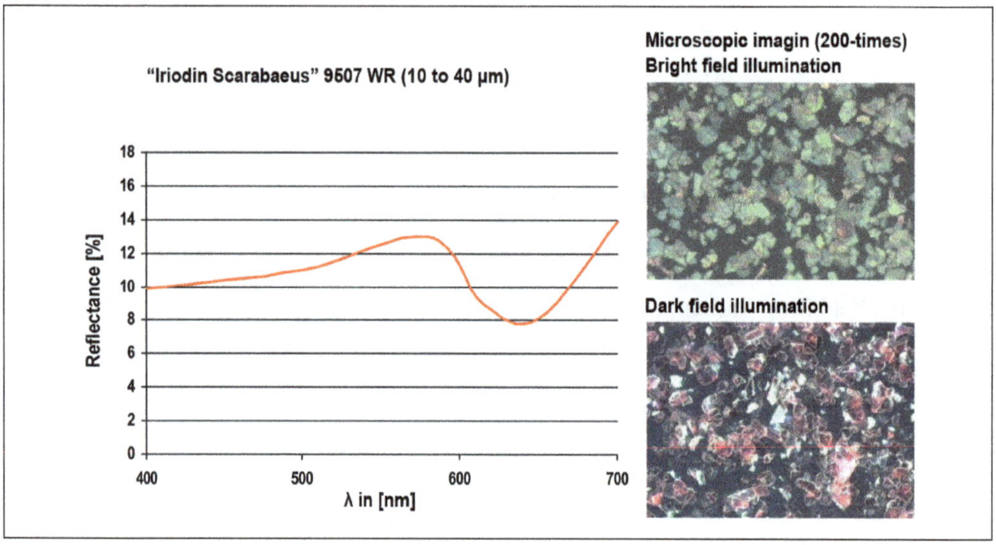


*Figure XII.11: Reflectance graph and microscopic image of a red/green pearlescent pigment.*

and as25°, whereas, at measuring angles of as45°, as75° and as105°, the pigment shows a quite steep absorption edge in the red region.

In general, iron oxide-coated pearlescent pigments yield more-saturated colours than corresponding titanium dioxide-coated pigments, as well as a steeper absorption edge.

### *Combination pearlescent pigments*

Among the ranks of combination yellow or gold pigments (titanium dioxide/iron oxide) is "moss green", a coating of titanium dioxide/chromium oxide (Merck KGaA). Bright field illumination reveals highly saturated green particles in addition to a small amount of other colours. The chromium oxide coating renders the particles highly opaque under dark field conditions. The reflectance curve over the visible wavelength range has the typical shape of chromium oxide.

## 2.4 *Miscellaneous effect pigments*

In this section, we consider the "more recent generation" of effect pigments based on various substrates, such as aluminium, synthetic substrate materials, and optimised, 3$^{rd}$-generation pearlescent pigments.

### 2.4.1 *Effect pigments based on aluminium coated with iron oxide*

The "Paliocrom" range (BASF SE) is based on aluminium flakes (see Section X.4). The substrates are cornflake or silver-dollar flakes which are coated with iron oxide by a special method. The desired colour is achieved by varying the thickness of the layer. This application exploits the high opacity of aluminium as a substrate.

Under bright field illumination, the particles exhibit a typical cornflake structure, with gold as the main colour. Some particles have a silver colour because they are uncoated or tend to

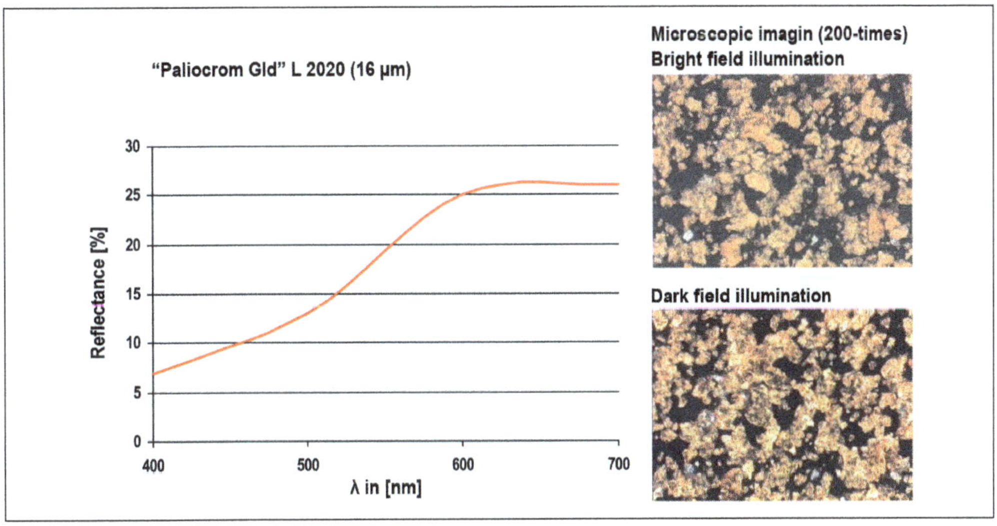


*Figure XII.12: Reflectance graph and microscopic image of an iron oxide coated aluminium flake.*

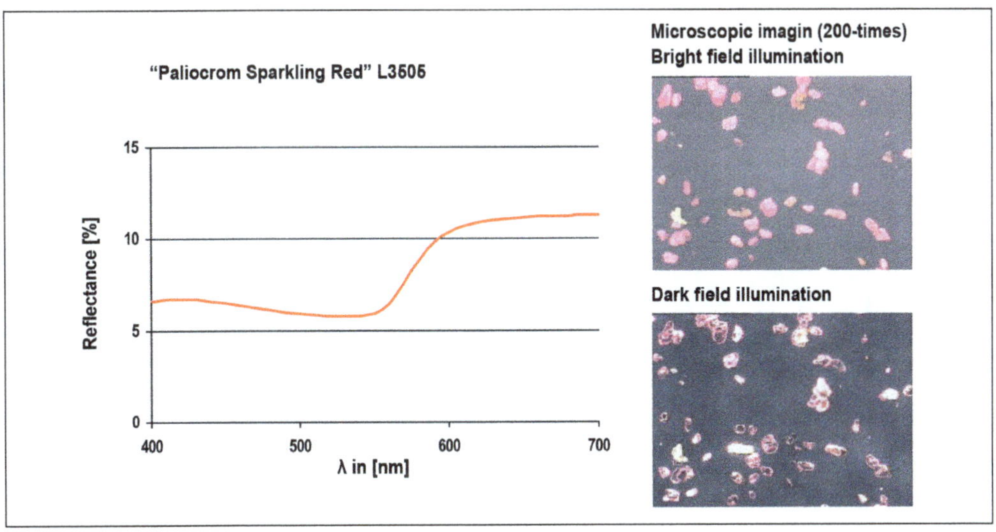


*Figure XII.13: Reflectance graph and microscopic image of an iron oxide coated aluminium flake, silver-dollar based.*

be redder due to a thicker coating of iron oxide. Dark field illumination yields no additional information, due to the high opacity of the substrate.

As mentioned above, the thickness of the layer determines whether gold, orange or red colours are obtained. A red pigment based on silver-dollar flakes has recently been introduced onto the market.

### *2.4.2 Effect pigments based on aluminium oxide platelets*

Based on synthetic aluminium oxide platelets and coated with either titanium dioxide or iron oxide layers, the "Xirallic" range (Merck KGaA) exhibits the same colour development as standard mica-based pearlescent pigments (see Sections X.4. and IX.4.1.2).

As a synthetic substrate is used, which is produced under controlled conditions, a very smooth surface of much greater brilliance is achieved. Coating with titanium dioxide yields

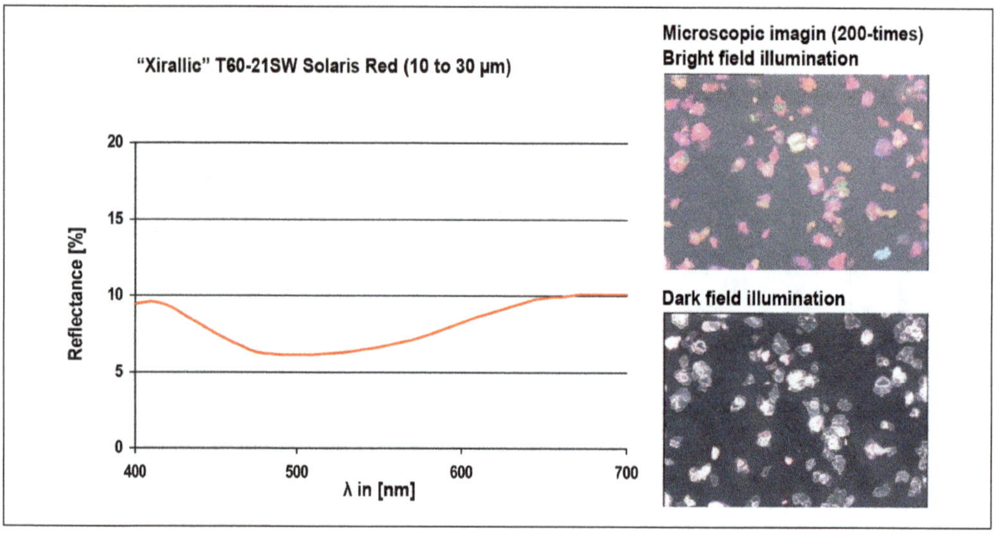


*Figure XII.14: Reflectance graph and microscopic image of a red aluminium oxide platelet coated with titanium dioxide.*

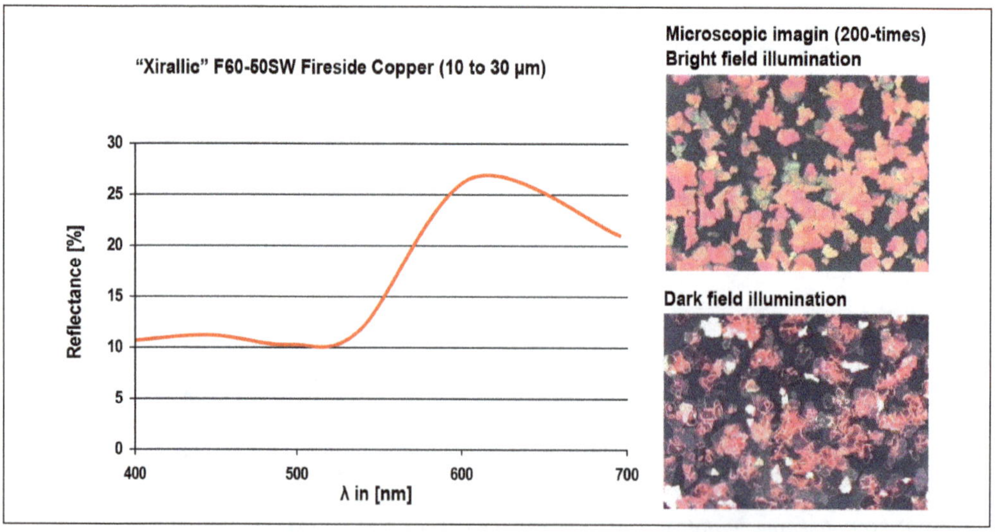


*Figure XII.15: Reflectance graph and microscopic image of a copper aluminium oxide platelet coated with iron oxide.*

the whole range from silver to interference green. The particle size is small, ranging from 10 to 30 μm. Similar considerations apply to the pigments coated with iron oxide. The particles have the characteristic "elephant trunk" shape, which stems from the production process. The red effect pigment shown here has the same colour under bright field illumination as the mica-based pigment, but has much higher transparency under dark field conditions.

The iron oxide-coated grade exhibits high opacity under dark field conditions. The reflectance curve has a steep absorption edge, with a characteristic downward bend between 600 and 700 nm.

### *2.4.3 Effect pigments based on silicon dioxide platelets*

Based on a synthetic silicon dioxide substrate (see Section X.4.1.4.), the "Colorstream" range from Merck KGaA) features pigments coated with titanium dioxide and iron oxide.

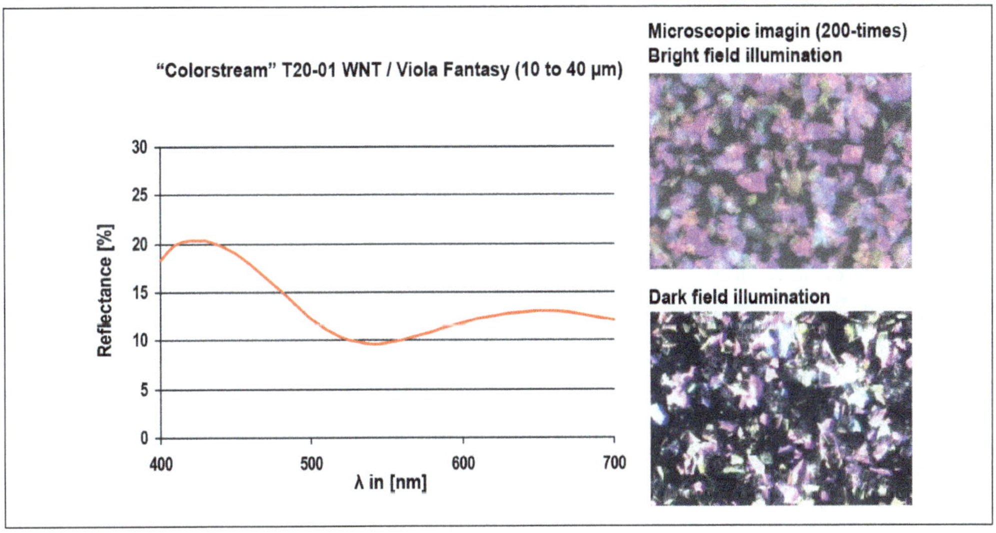


*Figure XII.16: Reflectance graph and microscopic image of a violet silicon dioxide platelet coated with titanium dioxide.*

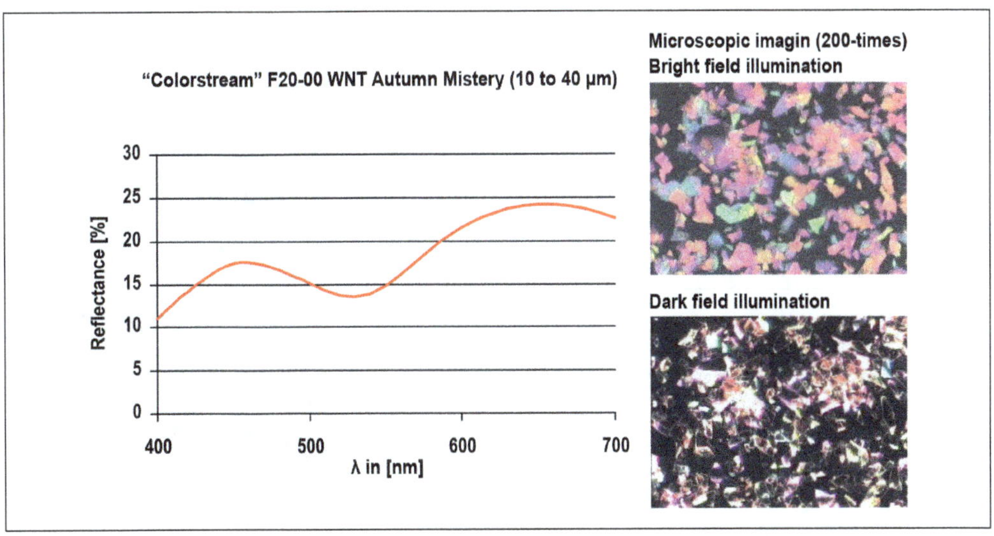


*Figure XII.17: Reflectance graph and microscopic image of a silicon dioxide platelet coated with iron oxide.*

The particle shape is totally different from that of the mica-based grades. Triangular or rhombohedral particles predominate, and have a glass-like, smooth surface.

Figure XII.16 shows the typical colour of a violet effect pigment under bright field conditions. The principal colour is violet, with some blue particles. Dark field illumination reveals transparency, in line with the nature of the substrate and the coating, along with edge scattering. The reflectance curve is very characteristic, with a maximum in both the short and long wavelength regions.

Iron oxide-coated silicon dioxide particles are mainly red, with some yellow and green particles, under bright field illumination. Under dark field conditions, the particles are mostly opaque. The reflectance curve has an absorption edge between 600 and 700 nm and a green contribution between 450 and 500 nm (see Figure XII.17).

### *2.4.4 Optimised pearlescent pigments with titanium dioxide coating*

Optimisation of 3rd-generation mica-based pigments targeted:

- particle shape, i.e. fewer edges and thus mainly rounded particles
- smoother surface, to minimise centres of scattering
- better control over coating thickness for each colour
- narrower particle size distribution combined with slightly smaller particles.

The outcome is highly optimised coloristic properties compared to those of the standard grades (see Section X.4).

Bright field illumination reveals the mainly rounded shape of the particle, and the uniform particle size. Dark field illumination yields high transparency and a much less structured surface (Figure XII.18).

### *2.4.5 Multilayer pigments with Fabry-Perot structure (OVIP)*

OVIP pigments (optically variable interference pigments) exhibit angle-dependent colour change across the entire spectrum.

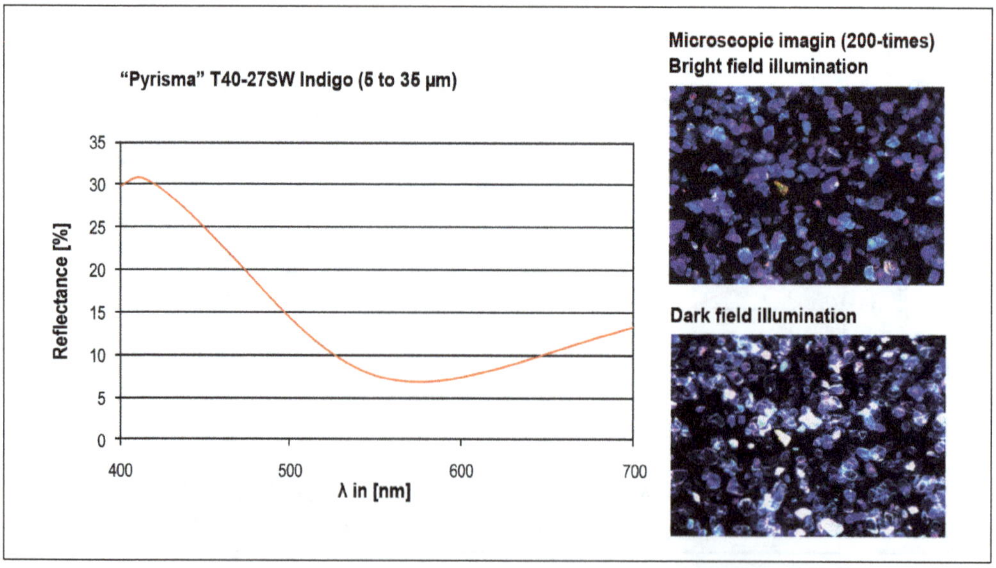


*Figure XII.18: "Prisma" Indigo as an example for the 3rd-generation of mica-based pigments.*

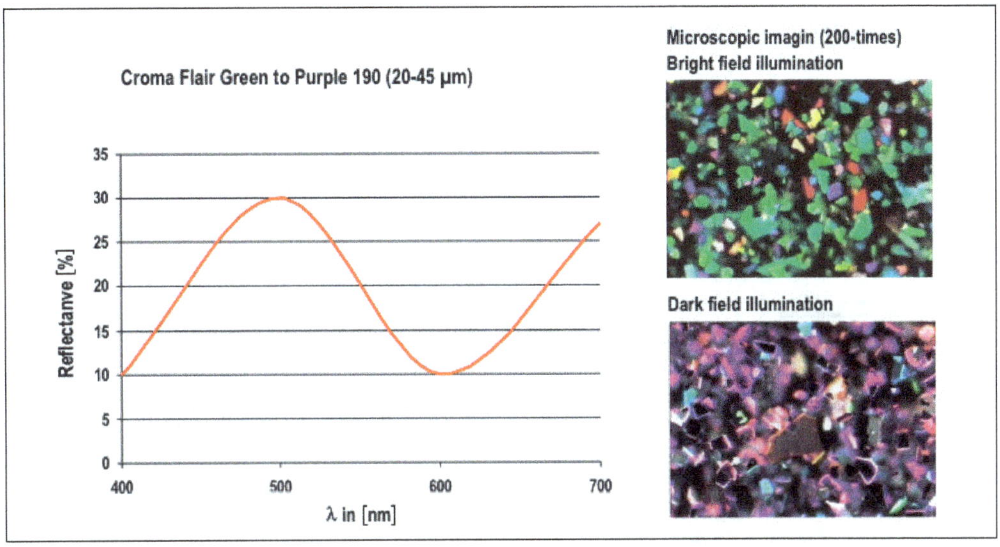


*Figure XII.19: "Croma Flair" Green of Flex as an example for a OVIP pigment.*

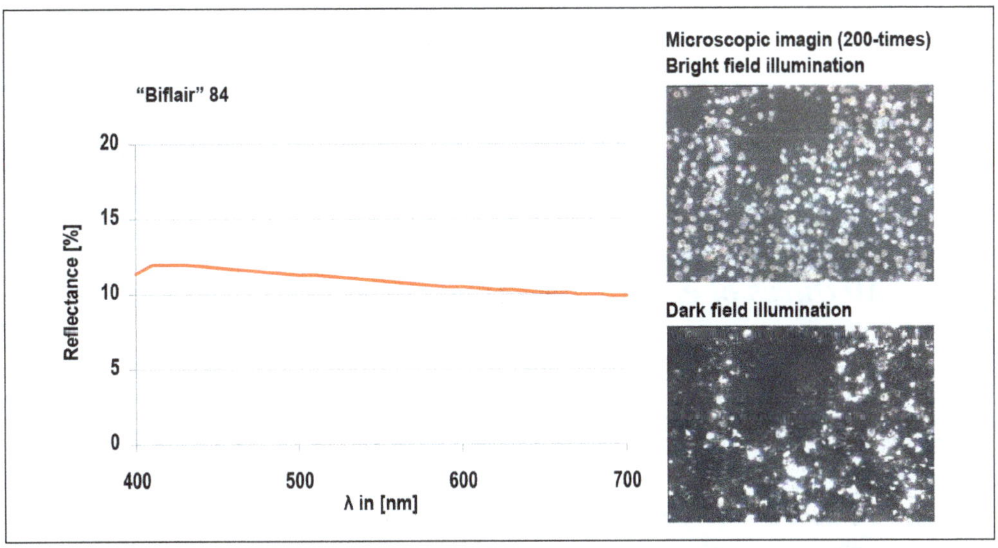


*Figure XII.20: "Biflair 84" as an example for a bismuth oxychloride pigment.*

Due to their extremely high prices, the pigments are used exclusively for premium applications in the automotive sector (see Section X.4.1.6) [7]. Under bright field illumination, the particles have a shape similar to that of the silicon dioxide-based "Colorstream" range.

Under dark field illumination, they exhibit a highly opaque, metallic aspect with pronounced light edges. The reflectance curves of all grades have a unique "fingerprint".

### 2.4.6 Bismuth oxychloride

Bismuth oxychloride pigments can easily be distinguished from other silver pearlescent pigments by their small particle size (7 to 14 µm) and hexagonal particle shape. They are mainly used in automotive interior paints because of their limited weathering fastness (yellowing). They are primarily used to achieve a chrome-like finish inside vehicles, e.g. for dashboards and steering wheels (see Figure XII.20).

# 3 Application examples

In this section, we look at several examples based on OEM paint formulations, illustrated with the aid of microscope images and reflectance curves.

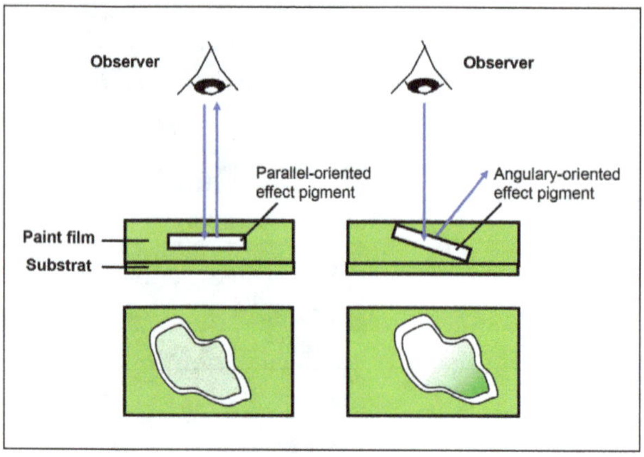


*Figure XII.21: Orientation of effect pigments, left: optimal, right: angular.*

## 3.1 Orientation of effect pigments in paint films

The orientation of effect pigments generally varies with the solids content of the binder system (low solids, medium-high solids, high-solids, and waterborne coats) and the method of paint application. The pigment loading in the effect pigment additionally influences the orientation of the particles.

The left-hand diagram of Figure XII.21 shows an aluminium flake in a green effect paint in which the particle is oriented horizontally and is uniformly coloured because of its position in the paint layer. On the right-hand side, the flake is oriented at an angle. The edge closer to the paint surface has a silver colour. This effect is even more pronounced at 500-times magnification.

## 3.2 Mixtures of effect pigments

The use of dark field illumination is essential when mixtures of effects are analysed.

Consider two different combinations:

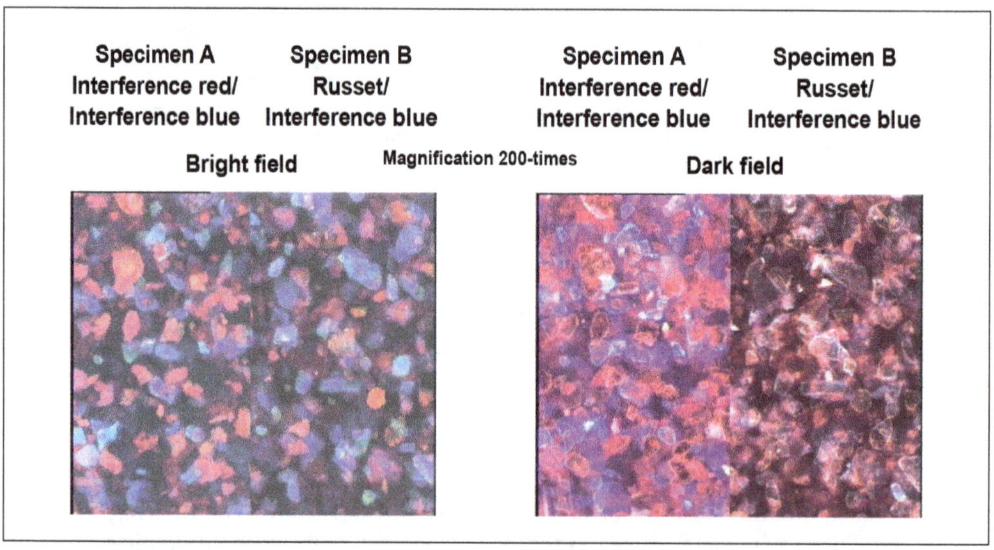


*Figure XII.22: Mixture of effect pigments, in each case left: sample A, right samle B with bright and dark field.*

**Specimen A**
Interference Red ($TiO_2$ coating)
Interference Blue ($TiO_2$ coating)
C.I. Pigment Red 122
C.I. Pigment Black 7

**Specimen B**
Russet ($Fe_2O_3$ coating)
Interference Blue ($TiO_2$ coating)
C.I. Pigment Red 122
C.I. Pigment Black 7

The combination ratios of the pigments in samples A and B are identical. Under bright field illumination (magnification 200-times) the two samples show a red and blue effect and have a coarse particle size and nearly identical colour. Dark field conditions clearly reveal the opacity of the red effect pigment used in sample B, whereas the blue effect pigments in both cases are transparent. This example shows the need to use dark field illumination for microscopic analysis of effect pigments.

## *3.3 Tinted clearcoat*

As an example, consider "Inferno Red", an OEM concept colour shade. The clearcoat includes a small amount of a red organic pigment (highly weather-fast). The amount is very low (0.4 to 1 %) and the pigment should be highly transparent for the purpose of boosting saturation.

A red organic pigment (C.I. Pigment Red 179) was used to tint the clearcoat. Figure XII.23 reveals a uniform red colour that does not influence the orientation or colour of the effect pigments in the base coat. This applies to the other colour regions.

## *3.4 Effect paints in combination with light-scattering pigments*

For many years, the trend for OEM colour effect shades leaned in favour of lighter/darker flop. That meant combinations of highly transparent pigments with effect pigments and carbon black. A lighter or a neutral flop is achieved by using highly opaque or light-scattering pigments, as seen in the current trend in automotive colours.

As an example, consider a styling concept based on:

- C.I. Pigment Red 254 (opaque version)
- silver-dollar flake
- pearlescent combination, gold
- C.I. Pigment Black 7

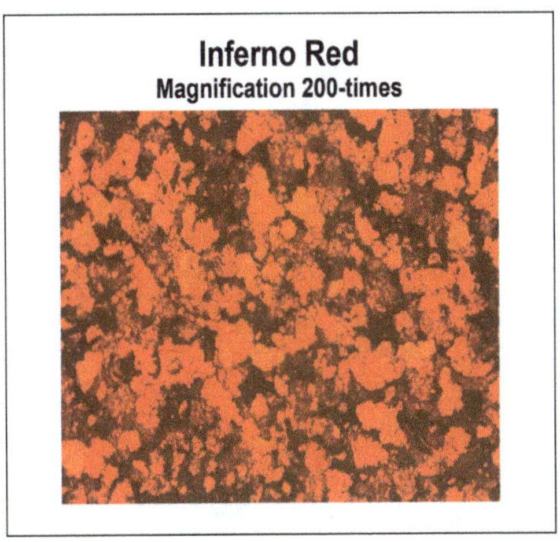


*Figure XII.23: Inferno Red as an example for an addition coloured clearcoat.*

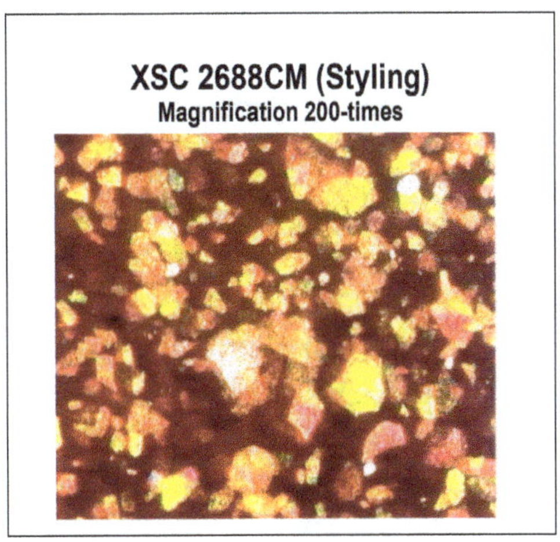


*Figure XII.24: Styling concept XSC 2688CM.*

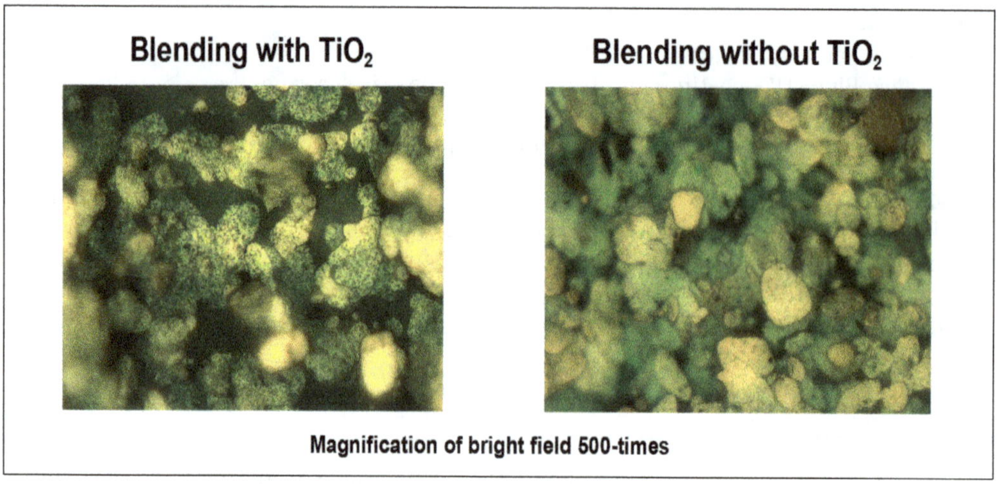


*Figure XII.25: $TiO_2$ in effect colours to achieve a bright or a neutral flop.*

Figure XII.24 shows this combination under bright field illumination. Due to the high amount of C.I. Pigment Red 254 [8], red dots can be readily detected on top of the aluminium flakes and the combination gold particles. Opaque yellow, white and orange pigments act similarly, provided that they exceed 5 % of the total pigmentation. Very often, opaque $TiO_2$ is used to achieve a lighter or neutral flop (see Figure XII.25). The microscopic pictures in Figure XII.25 show a green effect shade, with $TiO_2$ in the left figure and without $TiO_2$ in the right figure. The left figure clearly has a high amount of greyish dots on top of the aluminium flakes under bright field illumination.

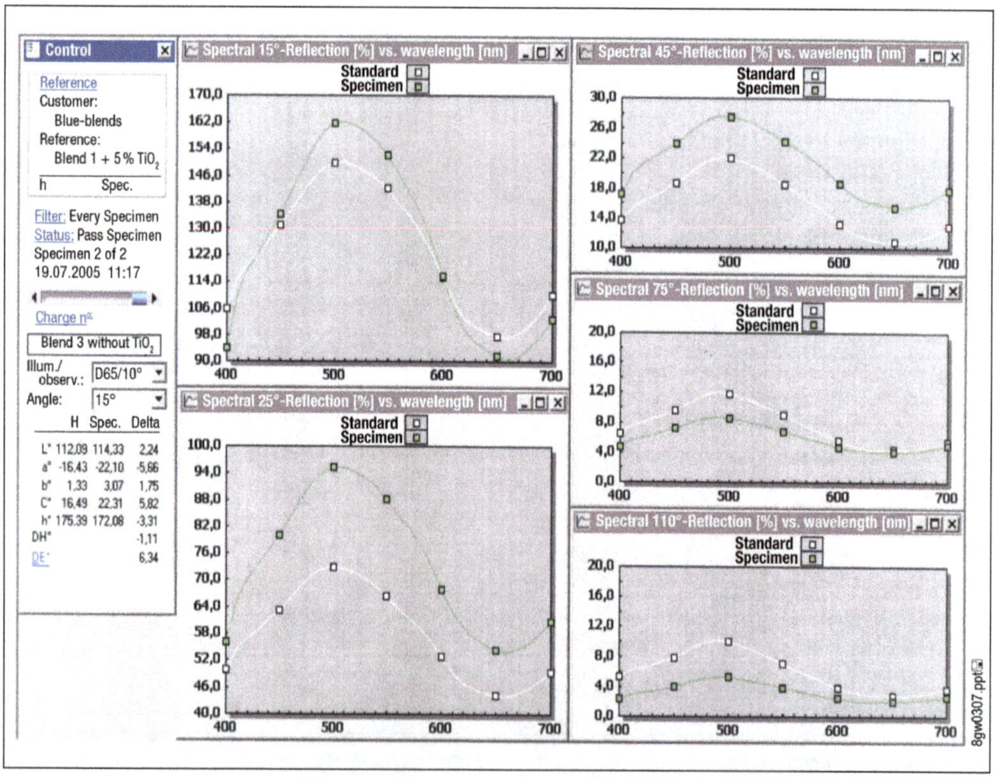


*Figure XII.26: Use of $TiO_2$ in effect coatings observed under fife angles.*

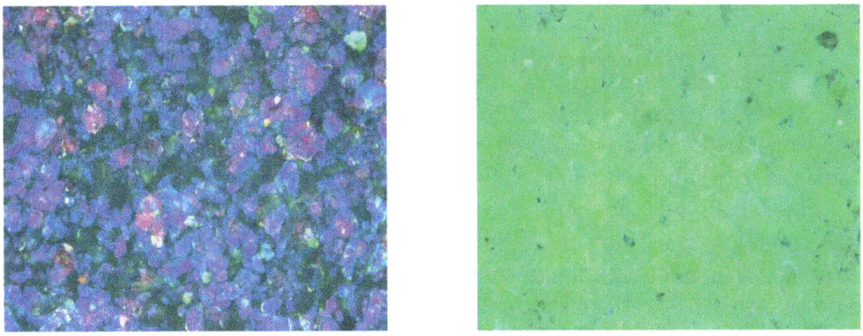


*Figure XII.27: Effect coatings on a coloured filler, "Indianapolis Green" of DuPont Exclusive Line.*

The microscope image on the right shows no dots at all on top of the aluminium flakes. The reflectance curve of the green effect shade containing $TiO_2$ exhibits distinctly lower reflectance compared to the colour without $TiO_2$ when measured at an angle of as15°/as25°/as45°. This confirms the well known rule that adding $TiO_2$ to an effect shade produces a darker colour at those angles (face-on view). (Adding white produces a darker colour!)

The highly opaque white pigment significantly reduces the reflectance of the aluminium flakes and creates a darker appearance. In contrast, at a viewing angle of as75°/as110°, the colour containing $TiO_2$ appears much lighter, which can be clearly seen in the measured reflectance curve. The light-scattering contribution of $TiO_2$ dominates at a low viewing angle.

### 3.5 Use of polarised light

Polarised light created by inserting a fixed analyser and a variable polariser into the light path under bright field illumination can be used to detect coloured primers applied under non-hiding effect shades (3-coat systems). Furthermore, micronised $TiO_2$ used for creating the "frost effect" can be detected in polarised light. It is identified by the "blue cloud" seen in the background of the microscope image.

Figure XII.27 clearly reveals the green filler under polarised light. The effect pigment employed is identified under bright field illumination as interference red mica. Not visible under polarised light on account of the high transparency and of the effect exhibited by the red pigment, the complementary colour under this aspect is of course green.

## 4 Systematic approach to colour-matching effect shades

The major problem associated with colour-matching effect shades is the change in colour that occurs with a change in viewing angle.

Recommended approaches include a colour-matching system comprising a multi-angle spectrophotometer and a microscope (for effect-pigment analysis), or a matching system based on single-angle measurement (d:8° as for solid colours), also in combination with microscopic analysis of the effect pigments and a multi-angle spectrophotometer for measuring the flop during the matching phase (see Chapter II). This will be described in the following sections.

## 4.1 Factors affecting the matching of effect shades

The binder system and/or its solids content greatly affects the orientation of the effect particles in the formulation: low-solids systems exhibit much better orientation (levelling) than high-solids and medium-high-solids systems. The orientation of the effect particles exerts a strong influence on the extent of the light/dark flop. When the same effect pigments are used in colour matching as were identified in the microscopic analysis, major problems need not be expected. The effect pigments should be incorporated into the paint in a low-shear stirring process (high shear causes the flakes to break).

Proper dispersion of organic/inorganic pigments, including carbon black, is essential because of the strong influence exerted on the colour strength, transparency and colour development of these pigments.

Additionally, the method of application (pneumatic, electrostatic) has a strong influence on flop, sparkling and colour.

## 4.2 Colour-matching system and systematic approach

The “PI Match Metallic System” employs single-angle measurement; therefore, as the angle-dependent change of colour and flop are not recorded, microscopic analysis is very important.

***Description of the PI-Match Metallic System***

- Measurement is carried out at a single angle (d:8°).
- The PI Match System is an “expert system”. A knowledge of the reflectance curves of the pigments used is essential, as it will speed up the process.
- Use of a microscope is essential, too. Proper identification of the effect pigments is necessary because of the single-angle measurement (flop behaviour). The measuring method is described in detail in Section II.

## 4.3 Preparation of calibration panels

The generation of sets of calibration panels for the chromatic and effect pigments is crucial. The first step is to define the type of binder system (solventborne/waterborne, etc.), and the carbon black and aluminium flake serving as reference. Then, the number of calibration panels and the mixing ratios must be defined. For reasons of reproducibility, automatic spray application is necessary and the applied dry film thickness should be in the usual range for automotive (OEM) applications. The pigment content of individual pigments is defined via their specific surface area, and the transparency and the colour space/strength are optimised via the dispersing method (grinding equipment, grinding conditions).

In critical colour regions, such as red pigments, either additional mixing ratios should be used or emphasis should be placed on the intended application for each pigment (mass tone or tinting pigment, etc.). The reason is the strong shift of up to 40° in colour position (hue angle) of red pigments upon addition of aluminium flakes/white pigment of between mass tone and pastel mixture to the bluer side (see Figure XII.28). In general, the ratios shown in Figure XII.28 can be used.

The measurement for saving in the system should be carried out by measuring against both a black and a white background; this obviates the need for a hiding application.

## 4.4 *Options and application of the colour-matching system*

The colour-matching system offers various options calculating a non-metameric pigment formulation by allowing fixed ratios of selected pigments. The reason is that most OEM pigmentations employ several effect pigments. The ratios must be determined visually under the microscope for later incorporation into the calculation.

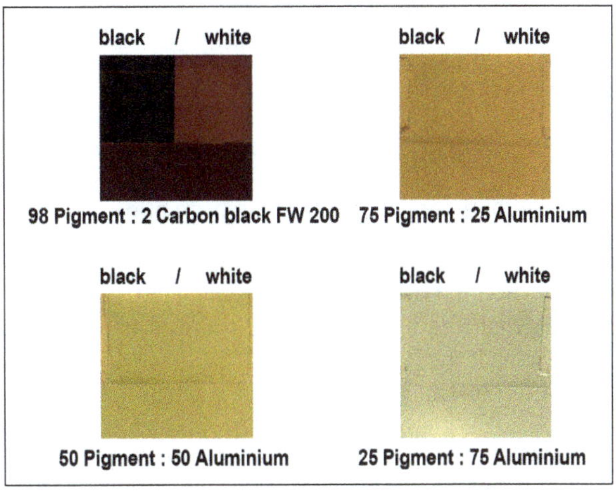


*Figure XII.28: Irgazin Yellow 2RLT (P.Y. 110) as an example with organic pigments for a calibration system of metallic coatings.*

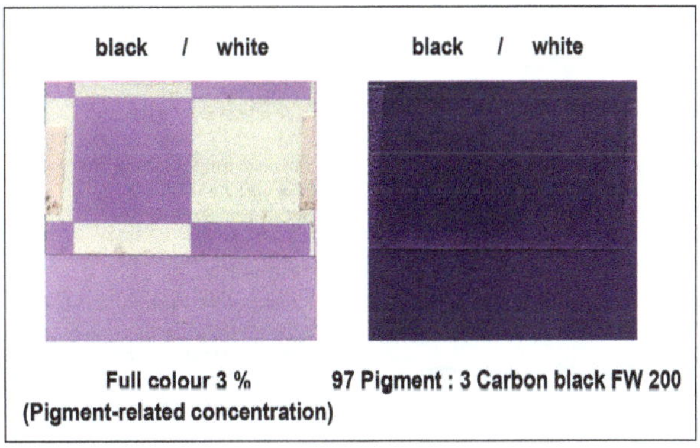


*Figure XII.29: Iriodin Pearlviolett 9219 WR as an example with mica pigments for a calibration system of metallic coatings.*

In general, there are two ways to calculate the recipe:

- **The "Expert Method":** The reflectance curve of the colour to be matched is used to identify the base pigments (from their contribution in the visible wavelength range) and these, along with the effect pigments identified under the microscope (library, both chromatic and effect pigments), are used to calculate a non-metameric colour match.
- "**Combination Method**": The maximum number of pigments (chromatic and effect) to be used to calculate the recipe can be specified. The system calculates all the possible combinations, sorted by their metameric behaviour under the chosen light sources.

***Options available in the PI Match colour-matching system***

- Fixing of ratios between selected pigments
- Selection of pigments (max. 9)
- Selection of combinations (min. 3/max. 8)
- Specification of dry film thickness
- Calculation of hiding power

This colour-matching system utilises the data for the saved pigments (sets of calibration panels). It is a "living, learning system" to which new pigments can be added when placed on the market.

***Number of pigments in the system***

At the moment, the system contains:

106 colour pigments (organic/inorganic), including carbon black and white ($TiO_2$)

- 30 types of aluminium flakes
- 95 different effect pigments
- 231 pigments in total

## 4.5 Colour-matching procedure

The systematic procedure for matching effect shades is described in Figure XII.30.

Basically, it starts with a microscopic analysis of the effect pigments, the results of which are incorporated into the subsequent colour-matching calculation process. After each spray-out, the result is checked under the microscope and a multi-angle spectrophotometer is used to check the flop. The details of the colour-matching method, e.g. measuring geometry, calculation method, are described in detail in earlier chapters.

The main points are:

- mount d:8° (deduction of surface gloss of 4°)
- 10 nm measuring increments over the visible wavelength range (400 to 700 nm)
- Kubelka-Munk theory
- Adjustment of the reflectance curve to minimise metamerism, optionally under two or three standard illuminants ($D_{65}$, A, F11).

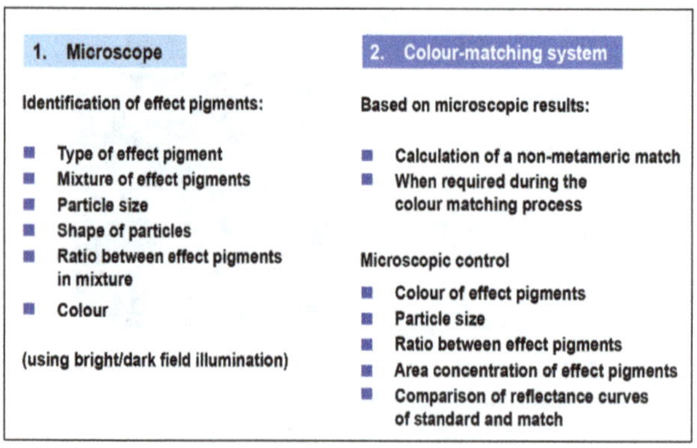


*Figure XII.30: Colour-matching procedure.*

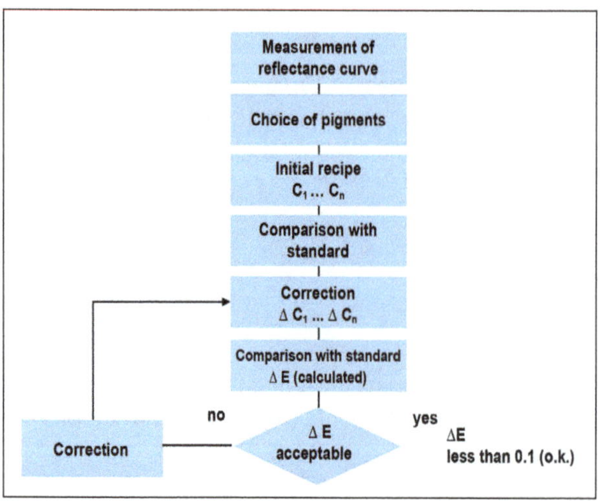


*Figure XII.31 illustrates the systematic colour-matching procedure.*

The starting formula is calculated using the colour-matching system and the final recipe is achieved via subsequent correction steps (always checked with the microscope and a flop control measurement as described above).

Consider the automotive colour "Jaune Soleil D-30" by way of example. The calculated formula shows information about

- dry film thickness (adjusted to 15 μm)
- calculated deviation from the "master standard" under two standard light conditions
- pigment/binder ratio
- total pigment concentration
- pigments used in the recipe
- weights of individual paint concentrates
- reflectance curves of standard/match

From the reflectance curve of the "master standard", it is obvious that a combination of a greenish yellow pigment and a yellowish red pigment must be the basic

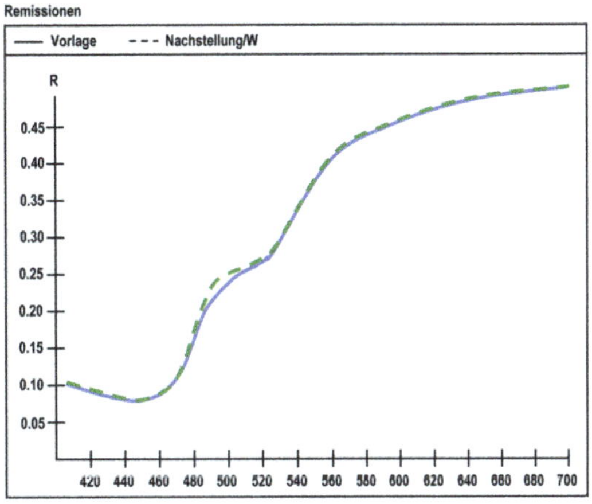

Recipe V-Jaune Soleil D-30 20.6.00 Datum 20.06.2000 11:44
Standard V-Jaune Soleil D-30 20.6.00 Anzahl Rezepte 1
Matching
System Metallic standard
Pigmentation Optimal dE
Film thickness 15.0 Lichtarten D65/10 A/10

Nr. 1

| | | D65/10 | A/10 |
|---|---|---|---|
| | dE Cielab | 0.202 | 0.06 |
| | dL* | 0.008 | - 0.01 |
| | dC* | 0.037 | 0.02 |
| | dH* | 0.199 | - 0.05 |
| | Contrast ratio CMS (1 t) | 1.00 | |

Pigment/binder ratio solids [%] 30.8
Pigment in liquid paint [%] 5.1

| Pigment | Pigment [%] | % in liquid paint | Weight (50 g paint) |
|---|---|---|---|
| 138 Fanchon Fast Yellow Y-5688 | 35.26 | 6.0 | 15.11 |
| 400 Hostaperm-Scharlach GO | 11.22 | 7.0 | 4.12 |
| 876 Farbruß FW 200 | 0.89 | 3.0 | 0.77 |
| 948 Iriodin Brillantgelb 9205 WR | 3.67 | 14.0 | 0.67 |
| 909 Mobilux R 187 | 24.48 | 18.0 | 4.20 |
| 900 Metallux 2153 | 24.48 | 15.0 | 4.20 |
| PE Klarlack | 138.38 | 0.0 | 20.93 |



*Figure XII.32: Metameric-free matching of a OEM paint.*

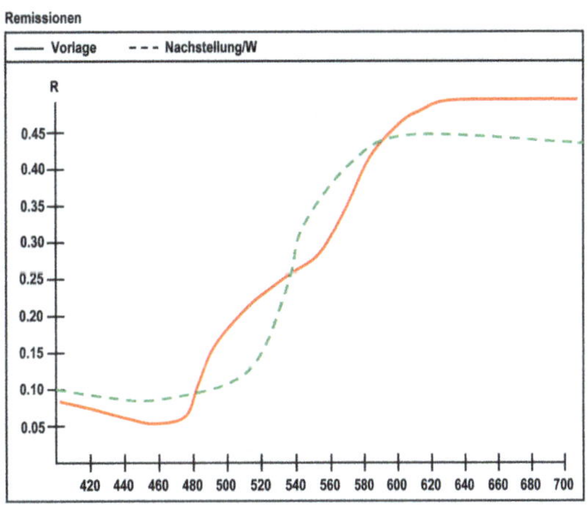

Recipe V-Jaune Soleil D-30 20.6.00 Datum 26.11.2001 07:12
Standard V-Jaune Soleil D-30 20.6.00 Anzahl Rezepte 1
Matching
System Metallic standard
Pigmentation Optimal dE
Film thickness 15.0 Lichtarten D65/10 A/10

Nr. 1

Rechnerische Schichtdicke

| | | D65/10 | A/10 |
|---|---|---|---|
| | dE Cielab | 5.241 | 4.74 |
| | dL* | -0.717 | - 0.82 |
| | dC* | -4.960 | -4.49 |
| | dH* | -1.533 | 1.30 |
| | Contrast ratio CMS (1 t) | 1.00 | |

Pigment/binder ratio solids [%] 45.2
Pigment in liquid paint [%] 7.4

| Pigment | Pigment [%] | % in liquid paint | Weight (50 g paint) |
|---|---|---|---|
| 876 Irgazin Gelb 2 RLT | 38.68 | 6.0 | 23.99 |
| 400 Hostaperm-Scharlach GO | 0.31 | 7.0 | 0.17 |
| 876 Farbruß FW 200 | 0.00 | 3.0 | 0.00 |
| 946 Iriodin Brillantgelb 9205 WR | 5.55 | 14.0 | 1.47 |
| 909 Mobilux R 187 | 27.73 | 15.0 | 6.88 |
| 900 Metallux 2153 | 27.73 | 15.0 | 6.88 |
| PE Klarlack | 48.44 | 0.0 | 10.61 |



*Figure XII.33: Metameric-free matching of a OEM paint.*

pigments to use (besides the effect pigments and carbon black) in order to suppress metamerism.

The step-like shape of the standard's reflectance curve is due to the absorption edge of the greenish yellow at approx. 480 nm and the yellowish red at 500 to 580 nm. If a single reddish-yellow pigment is used instead to calculate a match, the outcome is a totally metameric recipe.

This example confirms the need for reflectance curve adjustment if an absolutely non-metameric match is to be achieved. Once the colour has finally been adjusted, multi-angle measurements are taken between standard and match so that the flop behaviour can be documented.

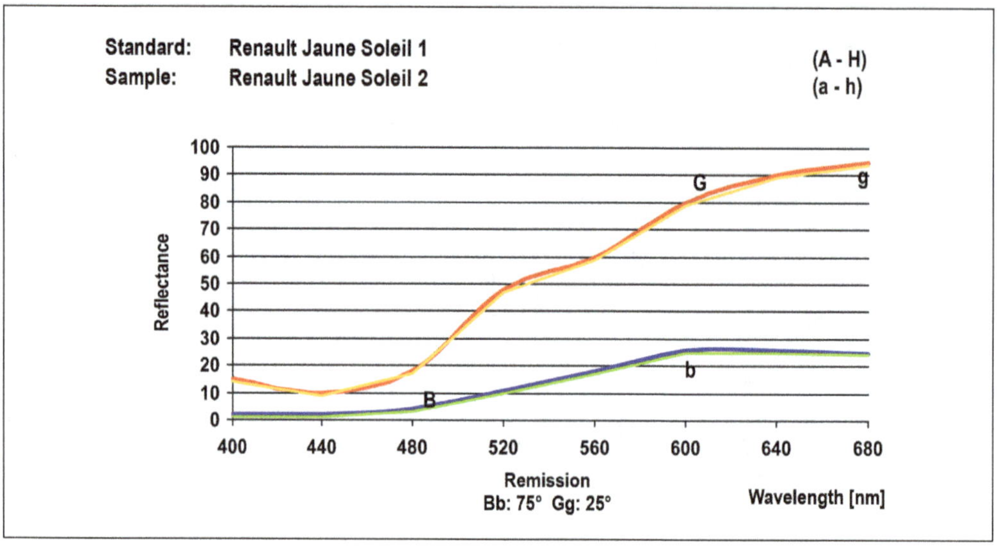


*Figure XII.34: Measurement with a multi-angle geniophotometer.*

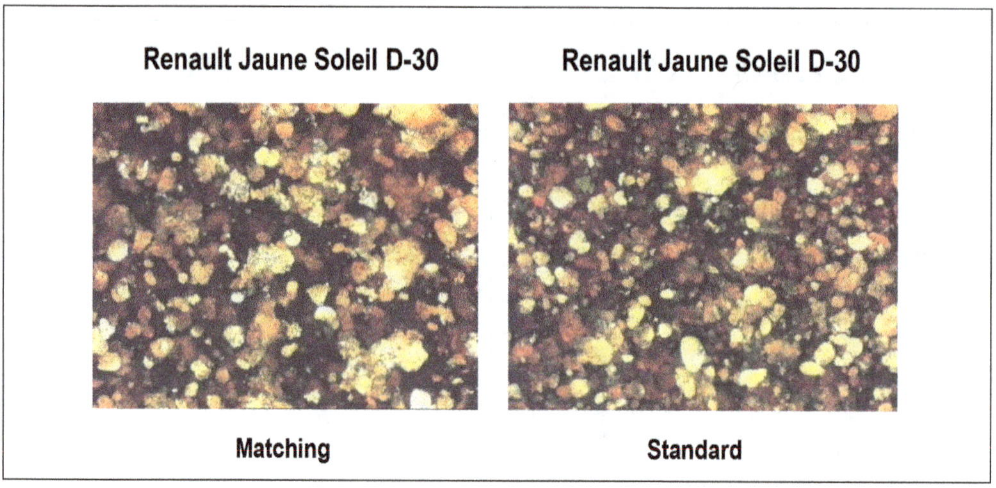


*Figure XII.35: Microscopic image of match and standard.*

In Figure XII.34, the comparison is based on measurements for two selected angles (25° and 75° from the specular). The results show good correspondence. The microscope images of the standard and match are quite similar.

# 5 Some useful hints for colour-matching effect paints

Here are some basic rules concerning pigmentation and flop adjustment of effect shades:

Dark flop

- Use highly transparent organic or inorganic pigments
- Coarse types of effect pigments (aluminium flakes, pearlescent pigments etc.)
- Choose silver-dollar types instead of cornflake types when choosing aluminium flakes

Light flop
- Opaque, light-scattering, organic/inorganic pigments
- Fine effect pigments
- Deorientation agents

Colour flop
- Blue with a green flop: Use phthalocyanine blues with a neutral flop (chlorinated types)
- Blue with a red flop: Use α/ε-phthalocyanine blues
- Effect pigments with extreme flop: "Scarabeus", "Cromaflair", "Colorstream" etc.
- Combination of highly opaque pigments with transparent effect pigments (complementary colour combinations, e.g. red/green or yellow/blue).

# *6 Literature*

[1] P. Wißling et al.: Metallic Effect Pigments, Vincentz Network 2006, Hanover, Germany

[2] I. Denne: Abrafati Congress, Sao Paulo 1995, Microscopical Evaluation of Effect Pigments

[3] G. Wilker: Abrafati Congress, Sao Paulo 1995, Colormatching of Effect Shades based on microscopical and spectroscopic method

[4] G. Pfaff: Special Effect Pigments, 2nd revised Ed., Vincenz Network 2008, Hanover, Germany

[5] R. Schmid, N. Mronga, V. Radtke: Optisch variable Glanzpigmente, Farbe und Lack 5/98

[6] F. Maile, Abrafati Congress 2001, New Effect Pigments for the Millennium

[7] G. Pfaff, A. Huber: Spezielle Effektpigmente auf Basis von Aluminiumoxidplättchen, Welt der Farbe, 9/2005

[8] W. Herbst, K. Hunger: Industrial Organic Pigments, 3rd revised Edition, Wiley - VCH Verlag 2004 Weinheim

# Authors

**Manfred Binder**, born in 1963, graduated as a mechanical engineer in 1989, majoring in metrology and control. He started his career as a design engineer for electronic microscopes. In 1991, he joined Infrared Engineering, where he was responsible for sales in southern Germany and Austria. Infrared Engineering developed a non-contact tristimulus colorimeter; that was his first encounter with colour measurement. In 1994, he switched to Kollmorgen Instruments. Here he worked as a sales engineer for laboratory colour measurement and colour matching systems in southern Germany, Austria, and Switzerland, and eastern Europe. The company changed its name to GretagMacbeth and then finally to X-Rite. In 2002, Manfred Binder assumed responsibility for applications involving the paper, glass, coil coating and plastics industry. Since 2012, he has been responsible for global in-line sales at X-Rite. He is a member of the DIN ISO Committees for paper, board and pulp and has held many seminars on colour, colour measurement and closed loop colour control around the world.

**Walter Franz** studied physics and micromechanics at the University of Ulm, Germany, and graduated with a degree in engineering. He joined the Carl Zeiss Company in Oberkochen, Germany, as Regional Sales Manager for Analytical Instruments. In 1984, he became head of the Datacolor International's sales office in Frankfurt. He was appointed Managing Director of Datacolor Germany in 1992 and then Director of Sales PCS for Europe, Middle East and Africa in 2002. In 2009, he became Director of Global Sales and in 2012 Director of Global Business Development at Datacolor until 2015. He is a member of Technical Committees at DIN (German Institute for Standardisation) and VDA (German Automotive Industry Association).

**Dr. Peter Gabel** studied physical chemistry at the Technical University of Braunschweig, Germany, where he also received his PhD. He joined Merck KGaA in 1986, where he worked in the quality-control department in Darmstadt. In 1991, he moved to the Merck plant at Gernsheim, Germany, where he became quality-control manager for pigments until 2015. He has worked for many years on the DIN Colour Standards committee, the Pigments and Extenders Standards committee, and the ASTM E12 Colour and Appearance committee.

**Dr. Stephan Gauss**, born in 1956, studied physics at the University of Tübingen, and obtained his PhD from the Institute for Industrial Physics at the Nuclear Research Centre in Karlsruhe. He was then delegated by Euroatom to occupy a senior position at an international research project on super-conducting magnets at the Oak Ridge National Laboratory in the USA. Thereafter, he worked in different roles on a number of research projects conducted at the Corporate Research department of Hoechst AG in Frankfurt, Germany. Since 1998, he has been head of the Colorimetry division and works in various capacities in quality of assurance of coloured pigments, pastes and colorants at Clariant AG. He is a member of several ISO and DIN work groups dedicated to colorimetry and pigment testing. He also actively participates in the German Society of Colour Science and in colorimetry training.

**Dr. Uwe Hempelmann** obtained his doctorate in chemistry at the University of Bielefeld, Germany. He started his career in pigments at Bayer AG in 1989, before joining Lanxess AG. There, he has held different positions in the research and the technical marketing department. He is now responsible for REACH, regulatory affairs and quality management. He sits on or chairs a number of committees of national and international standardisation bodies concerned with pigments and extenders. He gives lectures as part of the colorimetry seminars, which are held at the Fraunhofer Institute in Stuttgart and are organised by Vincentz Network.

**Dr. Rainer Henning**, born in 1949, obtained his PhD in chemistry from the University of Dortmund in 1977. At BASF Coatings GmbH in Münster, he learned about paint formulations for applications in plastics and effect coatings. Since 1995, he has worked in Technical Marketing, Pigments for Automotive Coatings, at BASF SE in Ludwigshafen, including coloristic properties of coloured pigments until end of 2015.

**Dr. Wilhelm H. Kettler** completed his academic studies of physics, astronomy, and mineralogy at the Ruhr-University of Bochum by specialising in experimental solid state physics. He joined the former Herberts GmbH in 1989, which was integrated into DuPont Performance Coatings GmbH in 1999, and headed the "Colour Technology" R&D unit until 2011. His research activities covered all domains of colour technology applications. He holds several patents in the fields of colour recipe prediction of glossy and matte-finished surface coatings, colour recipe correction, and special applications of colour technology. In 2011, Dr. Kettler joined X-Rite Europe GmbH, where his current activities are mainly focused on the development of algorithms to characterise and assess texture features of gonioapparent surface coatings. He has been a longstanding member of various DIN committees, the German Colour Association (DfwG), and the German Physics Society (DPG).

**Dr. Hans Joerg Kremitzl**, born in 1966, graduated from the Department of Chemistry at Friedrich Alexander University Erlangen-Nuremberg, Germany, in 1995. In 1997, he joined Eckart, a leading manufacturer of effect pigments and now part of the Altana Group. Initially, he was responsible for the chemical stabilisation and technical optimisation of gold bronze pigments. Later, his remit was extended to the full range of metal effect pigments. After 2007, he managed the Quality Coordination department, dealing with test methods and measuring equipment. Since 2014, he has been head of Chemical Process Development within the pilot plant.

**Dr. Gerhard Rösler**, 1950–2012, studied electrical and electronic engineering/cybernetics at the Technical University of Munich. He launched his career at Johne+Reilhofer. In 1979, he obtained his doctorate at the Ruhr-University of Bochum. He subsequently built up the new Electro-Optics department at Johne+Reilhofer, which was acquired by Macbeth in 1989 and renamed Kollmorgen Instruments. A company merger then led to the formation of GretagMacbeth in 1997. His work focus shifted to colorimetry in production. The fusion with X-Rite came in 2006. There he was in charge of global on-line colorimetry in production. He was President of the German Colour Science Society (DfwG), chairperson of several DIN standards committees for colour and sat on various committees at CIE, ASTM, VDA. His seminars and presentations on modern colorimetric methods for industry proved hugely beneficial to users.

**Sandra Weixel,** born in 1967, graduated in 1991 from the University of Applied Sciences in Kaiserslautern with a degree in textile engineering. From 1991 to 1995, she was employed at Datacolor International as an applications specialist for colour measurement in the textile and paint industry. During this time she was responsible for instrument installation and trouble-shooting. In 1995, she joined BYK-Gardner GmbH in Germany, where she is product manager for Colour and manager of the Applications department. Her activities here are focused on marketing portable and bench top colour instruments on a global basis. She is also an active member of DIN/ISO committees on coatings and colour topics and gives technical presentations to various paints/plastics and colours societies in Europe, Asia and the USA. Aside from these educational activities, she has always worked closely with the major paint and plastic companies to define their colour and appearance needs for new products.

**Gerhard Wilker** was born in 1947. After an apprenticeship as a laboratory technician at Farbwerke Hoechst, he studied chemical engineering at Darmstadt University of Applied Sciences, Germany. He joined the Pigments Application department of Hoechst AG in 1969. In 1989, he became head of Pigments for Industrial Paints and later held the same position in the Automotive Paints section. When the pigment activities of Hoechst AG were taken over by Clariant in 1997, he remained in this position, but took on worldwide responsibility for the technical side. He gave various technical presentations at International Congresses in many countries, and retired in 2010.

# Index

## Symbole



## A



## B



## C

# D

## P



## Q



## R

## V



## W



## Y

www.ingramcontent.com/pod-product-compliance
Lightning Source LLC
LaVergne TN
LVHW081522300926
849161LV00006B/1268